Fusion Visual Effects with

DAVINCI RESOLVE 16

by Damian Allen, Tony Gallardo, and Dion Scoppettuolo

Fusion Visual Effects with DaVinci Resolve 16

Damian Allen, Tony Gallardo, and Dion Scoppettuolo

Blackmagic Design

www.blackmagicdesign.com

To report errors, please send a note to training@blackmagicdesign.com.

Series Editor: Patricia Montesion

Editor: Dan Foster

Technical Review: David Hover

Cover Design: Blackmagic Design

ISBN 13: 978-1-7342279-5-6

Contents

Foreword

Welcome to **Fusion Visual Effects with DaVinci Resolve 16**

DaVinci Resolve 16 is the only post-production solution that brings editing, color correction, audio post, and visual effects together in the same software application! The Fusion page in DaVinci Resolve includes over 250 tools for advanced node-based visual effects compositing and motion graphics. That means you can switch between creative editorial, color grading, final audio mixing, and visual effects instantly, without having to explore or translate files between different applications!

Best of all, DaVinci Resolve 16 is absolutely free! Plus, we've made sure that the free version of DaVinci Resolve includes more features than any other paid editing system. That's because at Blackmagic Design, we believe everybody should have the tools to create professional, Hollywood-caliber content without having to spend thousands of dollars.

I hope you'll enjoy using DaVinci Resolve 16, and we can't wait to see the amazing work you produce!

Grant Petty
Blackmagic Design

Who This Book Is For

This hands-on training guide is designed for DaVinci Resolve editors, colorists, visual effects newcomers, and experienced compositors or motion designers who want to create visual effects and motion graphics in DaVinci Resolve 16. The book is divided into three parts:

Part I: Titling and motion graphics

Part II: Visual effects creation

Part III: 3D compositing

You'll start with an introductory composite that will give you a quick overview of the Fusion page interface and how nodes work. Each subsequent lesson builds your skills in the fundamentals of visual effects and motion graphics. You'll cover a variety of genres, techniques, and technical best practices including title design, hidden effects, green screen compositing, visual effects color management, and much more. Finally, you'll work in Fusion's powerful 3D environment creating broadcast design graphics, 3D particles, and a 3D-tracked composite. Although the last lesson of this guide requires DaVinci Resolve 16 Studio, you can go through Lessons 1 through 11 using the free download of DaVinci Resolve 16 at: blackmagicdesign.com

Getting Started

Welcome to An Introduction to **Fusion Visual Effects with DaVinci Resolve 16**, an official Blackmagic Design certified training book that teaches professionals and students the art of titling, motion graphics, visual effects compositing, and broadcast design in DaVinci Resolve 16. Editors will find clear workflow-driven lessons, while seasoned compositors will quickly learn Fusion's powerful node-based interface to accomplish incredible Hollywood caliber visual effects.

As you step through the lessons, you'll gain experience with Fusion's title animation tools, particle effects, blue and green screen keyer, powerful planar tracking capabilities, 3D compositing environment, and more! Best of all, you'll discover that there is no longer a need to send shots out to another application, because with DaVinci Resolve 16, fantastic visual effects and 3D motion design is simply a click away from editing.

This guide takes a practical, hands-on approach using real-world techniques for various compositing and broadcast design jobs including credit rolls, split screens, sky and sign replacement, 3D graphics, and 3D set extension workflow. Beyond the artistic side of visual effects and motion graphics, you'll also learn the fundamental science behind the techniques you are using, including compositing in linear light, managing premultiplied alpha channels, and maintaining resolution independence. Along the way, you'll find practical tricks and tips used by professional visual effects artists and motion designers to enhance the final outcome of your projects. As you complete each lesson, you'll have opportunities to perform optional practice exercises and answer sample test questions to test your comprehension of the techniques.

After completing this book, you are encouraged to take the 50-question online proficiency exam to receive a Certificate of Completion from Blackmagic Design. You can take the exam online at www.blackmagicdesign.com/products/davinciresolve/training.

About DaVinci Resolve 16

DaVinci Resolve is the world's fastest growing and most advanced editing software. It also has a long history of being the world's most trusted application for color correction. With DaVinci Resolve 16, Blackmagic Design has added a complete 2D and 3D visual effects compositing and motion graphics environment that enables you to complete even the most challenging of projects using only one piece of software!

What You Will Learn

In these lessons, you'll work with multiple projects and timelines to learn fundamental and *practical* techniques used in a wide range of visual effects and motion graphics genres. You'll acquire real-world skills that you can apply to real-world productions.

Lesson 1

A quick start guide that lets you explore the user interface by creating a simple but highly realistic sci-fi composite. This lesson is meant to get you comfortable with the interface and workflow since it touches on all the essential tools you will use throughout this guide.

Part I

Part I includes three lessons covering titling and 2D motion design. You'll learn to create smooth and highly customizable credit rolls, explore more creative title animation using Fusion's unique Follower modifier, and create motion paths that can be published and shared between any number of elements. Each lesson touches on some aspect of animation, including Fusion's powerful keyframe spline editor, procedural modifier-based animations, and simple expressions that can create incredibly natural movement.

Part II

Part II uses four lessons to cover the most common 2D visual effects techniques that you can use on a broad range of jobs. You'll uncover various techniques using Fusion's point and planar tracking tools so you can realistically integrate objects into a shot. Using the flexible vector-based paint tool, you'll remove objects to create *hidden effects* that viewers never even know are there. Finally, you'll learn how to approach classic green/blue screen compositing that epitomizes visual effects for most people.

Part III

The final part of this guide uses four lessons to explore 3D compositing and motion graphics. You'll start by exploring how to set up, navigate, and move around in a 3D compositing environment. Then, you'll introduce various 3D objects as you design a broadcast graphic, complete with extruded 3D text, shiny custom chrome materials, animated camera, and multiple light sources. Some of Fusion's most useful and fun tools, the particle tools, are delved into in Lesson 11, as you create billowing realistic smoke for a 3D composite in a music video. The last lesson in this guide requires DaVinci Resolve 16 Studio, as it guides you through setting up, optimizing, and making use of the integrated 3D camera tracking tool for simple set extensions.

The Blackmagic Design Certified Training Program

Blackmagic Design publishes several official certification books as part of the Blackmagic Design Learning Series. They include:

- The Beginners Guide to DaVinci Resolve 16
- Advanced Editing with DaVinci Resolve 16
- Color Correction with DaVinci Resolve 16
- Fairlight Audio Post with DaVinci Resolve 16
- And more to come

Whether you want to learn more advanced editing techniques, color grading, or audio post production, certified training has a learning path for you.

After completing this book, you are encouraged to take a one-hour, 50-question online proficiency exam to receive a certificate of completion from Blackmagic Design.

For more information on additional books in this series and the Blackmagic Design certification exam, visit www.blackmagicdesign.com/products/davinciresolve/training.

System Requirements

This book teaches DaVinci Resolve 16 for macOS and Windows. If you have an older version of DaVinci Resolve, you must upgrade to the current version to follow along with the lessons. Fortunately, DaVinci Resolve 16 is a free upgrade from previous versions of DaVinci Resolve.

The Fusion page uses a RAM-based playback system. A minimum of 32 GB of RAM is recommended to achieve a comfortable level of performance.

Download DaVinci Resolve 16

You can download the free version of DaVinci Resolve 16 from the Blackmagic Design website:

1. Open a web browser on your macOS, Windows, or Linux computer.
2. In the address field of your web browser, enter **www.blackmagicdesign.com/products/davinciresolve**.
3. On the DaVinci Resolve landing page, click the Download button when it appears.
4. Follow the installation instructions to complete the installation.

> **NOTE** DaVinci Resolve 16 Studio is required for Lesson 12.

When you have completed the software installation, follow the instructions in the following section to download the content for this book.

Acquiring the Lesson Files

You must download the zipped R16 Fusion lessons folder to acquire the media files you'll use while performing the exercises in this book. After you download and save the compressed files to your hard disk, extract the zipped files and copy the R16 Fusion lessons folder to your Documents folder.

To Download and Install the Lesson Files:

When you are ready to download the lesson files, follow these steps:

1. Connect to the Internet and navigate to:
   https://www.blackmagicdesign.com/products/davinciresolve/training

   Scroll down the page and click the link for the Fusion lesson files.

   The zip file is roughly 5 GB in size, and depending on your Internet connection, should take roughly five minutes to download to your computer using a standard broadband connection.
2. After downloading the zip file to your computer, open your Downloads folder, and double-click the zip file to unzip it (if your computer doesn't unzip the file automatically).
3. Drag the **R16 Fusion lessons** to your Documents folder.

   Your lessons and media are ready to go.

Acknowledgments

We would like to thank the following individuals for their contributions of media used throughout the book:

- Nguyen-Ahn Nguyen, for *Hyperlight*, a short film produced and directed by Nguyen-Ahn Nguyen. Property of Nguyen-Ahn Nguyen.
- Rafa Garcia, for the VAN clip. Directed and edited by Rafa Garcia. Property of Rafa Garcia Films.
- Lukas Colombo, for the Steve Val: *Dark Matte* music video. Visual effects supervisor Nic Torres. Property of Moai Films.
- HDRHaven.com for vignaioli_night_2k.exr.
- Brian J. Terwilliger, for Living in the Age of Airplanes.

Lesson 1

Quick Start: Learning the Fusion Page

Think back to the first time you saw the spectacular special effects of a big Hollywood blockbuster. Almost everyone remembers experiencing their first exciting captivation by cinematic magic.

Now, remember how you felt when you were asked to create a special effect on a project for the first time. You probably had a very different experience. Whether you're an editor, motion designer, visual effects artist, or a colorist, your first attempt at creating special effects was probably utterly befuddling. Where do you begin?

Throughout this training guide, you will build increasingly more sophisticated effects, one by one. After tackling several different shots and stringing together several FX techniques, you will have developed solid starting points for common visual effects tasks.

Time

This lesson takes approximately 40 minutes to complete.

Goals

The Fusion page is one of the five main functional pages in DaVinci Resolve. It is a fully built-in version of Blackmagic Design's Fusion software, a powerful 2D/3D visual effects compositing and motion graphics application. With Fusion built into DaVinci Resolve 16, you can instantly jump from editing to visual effects to 3D motion graphics and back to your edit timeline without having to import and export intermediate files and move between different applications.This type of workflow has only been dreamed about until now.

In this first Quick Start lesson, you'll explore the interface and general concepts of working in the Fusion page in DaVinci Resolve 16. Most of the tools and techniques you use here will repeat in more depth in later lessons, so this lesson will help familiarize you with a general effects workflow and the resources within DaVinci Resolve's Fusion page.

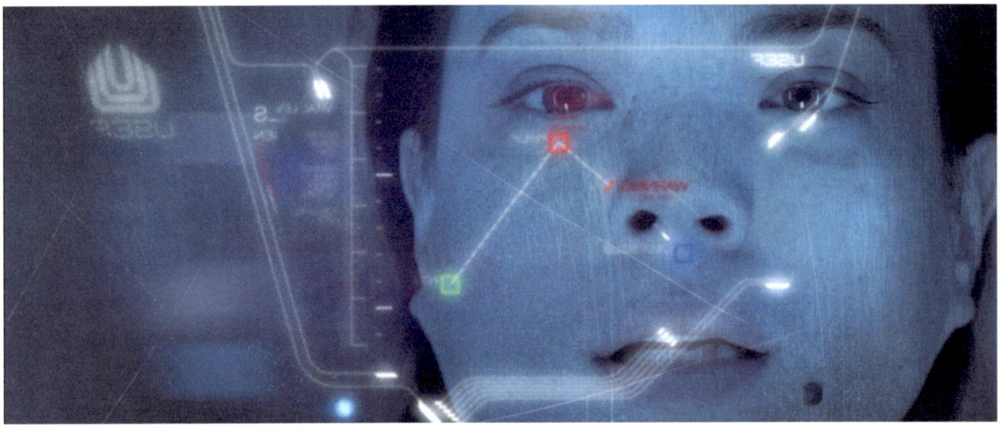

Completed composite for Lesson 1.

Exploring the Interface

In most editing systems, you put together your rough cut and then refine your edited versions in the edit page timeline. If you need compositing or motion graphics work, you export frames, open up different software, import the frames, and then render out the results for importing back into the edit timeline. In this first exercise, we'll open a project that will allow a quick tour of how DaVinci Resolve and the Fusion page simplify that entire process. Let's start by restoring a project and quickly getting into the Fusion page.

1 Open DaVinci Resolve, right-click in the project manager, and choose restore project archive.

2 Navigate to the R16 Fusion lessons folder, and restore the **Quickstart.dra** file.

   This project includes a single timeline that includes just one shot that requires visual effects.

3 In the edit page timeline, move the playhead over the close-up clip of the actress.

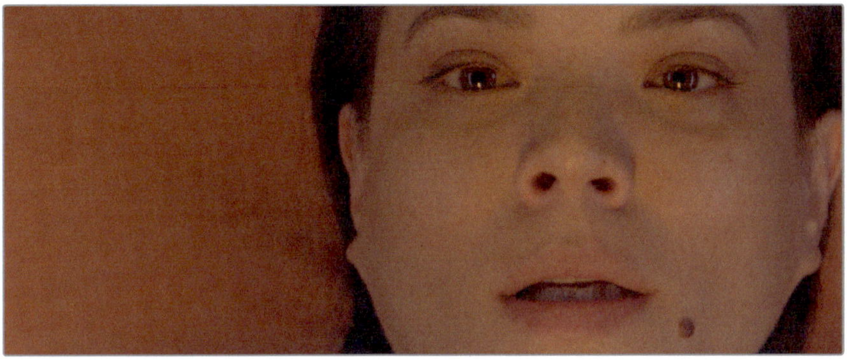

4 Click the Fusion page button, or press Shift-5.

That's all it takes to bring a single shot into the Fusion page, where you can apply effects.

But before you start creating those visual effects, let's get familiar with the Fusion page.

The page is organized into four main sections. Across the top, the two viewers display the images you are working on. Below the viewers, a toolbar includes the most commonly used effects, or *tools*. The lower work area, most often displayed as the node editor, is the heart of the Fusion page where you construct your effects. Finally, the inspector is to the right.

The toolbar provides buttons for adding commonly used effects or tools to the node editor.

The left and right viewers can show different images or effects from your composite.

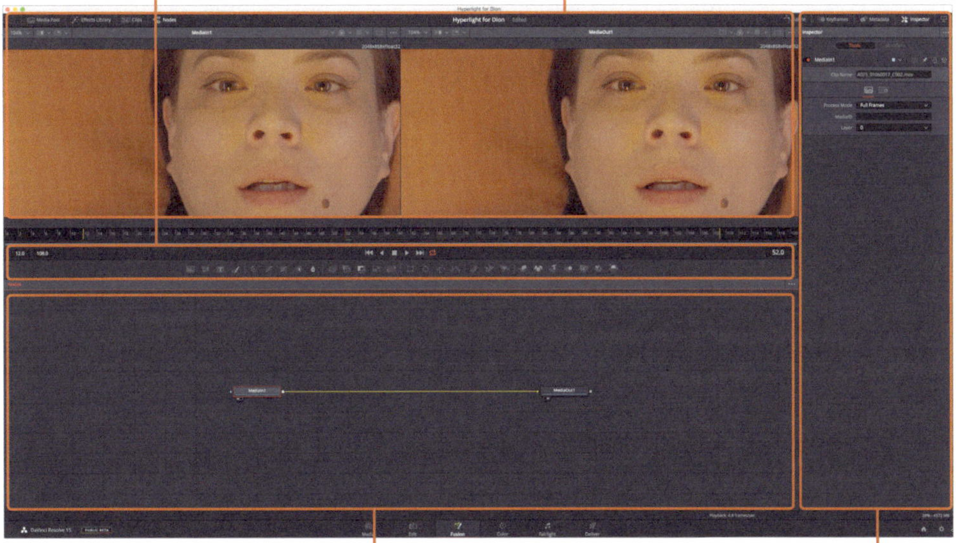

The work area can show any combination of the node editor, keyframes editor, or spline editor.

In the inspector, you can display and manipulate the parameters of any selected effect or tool in the node editor.

By default, the work area displays the node editor. Fusion does not use a timeline for compositing or applying effects, as does the edit page. Instead, Fusion uses a node tree in which each image or effect is a *node*, as represented by a rectangular icon in the node editor.

Nodes connect in a flowchart-like style, which makes it easy for you to see the entire structure of a composite and quickly make changes. Working with nodes in Fusion is much faster than hunting through nested stacks of precomposed layers and filters, as you might have to do with a layer-based interface. The node tree flows from one node to the next: One node applies an effect to an image and then passes it to another node for additional processing or output.

5 With the mouse pointer located over an empty gray space in the node editor, hold down the middle mouse button and drag to pan the node tree into the center of the panel.

Every clip or image file that you bring into the Fusion page is represented by a *media in* node in the node editor. The current media in node represents the clip from the edit page. The *media out* node represents the image sent back to the timeline on the edit page.

6 In the node editor, select the media in 1 node, and press the 1 key to display the image in viewer 1 to the left.

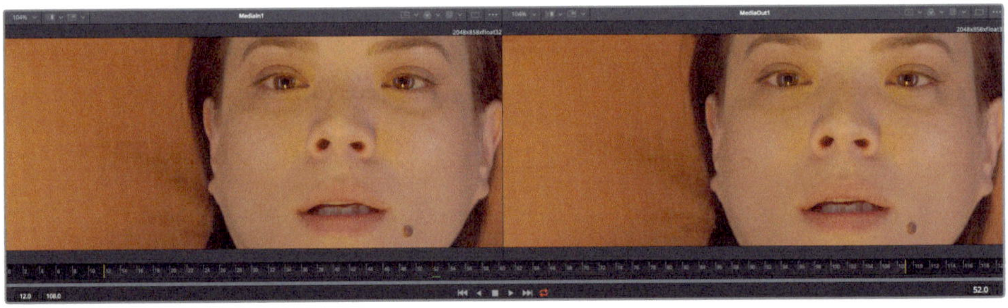

Viewer 1 Viewer 2

TIP The terms *node* and *tool* are used interchangeably to refer to an image-processing operation.

You can rename nodes to describe their function or image.

7 Select the media in 1 node. Press the F2 key, and rename the media in 1 node **ACTRESS**.

8 Select the media out node. Press the F2 key, and rename the media out node **OUTPUT**.

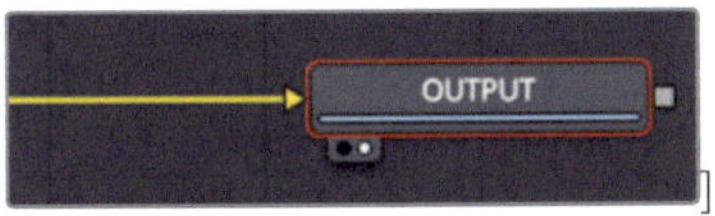

Under the viewers, a time ruler shows the duration of the current effect. Dragging the red playhead in the time ruler moves to different frames just like when you drag the jog bar under the source or timeline viewer in the edit page.

> **TIP** By default, the time ruler and all time fields on the Fusion page display frame numbers. To display timecode, choose Fusion > Fusion settings, and in the defaults panel, configure the Fusion page to do so.

The time ruler shows the entire source clip length, and the yellow lines indicate the *render range*, which is that portion of that clip actually used in the edit or cut page timeline.

9 Drag the playhead slowly through the render range from the first yellow line on the left to the second yellow line on the right.

As you drag the playhead through the render range, the current time display (to the right of the time ruler) displays the current frame number. To the left of the time ruler, you can see the render range start and end frames.

Under the time ruler, you can drag a two-handled scroll bar to zoom into the render range. This capability helps you when a source clip is very long and you are using only a small portion of it in your timeline.

10 At the far left edge of the two-handled scroll bar, drag in toward the center of the time ruler until the yellow line is near the beginning.

11 At the far-right edge of the two-handled scroll bar, drag in toward the center of the time ruler until the yellow line is near the end.

TIP You can change the render range to loop over a shorter area of a clip. Changing the render range does not change the composition length rendered in the edit page timeline.

As the playhead moves, a green line appears along the time ruler to indicate frames cached into RAM for smoother playback. The more RAM available in your system, the longer the cached region for RAM playback can be.

TIP You can assign more RAM for Fusion RAM playback in the Preferences panel. The amount of RAM assigned to Fusion RAM playback is released when you use other pages in DaVinci Resolve.

Adding the First Effect

To understand the basic principles of working with nodes instead of layers, let's add a simple effect to this clip. You'll find most common tools, or effects, that you will want to use in the toolbar.

1 Hover your mouse pointer over the first tool in the toolbar to display a tooltip with the name of the tool.

A thin divider separates each of the six toolbar categories. From left-to-right, the categories are: generators, color, compositing/transforms, masks, particles, and 3D.

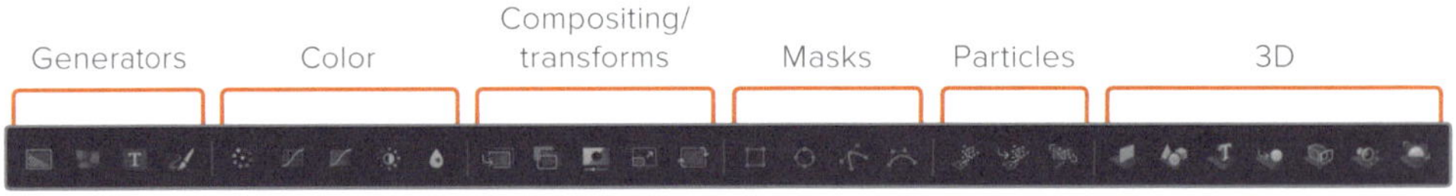

You build effects by connecting tools between the media in and media out nodes. In this example, you'll add a color corrector tool from the color category.

2 Move your mouse over the first tool after the first divider in the toolbar.

3 Drag the color corrector tool over the connection line between the ACTRESS node and the OUTPUT node.

4 When half of the line turns blue, release the mouse button to insert the color corrector tool as a new node.

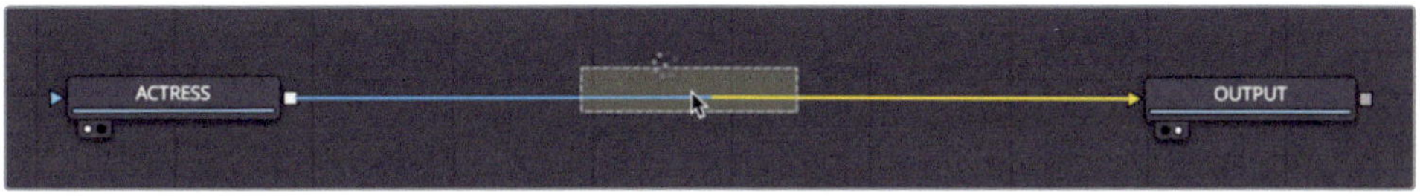

5 In the node editor, select the color corrector, and press the 2 key.

The color corrector results appear in viewer 2, while the original clip remains visible in viewer 1.

When a node is selected in the node editor, the parameters for adjusting that node appear in the inspector.

6 To give this shot a colder, outer space feel, in the inspector drag the color indicator toward blue until the ACTRESS has a light blue tint in viewer 2 .

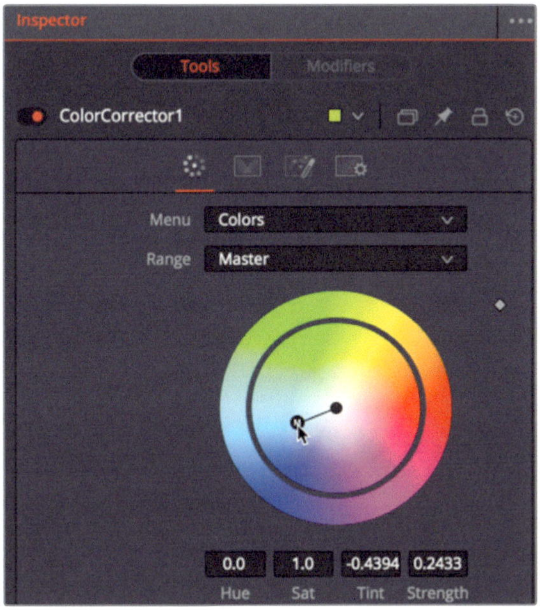

While the results are shown in viewer 2, viewer 1 still displays the original, unaltered ACTRESS image. By using the two viewers to see different aspects of your effect, you can more precisely compare and modify your adjustments.

Any effect you create in the Fusion page is also instantly visible in the timeline in the edit page.

7 At the bottom of the window, click the edit page button, or press Shift-4.

The edit page smart cache renders any Fusion effect in the background as you continue to edit.

TIP If you use DaVinci Resolve's user cache instead of smart cache, you can right-click over any Fusion clip in the edit page or color page timeline and choose cache Fusion output > On to manually cache the Fusion effect.

That's just the start of creating this effect; now let's return to the Fusion page.

8 At the bottom of the window, click the Fusion page button, or press Shift-5.

It's easy to see how the node tree functions like a flowchart with just a single image. The image comes in at the left, the effect is applied, and at the right, the modified image is sent back to the edit page. However, most visual effects use more than a single image, so next we'll learn how to bring in additional images that don't exist in the edit page timeline.

Adding Clips from the Media Pool

On the most basic level, visual effects are about combining two or more images to make a new image. Although you currently have only a single clip from the edit page timeline in your composite, you can access any clip from your project in the media pool. For the creation of this sci-fi shot, you will add a computer-generated head-up display (HUD) to make it appear that the actress is looking at a computer display.

1 In the upper-left corner of the interface, click the media pool button.

This Fusion page media pool is the same media pool you would find on the edit page and includes the same bins and clips.

NOTE Your bins and node editor may appear different than the pictures displayed in this book due to screen resolution and layout.

2 From the master bin, drag the HUD clip to an empty area of the node editor.

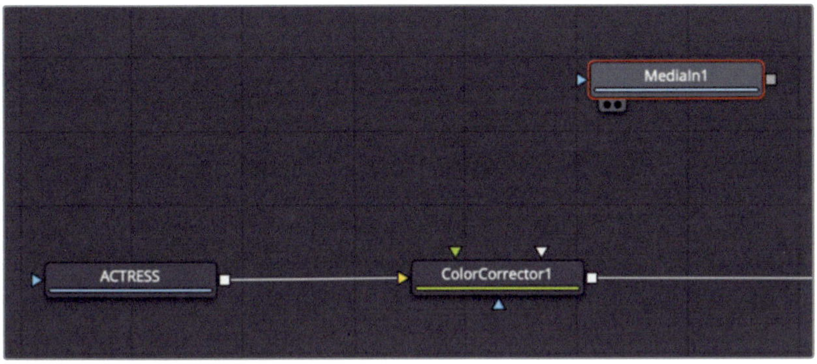

A new media in 1 node is created to represent the clip. The clip is not yet connected to your composite; it is just added to the node editor. Before connecting a clip to your composite, you can learn a lot about it by displaying it in the viewer.

3 Select the media in 1 node, and press F2 to rename the node to **HUD**; then, press 1 to display it in viewer 1.

In the lower-left of the HUD node, a small white dot, called the *view indicator button*, is highlighted to indicate that the node is displayed on viewer 1.

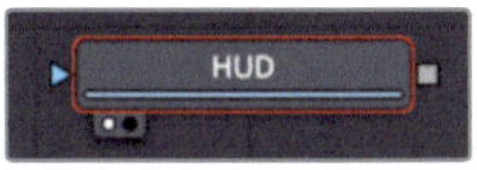

On the OUTPUT node, a second view indicator button is highlighted to the right of the first to indicate that this node is displayed on viewer 2. By displaying the OUTPUT node, viewer 2 will show you the final composite output destined for the edit page timeline.

> **TIP** Switching from the Fusion page to the edit page and back to the Fusion page will switch viewer 2 to display the media out node, regardless of what is selected before you leave the Fusion page.

Viewer 1 displays the HUD graphic with a checkerboard background to indicate that this computer-generated graphic has a transparent area created by the inclusion of an alpha channel.

> **TIP** An alpha channel is a fourth channel that accompanies the red, green, and blue channels of an image. Alpha channels determine which parts of an image are opaque and which parts are transparent.

4 Above viewer 1, click the color controls button; or click in the viewer and press the A key to view the alpha channel in the viewer.

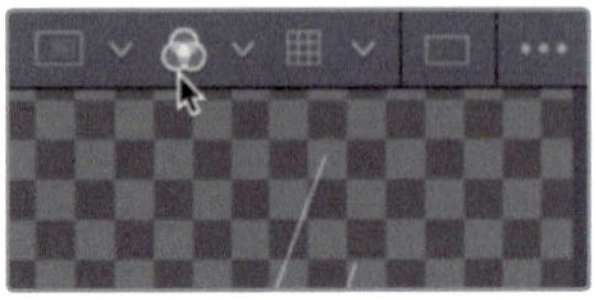

The color controls button now shows the alpha channel of the currently viewed clip.

5 Click the color controls button, or press the A key again, to once again display the full color image.

6 Click the media pool button to close the media pool and provide more room for the node editor and viewers.

With an additional image element added to the node editor, you are ready to begin blending two images together.

Understanding the Merge Node

The merge node is one of the most important and commonly used tools on the Fusion page. It composites, or blends, two images. You can add the merge node—as you would add other effects—by dragging it from the toolbar and inserting it in the node editor between the color corrector and OUTPUT nodes.

1 Directly after the second divider in the toolbar, drag the merge node into the node editor, and hover it over the connection line between the color corrector and OUTPUT nodes.

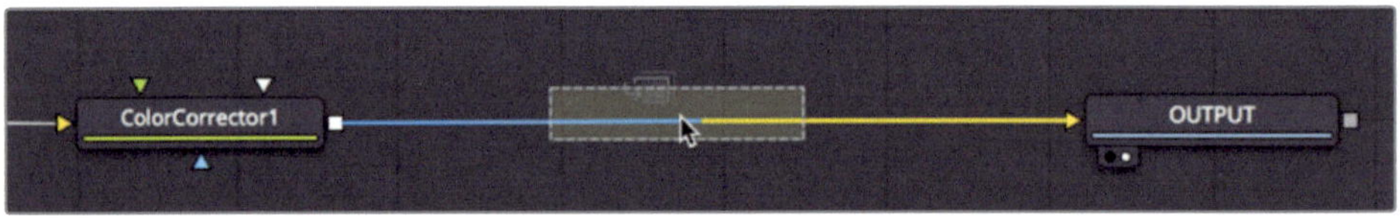

2 When half of the line turns blue, release the mouse button to insert a merge node between the color corrector and OUTPUT nodes.

> **TIP** Shift-dragging a node extracts it from the node tree and lets you reconnect it at a new location. Alternatively, selecting a node and pressing Delete or Backspace entirely removes that node from the node editor.

The merge node is now connected into the node tree, and you can use it to composite the HUD graphic over the shot of the actress. Before you combine the two images, let's explore the merge node in more depth.

A merge node has three inputs, one of which is for a background image (yellow triangle). You can think of this as a clip on V1 in the edit page timeline. A second input (green triangle) represents a foreground image, similar to a clip on V2 in the edit page timeline. A third input (blue triangle) is used exclusively for masking. Like all nodes, the merge node also includes an output (white square).

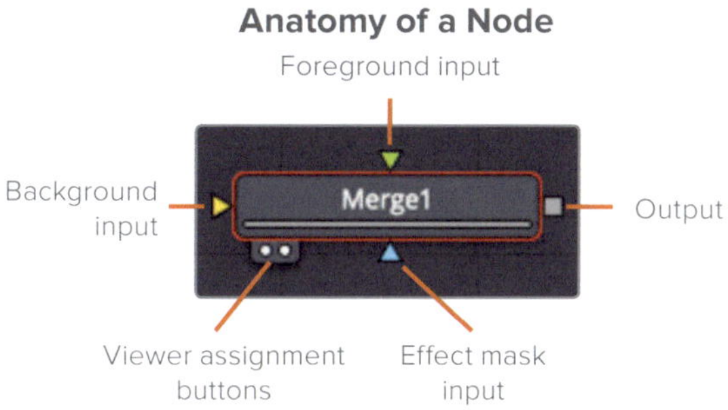

When you insert a merge node, it automatically connects the incoming connection line to the background input. It is up to you to manually connect the foreground image. In this node tree, the shot of the actress via the color corrector is connected to the yellow background input of the merge node.

3 To connect the HUD graphic into the foreground input of the merge node, drag the square output from the HUD node into the green foreground input of the merge node.

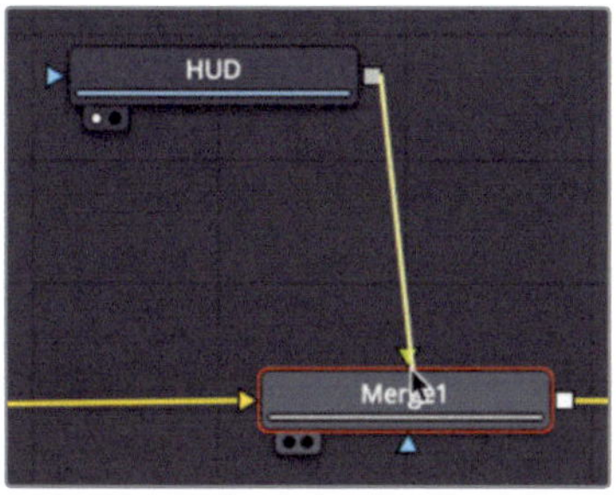

By adding the HUD graphic into the foreground input, you composite that graphic on top of the actress using the graphic's built-in alpha channel transparency.

TIP When using a third display, such as a full-screen broadcast monitor connected via a Blackmagic Design UltraStudio or DeckLink card, pressing 3 on your keyboard displays the node on the third monitor.

4 In the node editor, drag the HUD node below the merge node.

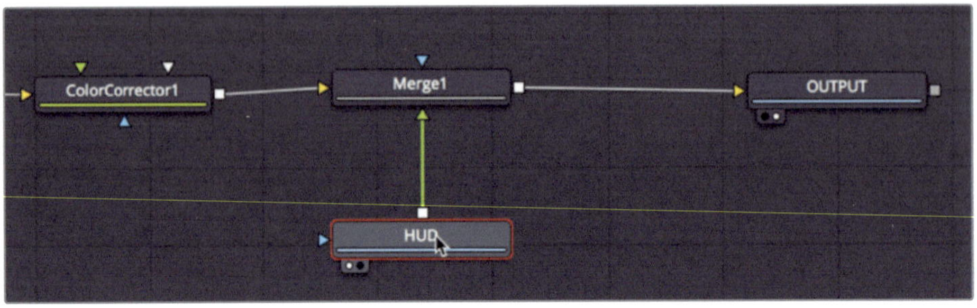

It is important to know that repositioning a node within the node editor in this way only visually organizes your node tree and has no impact on the compositing results. Only the connections made from node to node determine the processing order of images.

5 Return the placement of the HUD node above the merge node.

The merge node is the fundamental building block of almost every composite you will make in Fusion . Although it allows you to connect only two images, understanding how those two inputs interact is crucial because they enable you to link multiple merges and create much more advanced visual effects.

Inserting and Adjusting Effects

Node order is the single most important concept to understand in any node-based compositing system. Where you insert a node in the tree and the order in which you connect nodes determines the results of your composite.

The frame size of the HUD graphic is slightly larger than the background, so some of it is cut off. You'll need to add a transform node—but in precisely the correct location—to resize the graphic without resizing the background.

1 Just before the third divider in the toolbar, drag the transform tool to an empty place in the node editor.

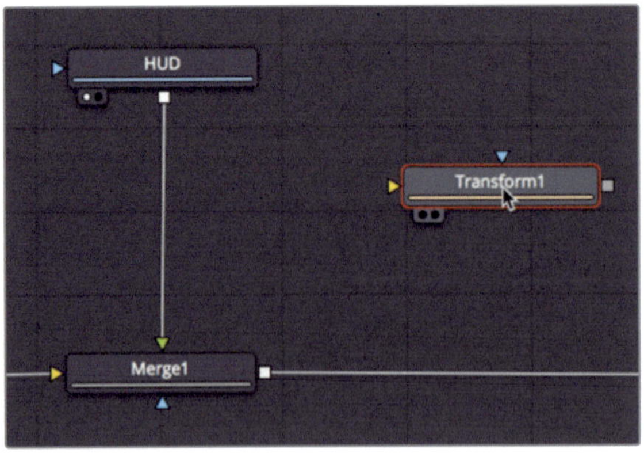

This transform tool must be placed into the node tree so it applies only to the HUD graphic. If you place it after the merge tool, this node will resize the entire shot.

2 Shift-drag the transform node between the HUD and the merge nodes. When the mouse pointer is over the connection line and the line turns blue, release the mouse button to insert the transform node.

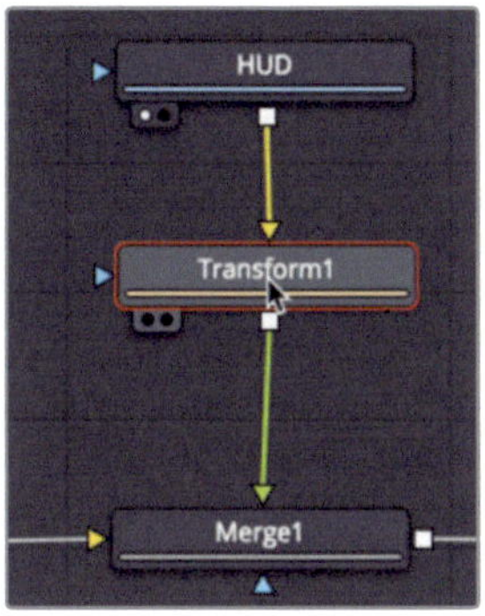

Although you have not made any changes, you can see the original graphic in viewer 1 and the transform adjustments in viewer 2 because it displays the OUTPUT node's results.

> **TIP** In the lower-left corner of the interface, the status bar displays basic metadata about any selected node.

When a node is selected in the node editor, the parameters for adjusting that node appear in the inspector with on-screen controls displayed in the viewer.

3 In the inspector, drag the transform size slider to the left to scale down the graphic until the logo on the left is no longer cut off.

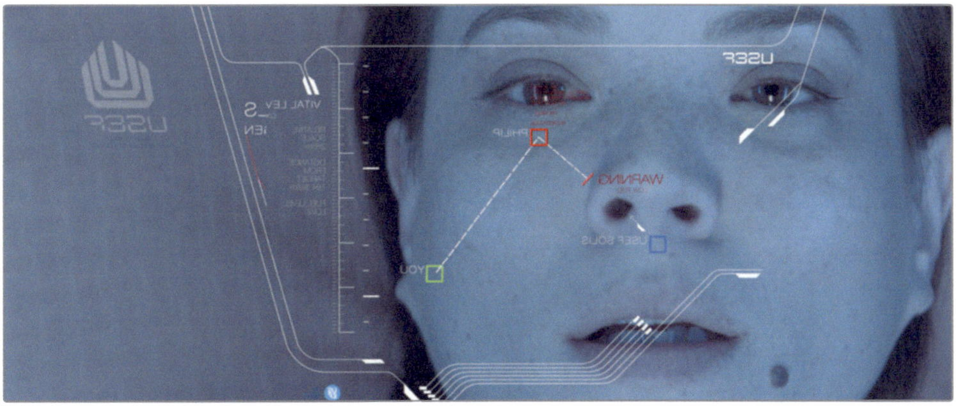

Adding additional effects to the same shot is as easy as clicking the tool in the toolbar.

4 Click the blur tool—located in the toolbar just before the second divider—to insert it directly after the transform node.

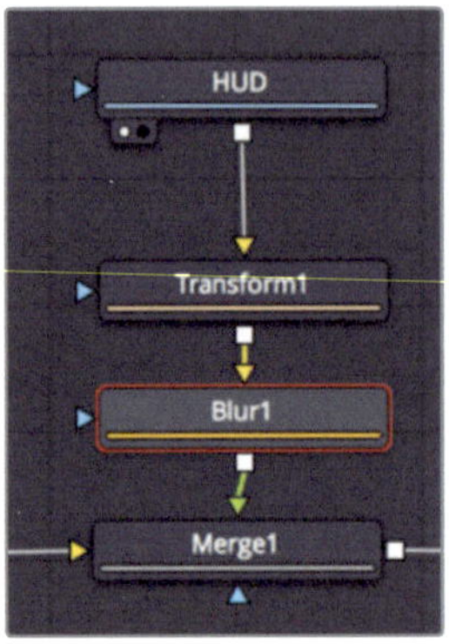

TIP When a node is selected in the node editor, clicking a button in the toolbar will insert the new tool directly after the selected node.

Often, computer graphics must be softened so they appear more natural when composited over live action. Next, you'll add a very small amount of blur to better blend the graphic with the live-action background.

5 In the inspector, drag the blur size slider to around 1.5 to slightly soften the graphics.

TIP It can be easier to enter values rather than dragging sliders when changing controls.

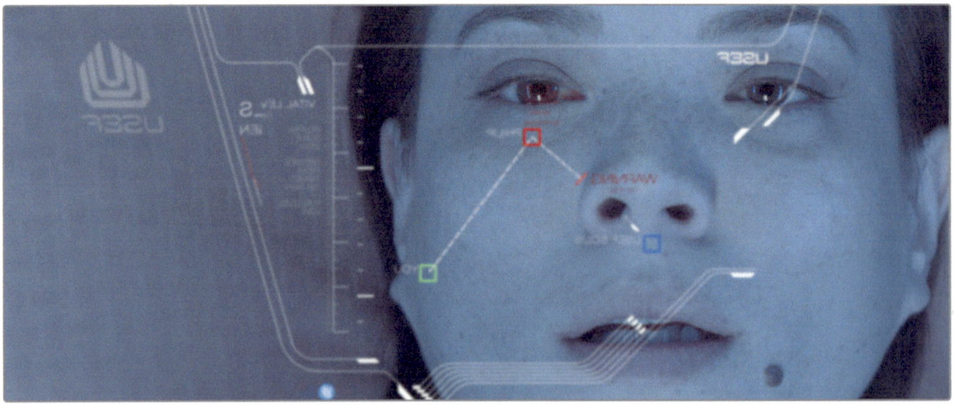

> **TIP** You can temporarily disable a node by selecting it in the node tree, and pressing cmd-P (macOS) or Ctrl-P (Windows).

Now, let's preview your composite.

6. In the transport controls, above the toolbar, press the first frame button to move the playhead to the start of the render range.
7. Press the spacebar to play the composite.

The results from the OUTPUT node are visible in viewer 2.

Chaining Merge Nodes Together

Merge nodes only process two images at a time: foreground and background. To add additional images, often called elements or plates, to a composite, you must chain merge nodes together using the output of one merge as the foreground or background of another merge. That's how you'll add a scratched glass element that we'll again get from the media pool.

1. In the upper-left corner of the interface, click the media pool button.
2. From the master bin, drag the glass clip to an empty area of the node editor.
3. Press 1 to see the image in viewer 1, and then press F2 to rename this node GLASS.

   This is the image we will composite over the top of the ACTRESS and the HUD.
4. In the node editor, select the merge 1 node.

   Instead of dragging from the toolbar, selecting a node allows you to click a tool in the toolbar and connect it into the selected node.
5. In the toolbar, click the merge tool.

The output of merge 1 is now connected to the yellow background input of merge 2. You can use the unconnected green foreground input on the merge 2 node to connect the scratched glass image.

6 Drag the output of the GLASS node to the green foreground input of the merge 2 node.

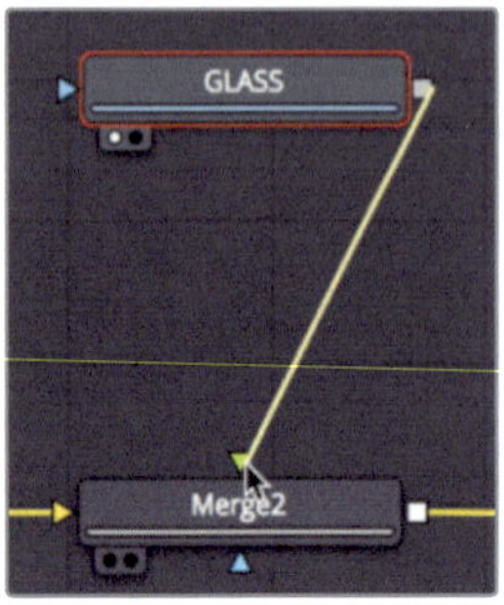

With the glass node connected, you can use the merge node to blend the images to create more realism.

7 In the node editor, select the merge 2 node; in the inspector, drag the blend slider to 0.3 to lower the opacity of the glass element.

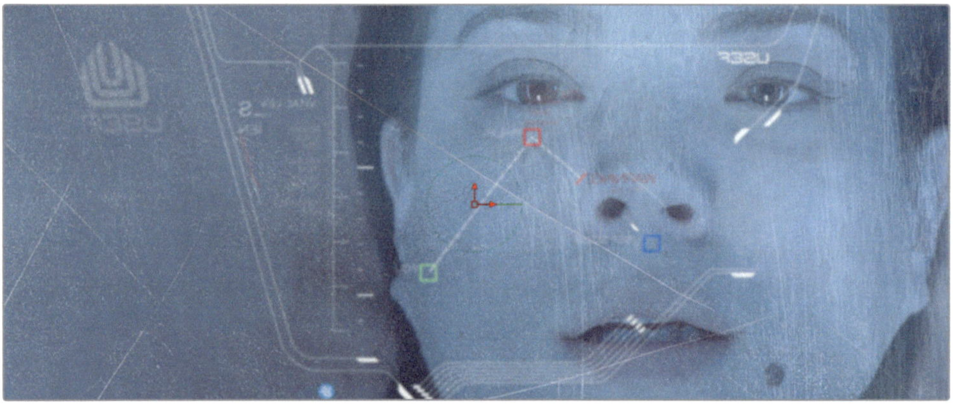

You can find more sophisticated blending operations in the apply mode drop-down menu. These settings use simple mathematical operations to blend images based on the color and luminance of the foreground and background.

Although you have a number of choices, the most common apply modes can be defined based on their overall effects on the composite:

- Darken colors: darken and multiply
- Lighten colors: lighten, and screen
- Increase color contrast: hardlight, overlay, and softlight

Because you want the scratches in the glass to be seen clearly, let's use an overlay mode that will composite the scratched glass with more contrast over the top of the actress.

8 In the apply mode menu, choose overlay.

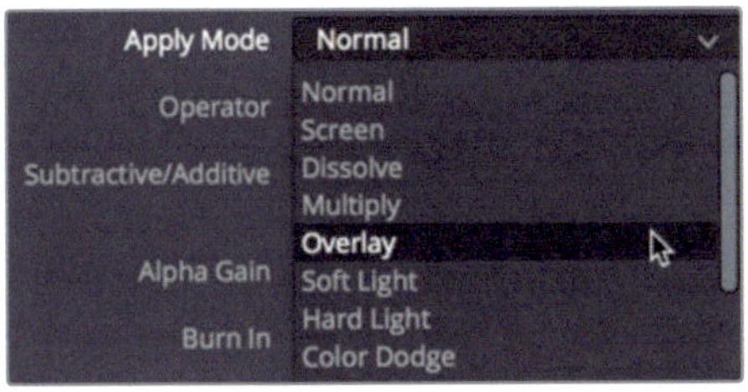

If you wanted to add even more image elements beyond the glass and HUD elements, you could continue chaining merges together. Adding new merge nodes before the existing ones would place new elements behind the current composite.

Masking Effects

Effects have an input and an output for connecting other nodes. These connections are color coded with a yellow triangle to represent the primary input, and a white square to represent an output. Nodes also have an effect mask input for masking use, as represented by a blue triangle. You can use the mask input to limit the area of the image affected by an effect. Let's create a subtle vignette that will darken the area around the actress' face.

1 In the node editor, select the color corrector node.

With the selected node being the one directly before the first merge node, any effect we add will affect only the actress.

2 In the toolbar before the second divider, click the brightness contrast tool to insert it into the node tree directly after the color corrector node.

Let's change the gamma in this node to darken the actress.

3 With the brightness contrast node selected, drag the gamma down to around 0.75.

You've darkened the entire image considerably. By adding a matte tool, you can limit where that darkness is applied.

4 With the brightness contrast node selected, in the mask category of the toolbar (located after the third divider), click the ellipse mask tool.

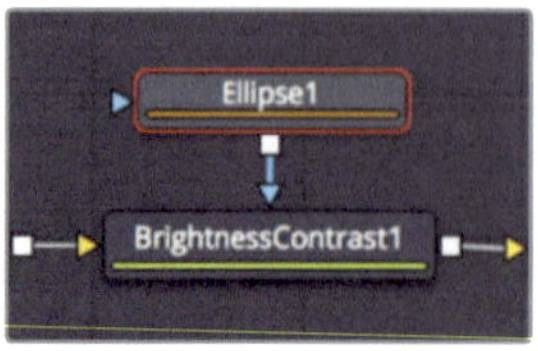

Because the brightness contrast node was selected, the ellipse tool automatically connects to it. The ellipse is a mask tool, so it automatically connects to the blue effect mask input.

In viewer 2, the gamma adjustment is now masked inside the ellipse shape. You must invert the ellipse matte so the darkness falls outside the shape.

5 In the inspector, select the invert checkbox to reverse the ellipse matte.

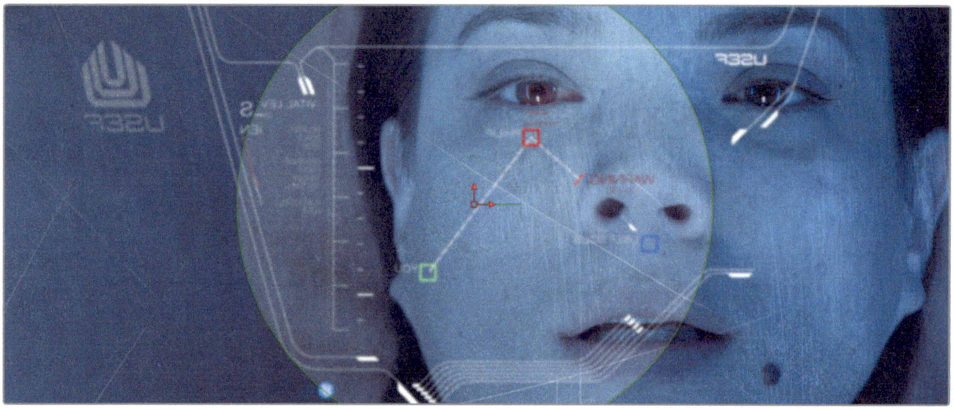

A few more adjustments will make the darkened vignette look even nicer.

6 In the inspector, refine the ellipse matte by increasing the soft edge slider to create a more gradual transition from the center to the darkened edges.

TIP A small gray dot appearing under any parameter slider indicates the default position of that parameter. Clicking the gray dot resets the slider to its default position.

You can use the on-screen controls to change the ellipse shape so it covers more of the wide aspect of the frame.

7 In viewer 2, drag the right edge of the ellipse to expand its width and cover more of the frame.

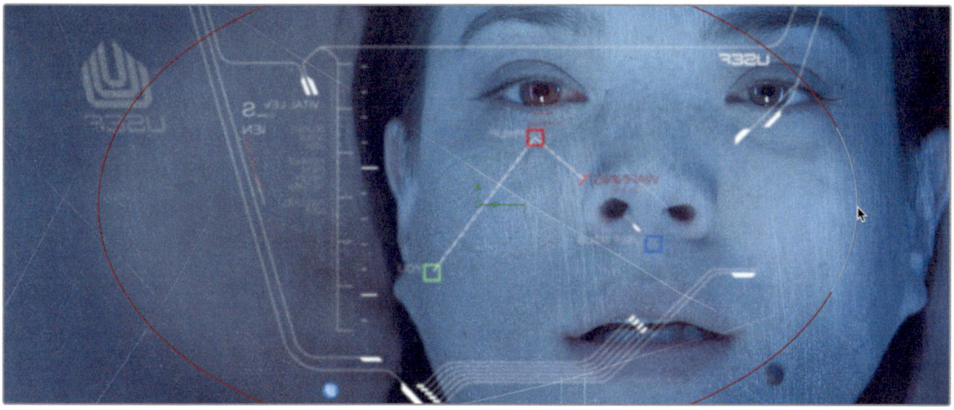

8 In the transport controls, above the toolbar, press the first frame button to move the playhead to the start of the render range, and press the spacebar to play through the composite.

The first time the composite is played, DaVinci Resolve loads the effect into RAM. A green line under the time ruler indicates the cached regions of the composite. When the entire composite, from the start of the render range to the end, is cached into RAM, the composite plays back in real time.

Adding Effects from the Effects Library

The Fusion page includes more effects than just those in the toolbar. It also includes over 250 tools that you can access in the effects library.

To add more realism to this shot, we should make the HUD appear more realistic and include some lens flares to simulate the effects of light from stars or suns in the area hitting the HUD glass.

Let's start with making the computer screen HUD glow.

1 In the node editor, select the blur1 that is connected to the merge 1 node.

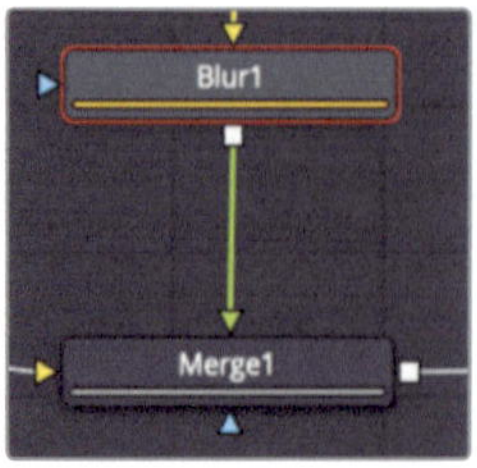

2 In the upper-left of the Fusion page, click the effects library button.

The library is organized by category. There are categories for everything from paint and particles to masking, image filters, tracking, and more.

3 In the effects library, select the tools > blur category and click the soft glow tool to add it to the node editor.

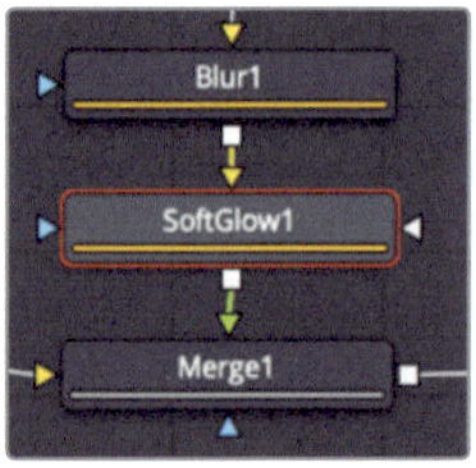

The soft glow gives the HUD a more realistic and dramatic appearance.

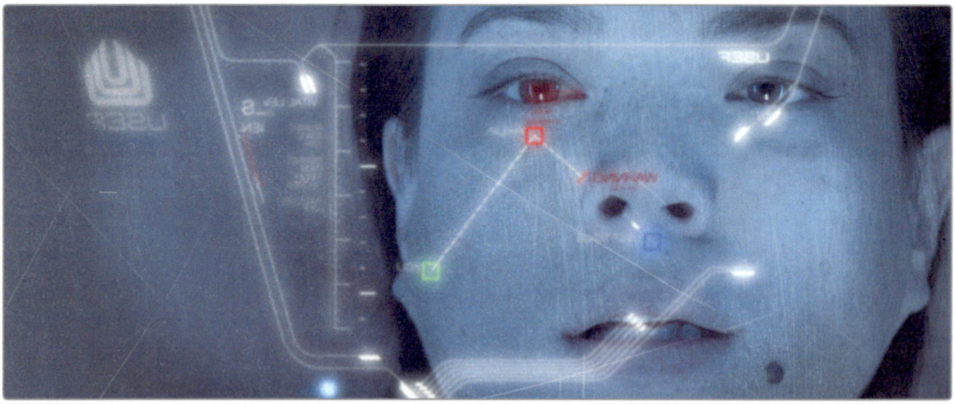

Now we'll add some light reflections.

4 In the effects library, click tools to close the category, and then click OpenFX > ResolveFX Light.

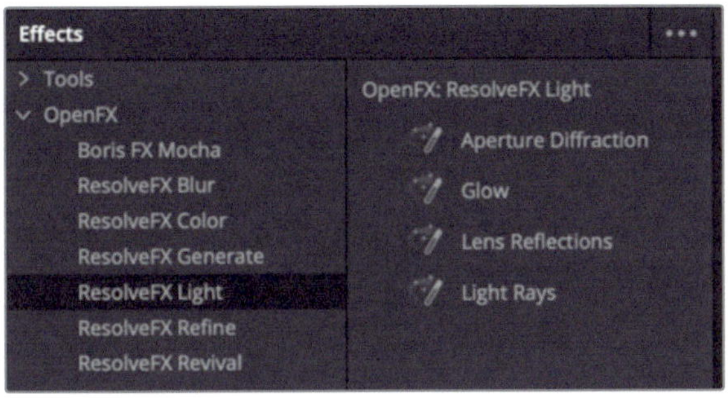

The OpenFX category contains the same filter effects you have in the effects libraries in the cut, edit, and color pages.

5 From the ResolveFX Light category, drag the Lens Reflections tool over the connection line between the merge 2 node and the OUTPUT nodes. When half of the line turns blue, release the mouse button to insert the Lens Reflections tool as a new node.

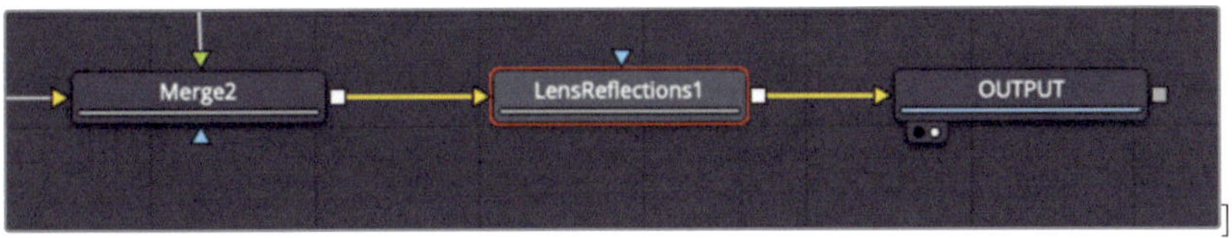

6 In the inspector, increase the Global brightness and Global blur values to suit your taste.

The Lens Reflections effect is used to simulate highlights reflecting off various optical elements within a lens to create flaring and scattering effects based on the motion of highlights in the shot.

This lesson was designed to introduce you to the general concepts and workflow for working in the Fusion page. You got a clip from the edit page and brought it into the Fusion page. In the Fusion page, you got nodes from both the toolbar and the effects library and inserted them into the node tree to create effects. Then, additional images were accessed from the media pool and composited using the merge node.

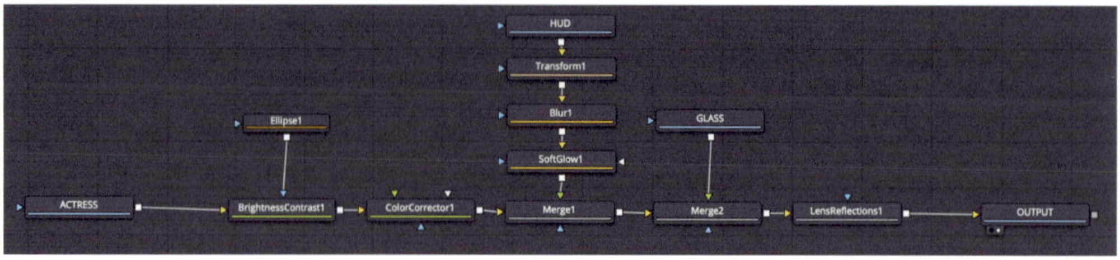

Completed node tree for Lesson 1.

Don't worry about the specific tools that you used because you'll have a chance to use many of them again during the upcoming lessons in this book. The important part is to acquire some level of comfort in moving around the Fusion page, connecting nodes, and modifying parameters.

Lesson Review

1. In the Fusion page, how can you display the output of a node in viewer 1?
2. True or false: To take a single clip from the edit page timeline into Fusion, the clip must be selected.
3. What node would you use to composite two images?
4. What is the yellow input on a merge node?
5. True or false? When you are on the Fusion page, you can disconnect the media out node because there is no use for it.

Answers

1 In the Fusion page, to display the output of a node on viewer 1, select the node and press the 1 key.

2 False. The playhead must be over the clip you want to take into the Fusion page, but the clip does not need to be selected.

3 A merge node is used to composite two images.

4 The yellow input on the merge node is for the background input.

5 False. The media out node is always the last node connected. It renders the node editor results back to the edit page timeline.

Part I
Titling and Motion Graphics

Lesson 2

The Art of the Credit Roll

Every scripted project, documentary, or reality program ends with credits.

A lot of effort and thought goes into deciding how those credits look, and the order in which the crew is listed. Since there are no rules that guide how credits are displayed, whoever is finishing the project must be flexible with the setup. One day all the text may be justified to the left; the next day it is in the center.

Although not principally designed for animating credit rolls, the flexibility inherent in Fusion's node-based interface turns out to be surprisingly good at producing credits. Building a credit roll will reinforce some of the things you learned in Lesson 1 while introducing you to text and some basic animation.

Time

This lesson takes approximately 40 minutes to complete.

Goals

Completed composite for Lesson 2.

Starting with a Fusion Generator

To start this project, you'll restore an archive that will be used during the next three motion graphics lessons that make up Section I of this book.

1. Open DaVinci Resolve. In the project manager, right-click, and choose restore project archive.
2. Navigate to the R16 Fusion Lessons folder you downloaded and restore **Fusion 16 Lessons part 1.dra**.

   When designing text, especially a credit roll, you don't necessarily need a video clip. You just need a blank canvas on which to create. For those situations, you can add a Fusion composition effect to your timeline and begin designing whatever you want.

   > **NOTE** The timelines bin includes a backups bin with timelines saved at various stages of each lesson and a completed bin with finished compositions. These bins are available for both reference and reverse engineering the node trees.

3. In the edit page, load the Part 1-START timeline and open the effects library, and then select the effects category.

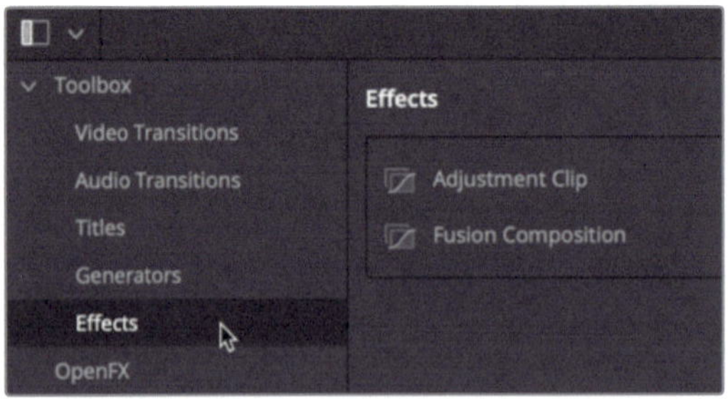

The effects category includes two effects that act like containers for other effects.

4 Drag the Fusion composition effect to the end of the timeline, making it the last clip.

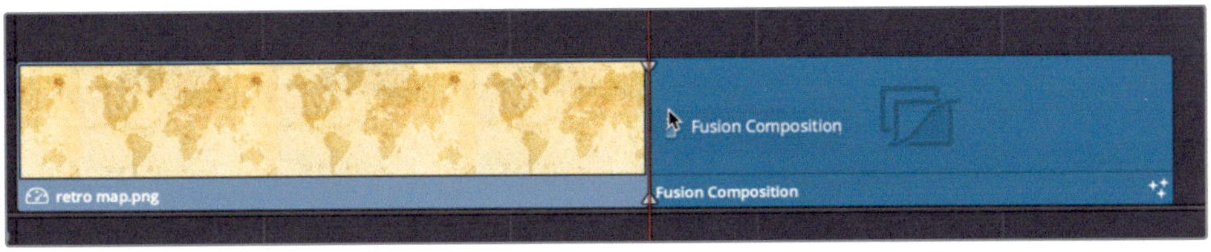

5 Right-click over the Fusion Composition and choose Change Clip Duration. In the dialog, enter **00010000** to create a 1-minute clip, and then click Change.

Admittedly, this is on the shorter side of credit rolls, but rather than have you sit through 5 minutes of credits, without the hope of seeing outtakes from the movie, this will work for learning purposes.

6 Position the playhead over the Fusion composition effect in the timeline, and click the Fusion button to go to the Fusion page.

A single media out node is the only node in the node editor. When using the Fusion composition effect, you have no media in nodes. It is up to you to add whatever content is required.

Adding a Text+ Node

When you need to create text in the Fusion page, the primary tool is the text+ tool. You use it for any 2D title designs, from dynamic opening titles to simple credit rolls. You'll add a text+ tool to the node editor and then begin adding and formatting the text for your credit roll.

7 Just before the first divider in the toolbar, drag the text+ tool into the node editor and connect it to the media out node.

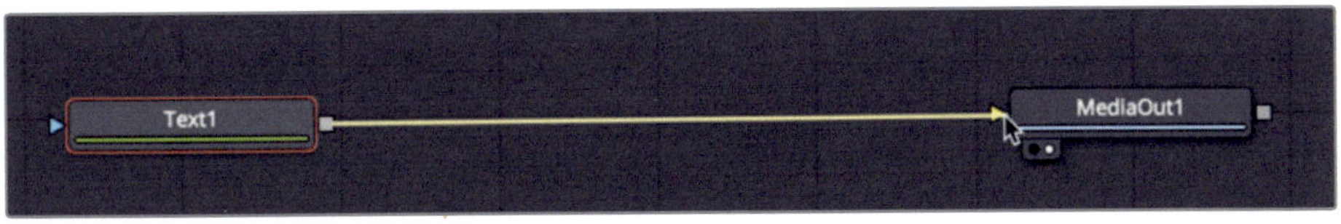

No matter what you are creating or how you started the Fusion composition, you always need to connect to the media out node to render back into the edit page timeline.

TIP You can also find the text+ tool in the edit page effects library or the cut page title library. Adding it to the edit page or cut page timeline provides the same text formatting controls you'll find here, in the inspector.

You'll need only one viewer for this lesson. To give yourself more room, you can use either viewer 1 or viewer 2 in a single viewer layout by clicking on the single viewer button in the upper-right corner of either viewer.

8 In the upper-right corner of viewer 1, click the single viewer button.

9 Select the text node in the node editor and press 1 to see it in the viewer.

10 In the inspector's styled text field, enter **DREAMLAND PRODUCTIONS**, and then press return or enter to create a line break.

11 Set the font to open sans, and set the font typeface to regular.

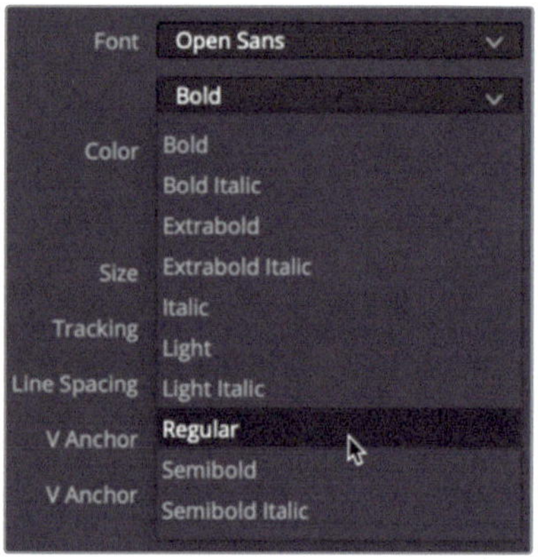

12 Change the text size to around 0.05.

TIP The text size is not based on font point size but is measured relative to the width of the frame. So, changing the resolution maintains the relative size of the text in the frame.

You can enter multiple lines of text and format them how you want. However, the easiest method for a credit roll is to write them in a word processor and then copy them into the text+ tool.

Aligning Text with Tabs

Credits can be hundreds of names long. Grouping your credits into categories is the only way to keep things organized. You can have all the information, but unless you carefully organize it, you won't be able to correct names or reorder the credits easily. To start, write your credits in a simple text app or word processor. You can then check spelling and easily share it with other people in the crew who might not be involved in post production. Once you have the text document, simply copy and paste it into the text+ tool.

1 On your computer's hard drive, navigate to the R16 Fusion lessons folder > Fusion files and open the CREDITS document in a text app or Notepad.

2 Select and copy the first two lines for the executive producer and the director.

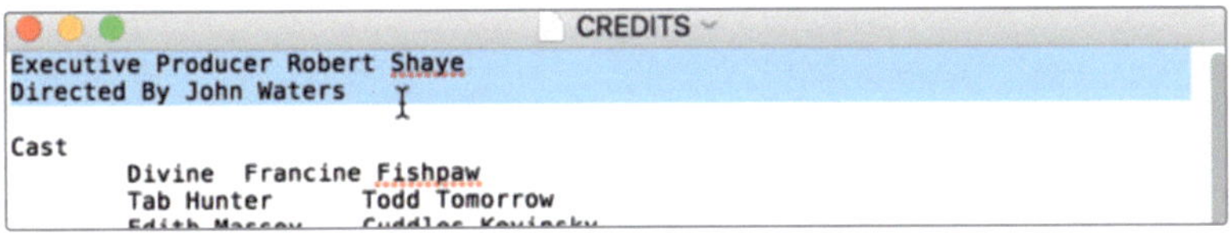

Once you have copied the first two names, you can return to the Fusion page and paste it into the text+ tool.

3 In the Inspector, click below DREAMLAND PRODUCTIONS to insert the cursor in the field, and then press cmd-V (macOS) or Ctrl-V(Windows).

The text is pasted into the styled text field and seen in the viewer. All words in the text+ tool start center justified; however, that can be changed using tabs. You'll leave DREAMLAND PRODUCTIONS center justified but add tabs to align the executive producer and director's credits.

4 In the text field, before the E in Executive Producer, insert the text cursor and press tab.

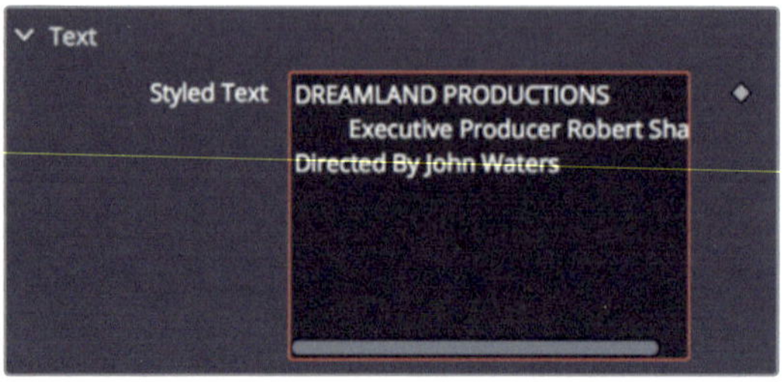

5 Insert the text cursor before the R in Robert and press tab again.

6 Do the same before the D in Director and the J in John, inserting two tabs.

The text in the viewer appears a bit worse in terms of layout, but that is only because you have not positioned the tabs.

7 At the bottom of the Inspector, click the disclosure arrow to reveal the tab spacing controls.

When you add tabs to the text, the tab spacing controls allow you to position the tabs in the frame and determine how the text aligns with those tabs. The number menu at the top of the tab spacing controls indicates which tab you are currently adjusting.

8 Drag the position slider or type in **-0.1**, or to be more precise enter it in the numeric field.

The tabs use a value of 0.0 as the center of the frame, -0.5 is the left edge and 0.5 is the right edge. So, tab 1 is now set slightly off to the left of the center. However, the text is still center aligned on that tab. Using the alignment slider, you can choose to align the text left, center, right, or any offset in between.

TIP Clicking the tab handle above the viewer cycles through left, right, and center alignment.

9 Drag the alignment slider all the way to the right, right justifying the text on the tab. This causes the ending of the words Executive Producer and Directed By to be aligned to the tab.

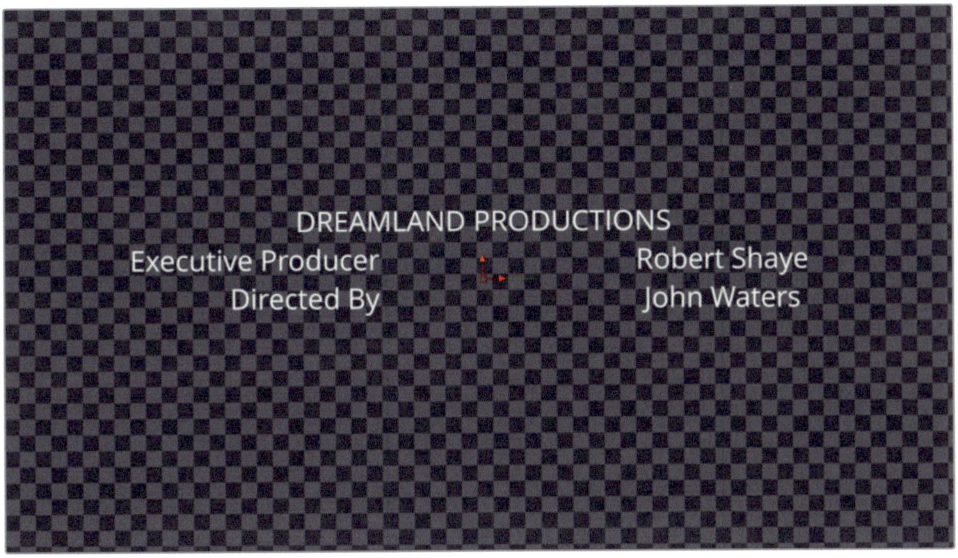

Now for tab 2, you'll do the same setup in the opposite direction.

10 In the tab spacing controls, change the Tab pop-up to 2 in order to control the position and alignment properties for tab 2.

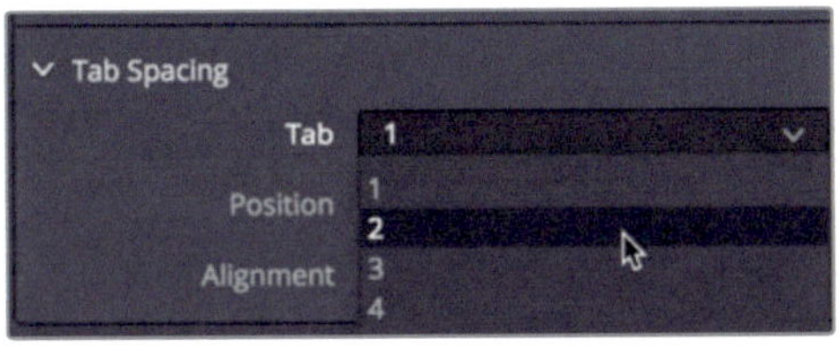

11 Drag the position slider or type in **0.1** to move tab 2 off to the right.

12 Drag the alignment slider all the way to the left, left justifying the text on the tab. This causes the start of both men's names to be aligned to the tab.

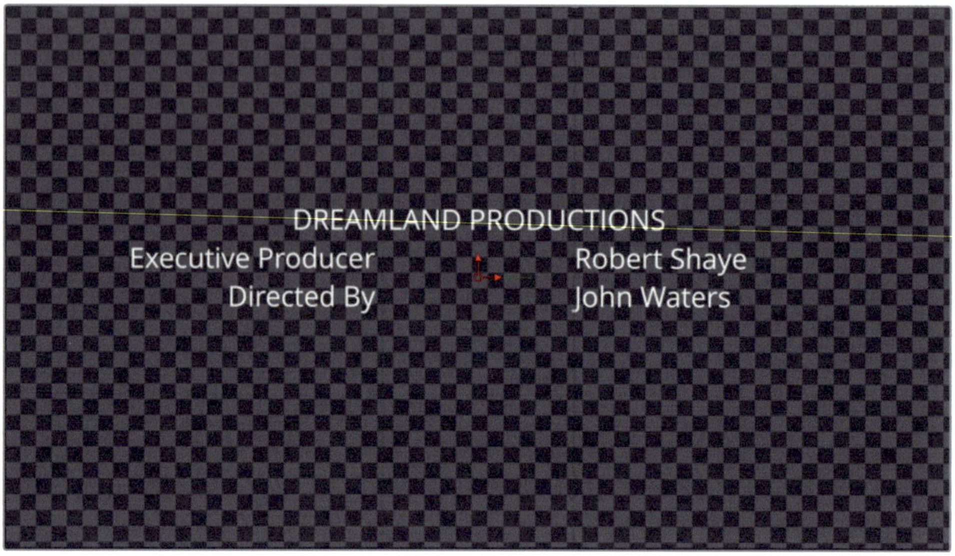

You now have the basic setup for your credit roll. Any text that you type or copy and paste into the styled text field is now center aligned in the frame unless they have tabs. If the text you type or paste includes tabs, the formatting is based on the tabs in the text+ tool.

13 Return to the CREDITS document, select the text starting from the word CAST and including up to the last actor. Press cmd-C (macOS) or Ctrl-C (Windows) to copy the text.

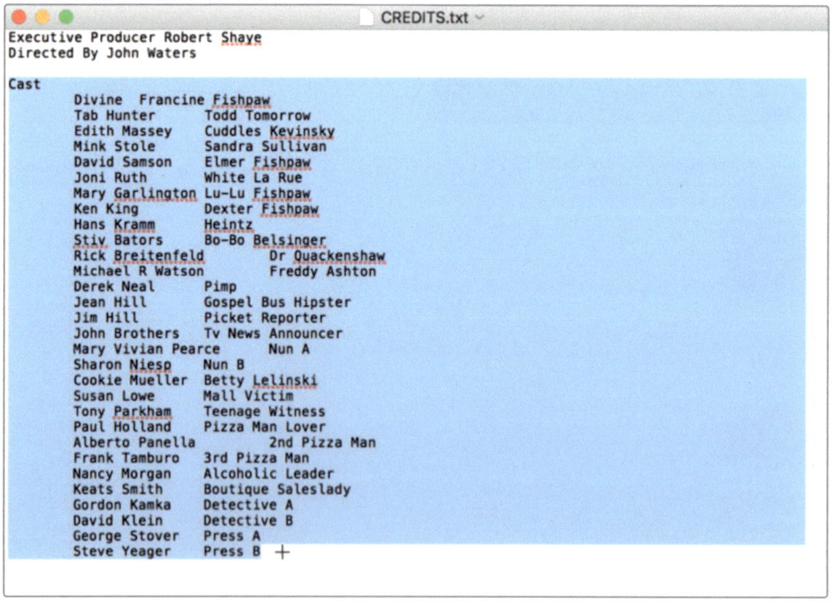
CREDITS.txt

```
Executive Producer Robert Shaye
Directed By John Waters

Cast
        Divine  Francine Fishpaw
        Tab Hunter      Todd Tomorrow
        Edith Massey    Cuddles Kevinsky
        Mink Stole      Sandra Sullivan
        David Samson    Elmer Fishpaw
        Joni Ruth       White La Rue
        Mary Garlington Lu-Lu Fishpaw
        Ken King        Dexter Fishpaw
        Hans Kramm      Heintz
        Stiv Bators     Bo-Bo Belsinger
        Rick Breitenfeld        Dr Quackenshaw
        Michael R Watson        Freddy Ashton
        Derek Neal      Pimp
        Jean Hill       Gospel Bus Hipster
        Jim Hill        Picket Reporter
        John Brothers   Tv News Announcer
        Mary Vivian Pearce      Nun A
        Sharon Niesp    Nun B
        Cookie Mueller  Betty Lelinski
        Susan Lowe      Mall Victim
        Tony Parkham    Teenage Witness
        Paul Holland    Pizza Man Lover
        Alberto Panella         2nd Pizza Man
        Frank Tamburo   3rd Pizza Man
        Nancy Morgan    Alcoholic Leader
        Keats Smith     Boutique Saleslady
        Gordon Kamka    Detective A
        David Klein     Detective B
        George Stover   Press A
        Steve Yeager    Press B
```

14 Back in the Fusion page Inspector, click below DIRECTED BY to insert the cursor in the styled text field, and then press cmd-V (macOS) or Ctrl-V(Windows) to paste the text.

TIP Before pasting the text, you may need to press enter or return to create a line break after the director's name if you haven't done so already.

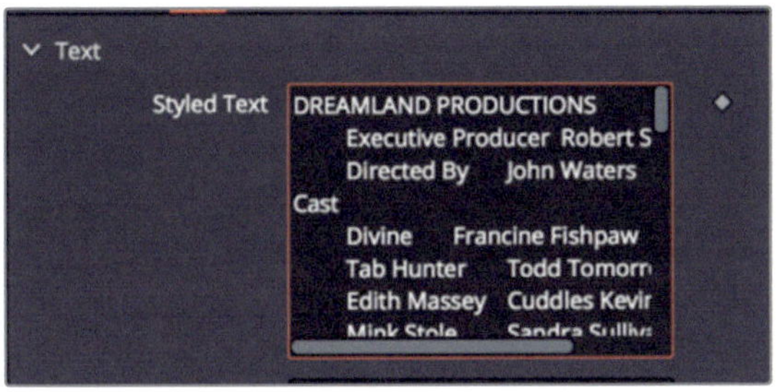

Since the text document already includes tabs, the pasted text aligns to the tabs in the text+ tool. The text layout in the viewer expands to include the new text, causing some of the text to fall outside the active frame area. Text that falls outside the active frame area can be seen as outlines, making it easy to read and format all the text in the credit roll.

Formatting a Single Line

If you change the font, size, or color of the text in the text+ node, it changes the entire text block that you have entered. To change the style of a single letter, word, or line, you must use a special modifier called character level styling.

1 Right-click in the styled text field and choose character level styling from the menu.

Modifiers are add-on functionality to the core capabilities of a tool. They are applied by right-clicking on the parameter you want to modify and choosing the appropriate modifier. The character level styling modifier allows you to override the text+ tool's global styling. Now, you can select the letters in the viewer that you want to style differently.

2 With the mouse pointer over the viewer, hold the cmd key (macOS) or the Ctrl key (Windows) and scroll the middle mouse wheel until you are able to see the outlines of the DREAMLAND PRODUCTIONS text above the frame.

You'll make the DREAMLAND PRODUCTIONS heading use a bold typeface.

3 Drag a selection rectangle around DREAMLAND PRODUCTIONS to select it, being careful not to select other letters.

4 In the inspector, click the modifiers tab.

When using a modifier, all adjustments for the modifier must be done in the modifier tab of the inspector.

5 Set the font typeface menu to bold and set the size to **0.06**.

You can still return to the main tools tab in the inspector to change the main body of the text. Text that has character level styling only updates if you select the text in the viewer and use the modifiers tab.

Inserting Logos and Graphics

Very often, credits are not limited to text, and you come across logos that you must add into the credit roll. Using a transform node and a merge, you can insert your logo or additional pages of text anywhere in the credit roll.

1 In the upper-left corner of the window, click the media pool button.
2 Select the credit roll bin and drag the DREAMLAND LOGO tiff into an empty area of the node editor.

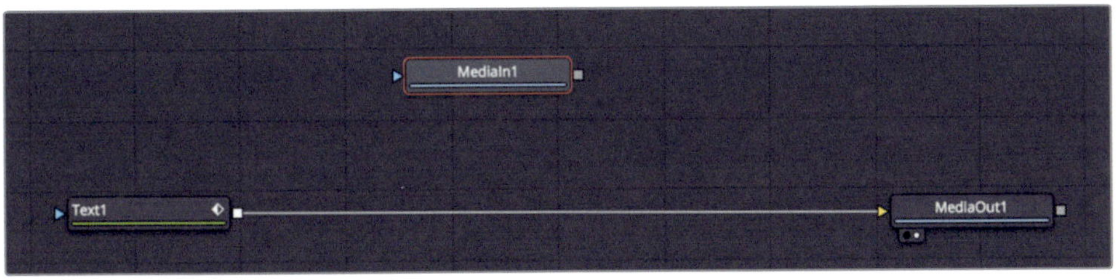

3 Just after the second divider in the toolbar, drag the merge tool over the connection line between the text+ and the media out node in the node editor.

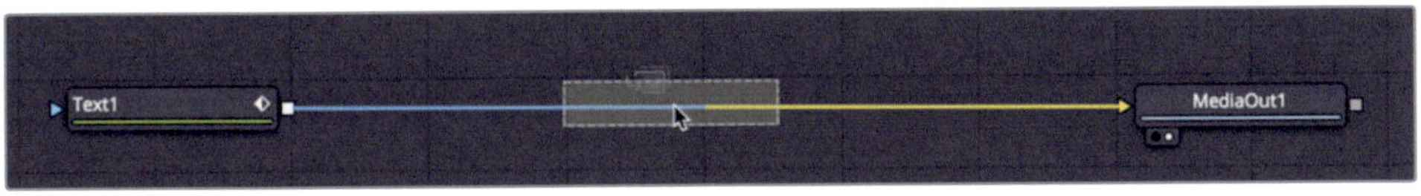

4 When half of the line turns blue, release the mouse button to insert the merge tool into the node tree.
5 Connect the output of the media in 1 node to the green foreground input of the merge node you just added.

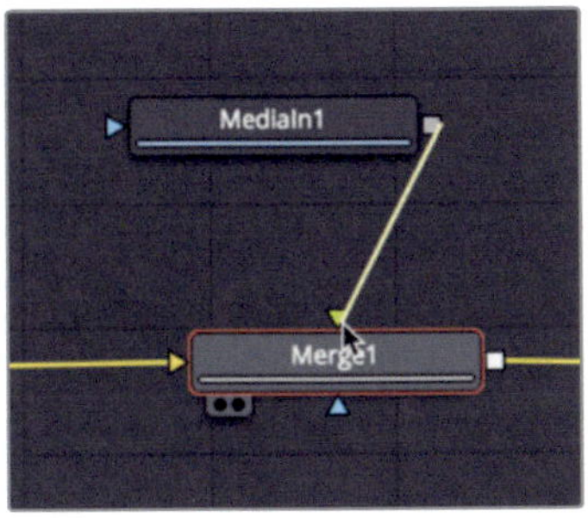

6 Select the merge node and press 1 to see it in the viewer.

Like all images you add to the node editor, its location is in the center of the frame. Media in nodes have no inherent positioning controls built into them. If you want to move or scale a media in node's image, you must use a transform node.

7 From the toolbar, drag the transform tool and insert it in between the media in 1 node and the merge node.

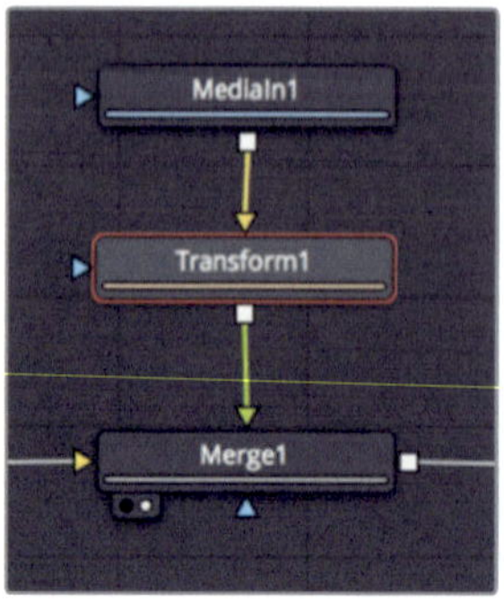

8 In the Inspector, drag the size slider down to about 0.5.

Assuming you want this logo to appear at the end of the credits, to position it accurately you'll need to see the logo in context with the cast text.

9 In the node editor, drag a selection rectangle around the text+ node, the merge node, the media in 1 node and its transform node.

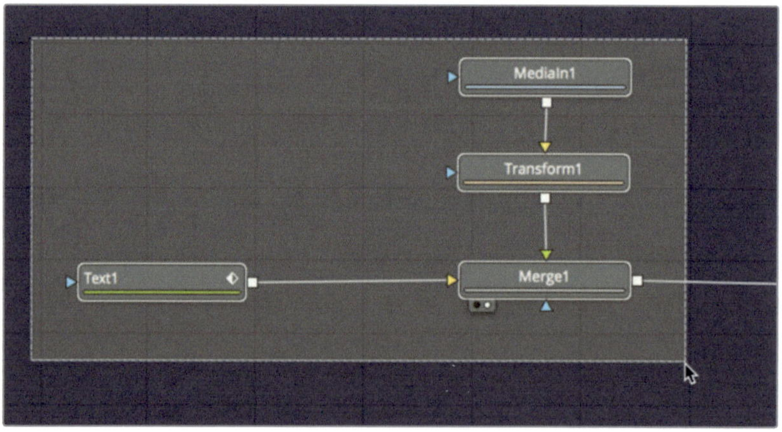

Now you can see the cast text and the logo in the viewer. Instead of scrolling the viewer back to see the last name of the cast text, you can get away with just panning the viewer up.

10 In the viewer, press the middle mouse wheel and drag up until you are able to see the bottom of the cast text, and a bit below where you will want the logo to be positioned.

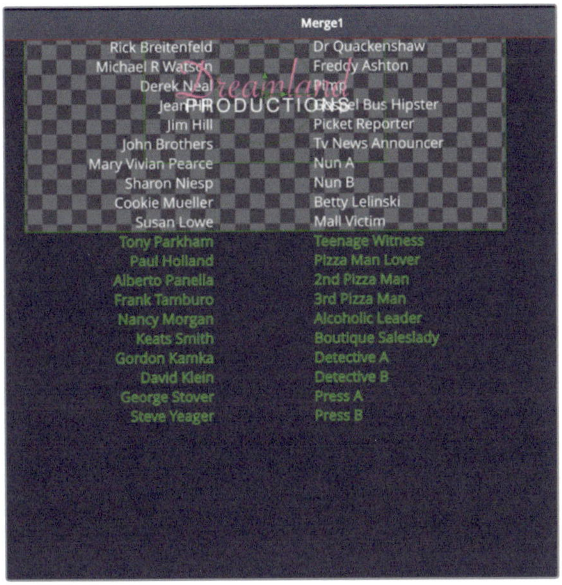

11 In the inspector, use the transform1 center Y controls to position the logo below the cast text.

You can continue adding new pages and logos to create longer credits, using additional text+ nodes for each department in the crew. Once you have everything in place, you can create the actual movement for the roll.

Making Credits Roll

Animating in Fusion is not very different from animating in the edit page. Although the Fusion page provides in-depth tools for refining animation, to get something moving you can just use the inspector.

1 Position the playhead at the start of the render range by dragging the red playhead or clicking the first frame button in the transport controls.

The easiest way to move the text and the logo is to add a transform tool directly after the merge node. This causes both the text and the logo to be controlled by the new transform node simultaneously. Changing the transform parameters causes any items connected to its input to follow.

2 From the toolbar, drag a transform tool and insert it in between the merge node and the media out node.

3 Press 1 to see the transform 1 tool in the viewer.

4 To fill the viewer with the entire frame, choose fit from the magnification pop-up in the upper-left corner of the viewer or click anywhere in the viewer and press cmd-F (macOS) or Ctrl-F (Windows).

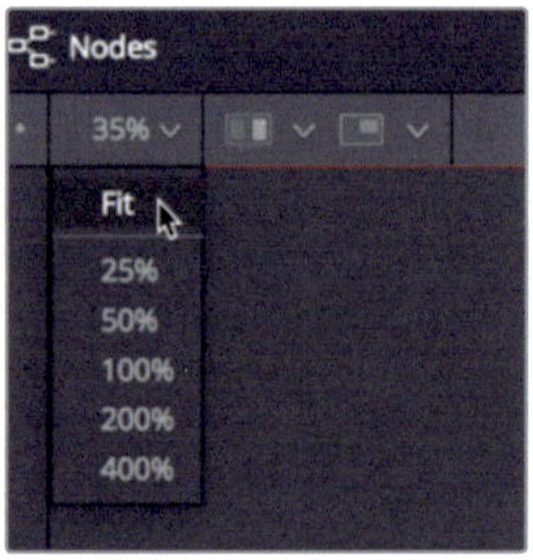

5 In the inspector, click the keyframe button to the right of the center X and Y controls.

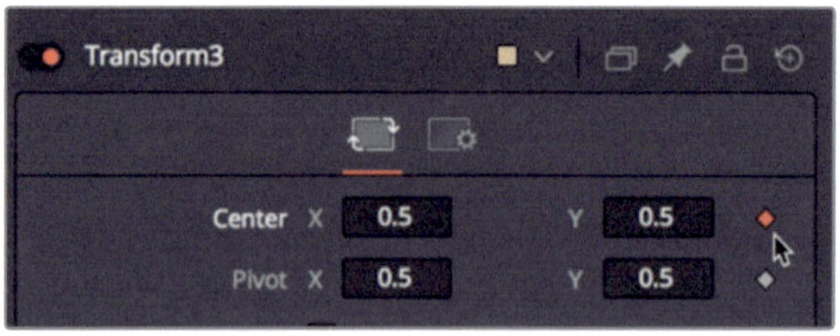

Just like the edit page, you have now enabled keyframing for the center X and Y position controls. You have also set a keyframe at the current playhead position.

6 In the inspector, position the mouse pointer over the center Y numeric field and drag to the left to lower the text in the frame. Keep dragging until the first line of the credit roll is out of the frame (around -1.3 or -1.4).

The easy method of finishing a credit roll is to move the playhead to the end, set a new keyframe, and scroll the text off the top of the frame. However, pesky motion artifacts typically result using that technique. No matter if you call it jitter, flicker, stutter, or strobing, fixing motion artifacts with credit rolls can seem hit or miss. Luckily, it is reasonably straightforward to correct this in Fusion.

Converting to Pixel Values

The only real way to achieve flicker-free rolling credits is to create a roll that moves two or three pixels per frame. It is essential that the roll not move at a sub-pixel rate; it must use integer (whole number) values. This restriction is not dependent on which software you use. Wherever you create your rolling credits, to avoid flickering, you cannot arbitrarily set keyframes because that can result in fractional position values and hence flicker.

The first step in making the roll smooth is making sure you position elements based on pixels. By default, Fusion uses normalized screen dimensions that go from 0 (bottom of the frame) to 1 (top of the frame). You must convert those coordinates to use pixel values.

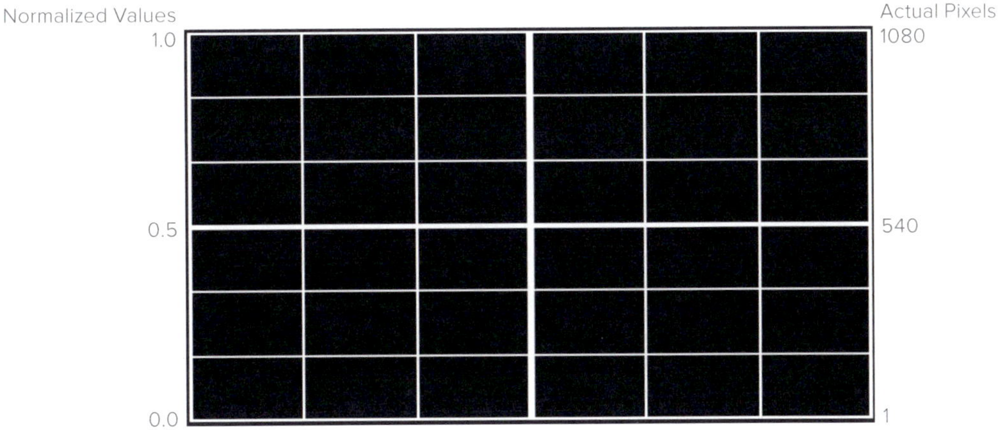

1 Select the transform 2 node.

2 At the bottom of the inspector open the reference sizing controls.

3 Click the checkbox for auto resolution.

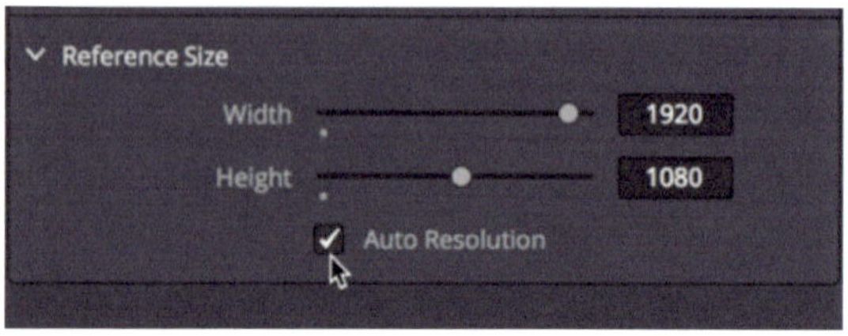

The auto resolution checkbox converts the values in the center X and Y field to use pixels instead of the normalized values of 0 to 1. With the resolution set to represent pixels, a value of 1 represents the bottom of the frame, and 1080 represents the top. To create the actual roll, you'll set the first keyframe to a whole number and set a second keyframe on the next frame, just 3 pixels higher. From there, you'll use Fusion's keyframe prowess to extrapolate the remaining duration of the roll.

Setting and Looping Keyframes

To create a smooth roll, you need two things. First, you set the initial keyframe to a whole number value. Second, you must consistently move 2 or 3 pixels per frame for the duration of the roll. Let's correct the first keyframe in the roll to a whole number value.

1 Go to the start of the render range and look at the center Y value. It probably has a value between -1300 and -1500 but with a few decimal points.

  This is the starting pixel value for the center Y position. You need to make sure this starts on an integer or whole number. Let's round off the starting center Y value.

2 In the inspector, enter **-1450** for center Y. This should be a large enough value to start the text off the screen and provides us with a nice whole number.

  Now, you'll set a second keyframe on the very next frame but only move the Y center value by 3 pixels.

3 On the far right side of the time ruler, enter **1.0** for the current frame. This will move the playhead one frame forward, to frame number 1.

TIP A Fusion Composition effect always starts with frame 0.

4 In the center Y field, double-click to select the entire field, and then enter (minus) -1447. Then press enter or return.

Setting these two keyframes moves the text three pixels higher in the frame. Now you need a way to repeat that pixel movement on each subsequent frame for the entire duration of the roll. For this, you'll use the spline editor.

5 In the upper-right of the Resolve window, click the spline button. Then, in the upper left, click the nodes button to hide the node editor, allowing more room to view your keyframes.

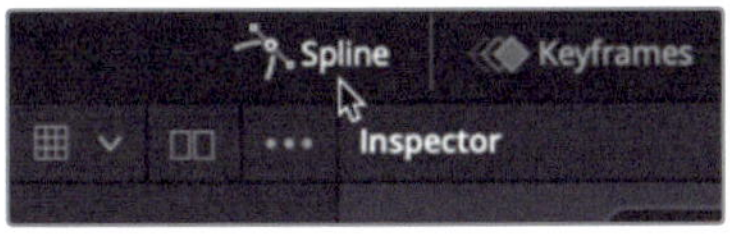

The spline editor is an incredibly deep panel, used to refine the motion between keyframes. You'll learn more about it in later lessons. For this lesson, all you need to be aware of to start is the list of animated parameters displayed on the left side and the toolbar displayed along the bottom.

6 Click the checkbox next to transform to display the transform animation line in the graph area.

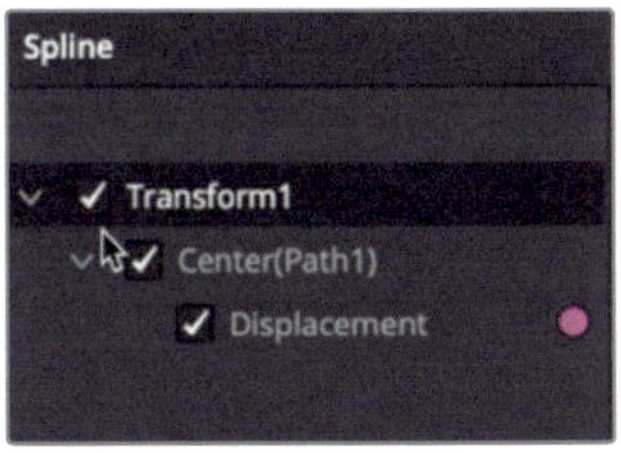

The animation line (or curve) displayed in the graph shows how values are interpolated between keyframes.

7 In the upper right of the spline editor, click the zoom to fit button to view the two keyframes in the graph area.

For this lesson, you do not need to move the existing keyframes in the graph area. You only need to make them repeat for the duration of the roll.

Along the bottom-left corner of the spline editor are several different tools for manipulating selected keyframes in the graph.

8 Drag a selection rectangle around both keyframes in the spline editor.

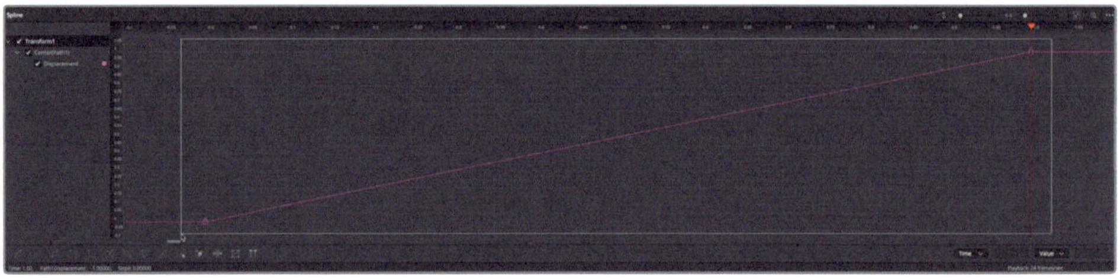

9 In the lower-left corner of the spline editor, click the sixth button from the right end of the toolbar to create a relative loop of the keyframes.

The relative loop button repeats the selected keyframes for the entire duration of the effect. Each repetition starts where the last one left off and adds on it so that the values increase steadily over time. In our case, each loop is one frame long and changes the Y value by three pixels.

This is just a quick sample of the powerful tools you'll discover in the spline editor that can make animation much more flexible and creative.

TIP Right-clicking over keyframes in the spline graph and choosing gradient extrapolation will provide similar results to a relative loop.

10 In the upper right of the Resolve window, click the spline button to close the spline editor. Then, in the upper left, click the nodes button to return to viewing the node editor.

11 Click the edit page button.

12 Wait until the credit roll clip has cached, and then press cmd-F (macOS) or Ctrl-F (Windows) and play the roll full screen.

The credit roll is smooth and long enough to fit all our text. Naturally, for the amount of text you have, you would need to modify the duration, but the relative loop will take care of extending the animation.

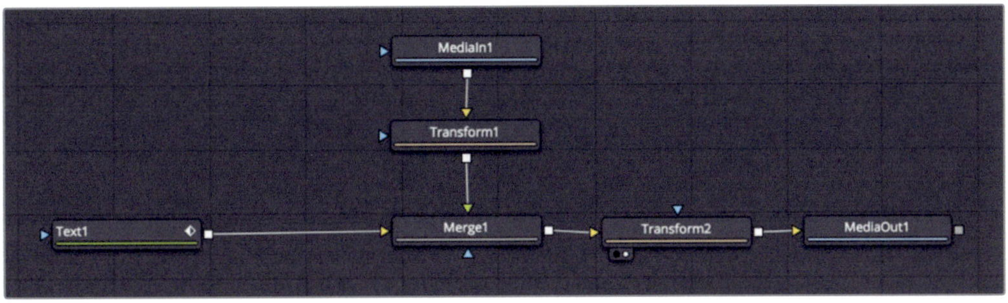

Completed node tree for Lesson 2.

Lesson Review

1 True or false: The Fusion composition effect is a way to start a Fusion composition without requiring a video clip.

2 True or false: The character level styling modifier is located in the effects library.

3 True or false: To change the position of a logo or graphic that has been imported from the media pool, select the media in node and use the position controls.

4 True or false: Tabs can only be left, right, or center justified in Fusion.

5 True or false: To set a keyframe, you can click on the diamond icon to the right of the control's name in the inspector.

Answers

1 True. The Fusion composition effect is a way to start a fusion composition without requiring a video clip.

2 False. The character level styling modifier is applied by right-clicking in the styled text fields and using the modifiers tab in the inspector.

3 False. Media in nodes have no inherent position controls. You must use a transform node.

4 False. The position slider in the tab spacing controls aligns the text to left, center, right, or any offset in between.

5 True. To set a keyframe, you click on the diamond icon to the right of the control's name in the inspector.

Lesson 3

Creating Animated Titles

Realistically, a credit roll has only a handful of design options since the aim is simply to convey information. When designing a main title, a tag line in a commercial, or lower thirds, you are not only conveying important information, but you must also add to the overall tone of the program and hold the audience's attention.

We briefly saw some of the functionality of the text+ tool in the previous lesson, but the text+ tool has a fantastic amount of control over the look, layout, and animation of your title designs.

In this lesson, you'll create an animated lower third and explore the rich feature set of the text+ tool. You'll also create *versions* to quickly try out and switch between different options for any tool. Finally, you'll save the lower third as a template that you can reuse directly in the edit or cut page.

Time

This lesson takes approximately 65 minutes to complete.

Goals

Completed title for Lesson 3.

Styling Text in the Edit Page

To create an animated lower third title, you can begin in the edit page and temporarily use a video background to position the text.

NOTE The timelines bin includes a backups bin with timelines saved at various stages of the lesson and a completed projects bin with finished compositions. These bins are both available for reference and reverse engineering the node trees.

1 Open DaVinci Resolve 16, and in the project manager, open the Fusion 16 Lessons part 1 project that you restored in the previous lesson.

2 From the timelines bin in the edit page, load the Part 1 -START.

3 Open the effects library and select the Titles category.

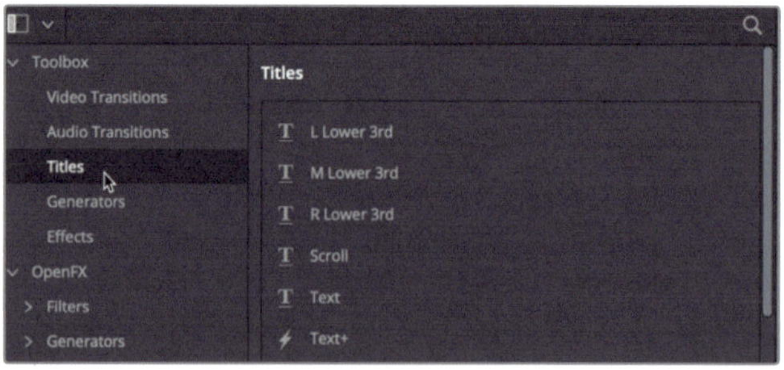

4 Drag the text+ template into the timeline directly above the first clip that contains our interview.

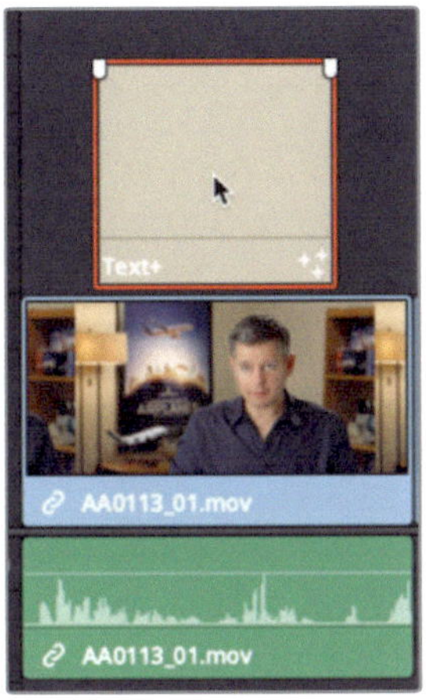

5 Select the text+ clip and open the inspector.

The text+ template in the edit page is the same text+ tool located in the Fusion page. You can begin creating titles in the edit page and then move to the Fusion page when you want to expand on the title animation.

6 In the inspector's styled text field, enter the name **Brian J. Terwilliger**.

7 Set the font to open sans and set the font typeface to bold.

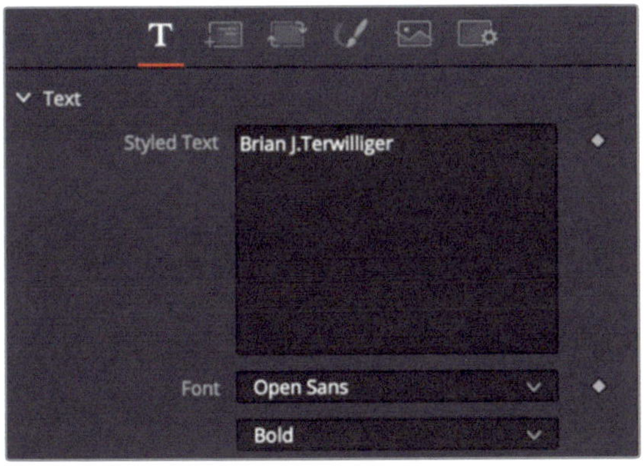

8 Change the text size to around 0.09.

9 At the top of the inspector, click the layout tab.

Beyond the familiar text formatting options on the text tab, the layout tab is used to position the person's name as a lower third title.

10 Drag over the center X and Y fields to move the text to the left side and lower third of the frame.

11 In the inspector, click the shading tab.

The shading tab is for styling the appearance of text. Instead of using a single solid color, you can convert the fill type to a gradient.

12 In the type drop-down menu, choose gradient.

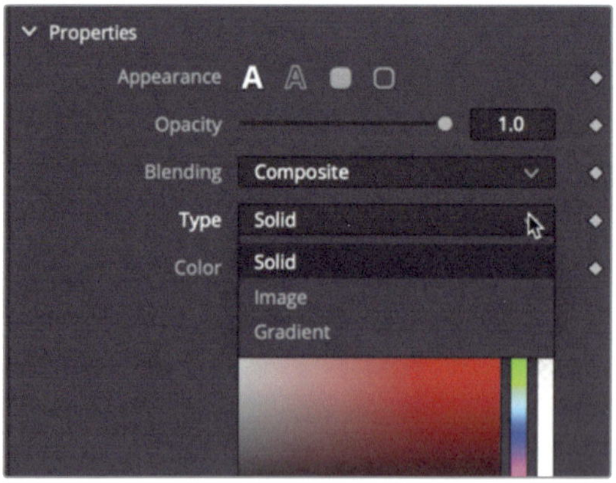

A gradient bar appears below the menu. In this gradient bar, you can assign the various colors that fill the text. The first white color stop on the bar selects the lower color of the gradient, and the white color stop on the right selects the upper color. The first color stop is selected by default, so you can set its color just by selecting it in the color swatch.

13 In the color swatch, select a dark gray color to set the lower gradient color.

14 To set the upper gradient color, click the white color stop on the right end of the gradient bar, and in the color swatch, choose a similar dark gray.

TIP You can drag any color stop to reposition the spread of the gradient colors.

15 To add a new color stop to the gradient, click directly on the middle of the gradient bar.

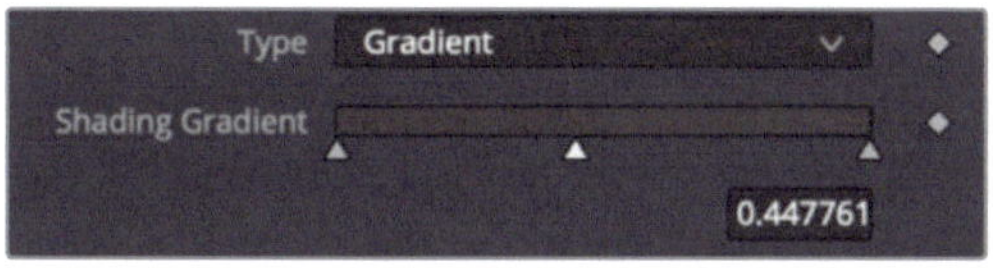

16 Drag in the color swatch to select bright white.

TIP To delete a color stop, drag it to the top of the inspector.

You can add the text+ title in both the edit and cut pages. Consider it the go-to title for designing lower thirds and main titles because it has so many options. Here, it allowed you to correctly size and roughly position the text against a background video track. The other benefit of the text+ title is that you can import it into Fusion and add some other elements along with some animation.

Moving Text to the Fusion Page

Moving text from the edit page into Fusion is the same as moving a video clip or the Fusion Composition effect: Just position the playhead and click the Fusion button.

1 Position the playhead over the text+ title clip, and click the Fusion button to go to the Fusion page.

   As a continuation from the previous lesson, you'll use a single viewer layout.

   The title appears in the Fusion viewer precisely as you had it designed in the edit page. The node is labeled template, so you know it is a title template from the edit page. Let's add a second line of text.

2 From the toolbar, drag the text+ tool into the node editor and leave it unconnected to any other node.

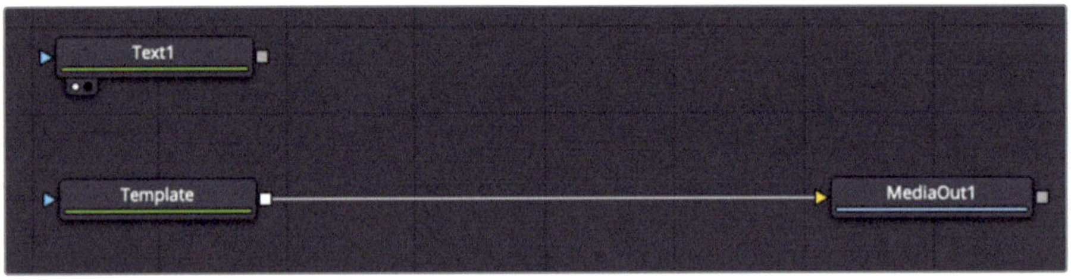

3 Press 1 to display it in viewer 1.

   We'll use two text+ nodes so we can animate the person's name and title independently.

4 In the inspector's styled text field, enter **PRODUCER.**

5 Set the font to open sans and set the font typeface to light.

6 Change the text size to around 0.07.

> **TIP** As a general rule, a change in font usually implies a change in meaning. Too many typefaces are distracting and can confuse the audience. Restrain yourself to using only one or two fonts in a single project. Use typefaces such as bold, light, or italics (as we have done here) to inject variety.

7 To make sure you stay organized, rename each node based on the text you've entered: Template becomes **BRIAN**, text 1 becomes **PRODUCER**.

   Now, you can connect both nodes to form a basic layout. If you drag one node's output to another node's output, they automatically create and connect to a merge node.

8 Drag the output of the PRODUCER node to the output of the BRIAN node to merge the two.

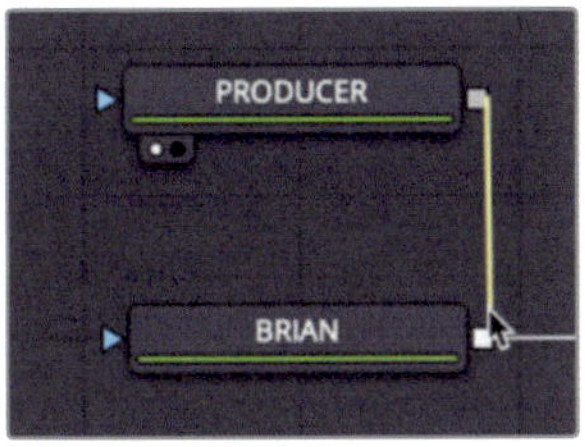

Dragging the PRODUCER node to the BRIAN node makes the PRODUCER node the foreground and the BRIAN node, the background in the merge node.

9 Select the merge node and press 1 to display it in the viewer.

The two text elements are seen in the viewer, but you still need to position the PRODUCER title under the interview subject's name.

10 Select the PRODUCER node, and in the viewer, drag the PRODUCER text under the name.

TIP You can use the center X and Y controls in the inspector's layout to be for more precision when positioning text.

11 In the inspector, modify the tracking slider so the PRODUCER text spreads to the same width as the person's name above it.

This is the general layout of the lower third title. In the following steps, we'll create some visual interest by animating the text.

Revealing Text with Mattes

Static text implies *"move along, nothing to see here,"* and the audience tunes out. Even minimal text motion adds interest to flat type to help keep an audience focused long enough to read the message. You can animate text in many different ways in Fusion, but you'll explore two of the most popular techniques in this lesson. First, you'll reveal the text with the use of a matte.

1 In the node editor, select the BRIAN node, and from the toolbar click the rectangle tool.

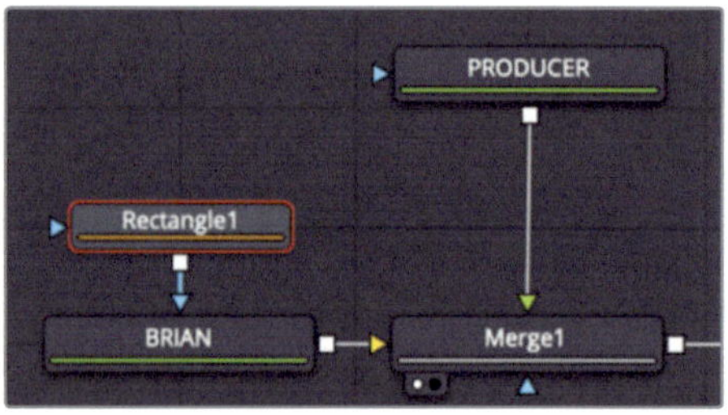

Clicking any matte tool connects the matte into the blue effect mask input of the selected node. You'll use this matte as a window of sorts. Positioning the text outside the rectangle hides it. Then, by animating the text's position, you can slide it into the rectangle to reveal the text.

The rectangle should be positioned over the text on the left side of the frame, and it must be large enough to cover the entire name.

2 In the node editor, select the rectangle node, and in the inspector, drag in the center x field until the bottom line of the rectangle is aligned with the bottom of the BRIAN text.

3 Change the height slider in the inspector until the rectangle mask fits tightly around the text. You may have to reposition the rectangle again to get it to fit nicely around the text.

4 Select the PRODUCER node and click the rectangle mask tool in the toolbar.

5 Drag it so the top line of the rectangle is aligned with the top of the PRODUCER text and use the height slider to fit it tightly around the text.

Now you can animate both text elements so they begin outside the rectangle and then slide into place.

6 Move to the start of the render range in the timeline.

7 Select the BRIAN node, and in the Inspector, click the layout tab.

The layout tab includes settings to place the text on a circle, a path, or just position it.

8 To the right of the center X and Y fields, click keyframe button.

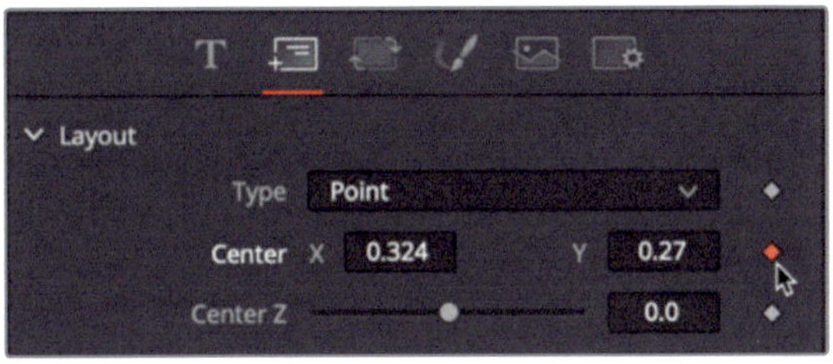

9 In the viewer, drag the center Y on-screen control down until the Brian text disappears below the rectangle mask.

10 Move the playhead to frame 15.

Here is where the Brian text will reappear on-screen.

11 In the viewer, drag the center Y on-screen control up until the text reappears, centered within the rectangle mask.

We'll create the same animation for the Producer text.

12 Move to the start of the render range.

TIP You can press cmd-left arrow (macOS) or Ctrl-left arrow (Windows) to move the playhead to the start of the render range.

13 Select the PRODUCER node, and in the inspector, click the layout tab.

14 To the right of the center X and Y fields, click the keyframe button.

15 In the viewer, drag the center Y on-screen control up until the text disappears above the rectangle mask.

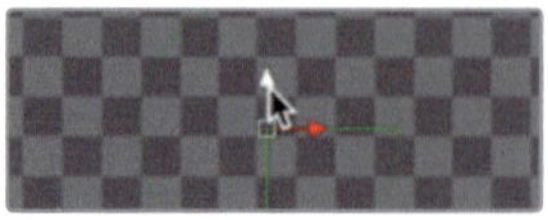

16 Move the playhead to frame 15.

17 In the viewer, drag the center Y on-screen control down until the text reappears, centered within the rectangle mask.

18 Click the play button to view the animation.

Both lines of text slide out from the center of the banner, revealed only within the rectangular mattes. Using a combination of position keyframes and mattes, you can create very complex animations with a straightforward setup.

Creating a Background Banner

The creation of graphical elements like background banners uses a slightly different approach than most motion graphics applications. You use a color generator as fill and a matte to cut the shape you want. We'll create a background banner with rounded corners for our text. This makes the title stand out more when placed over video.

1 From the toolbar, drag a background node into an empty area of the node editor.

The background node is a color generator. It will create the background color of the banner background.

2 Press 1 to see the default back color in the viewer.

The background node allows for both solid color and gradient backgrounds. For this banner, you will create a vertical gradient.

3 In the inspector select vertical from the type menu.

The vertical gradient has top and bottom color swatches where you select the two colors of the vertical gradient.

4 In the top color swatch, select a light gray blue color, and in the bottom swatch, select a darker gray blue.

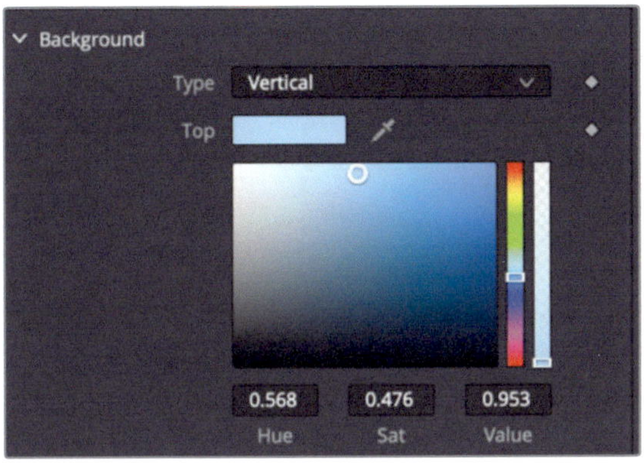

Next, you can add a mask to form the shape of the banner.

5 With the background node selected, click the rectangle mask tool in the toolbar.

The Rectangle mask crops the color into a smaller rectangle centered in the frame.

6 In the inspector set the width to 0.6, the height to 0.3, and the corner radius to 1.0.

7 Rename the background node to BANNER.

To position this correctly behind our text you must merge the banner using the text as the foreground.

8 Drag the output of merge 1 to the output of the banner node.

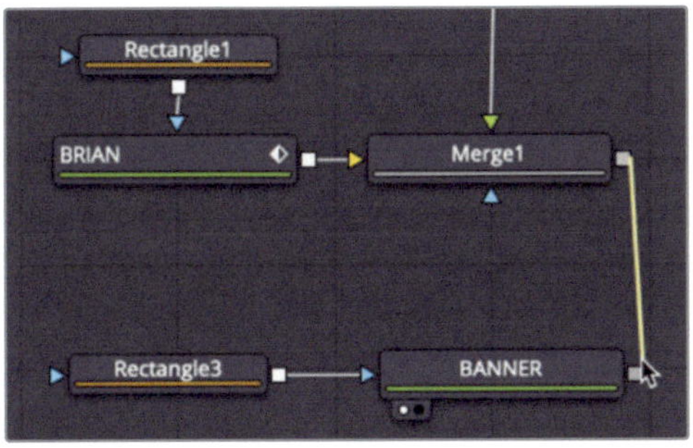

9 Select the newly created merge 2 and press 1 to see it in the viewer.

Now that we can finally see all our elements in the viewer, you'll need to position the banner under the text. You can do this by positioning the rectangle mask.

10 Select the rectangle 3 node and use the on-screen controls in the viewer to position it directly behind the text.

With the banner in place, we can now see it looks a little plain. In the next exercise, you'll add a border around the banner to enhance its appearance.

Copying and Instancing Nodes

When creating graphical elements, copying Nodes is a fast method of generating elements with similar settings. The copied nodes are entirely independent of the original. You can also create an instance of a node, which duplicates the node but retains a live link to the original. Changing one changes the other. We'll use both to create an outline around our banner shape.

1 Select the banner node and press cmd-C (macOS) or Ctrl-C (Windows).

2 Click directly under the banner node and press cmd-V (macOS) or Ctrl-V (Windows) to paste the new background node with the same gradient color.

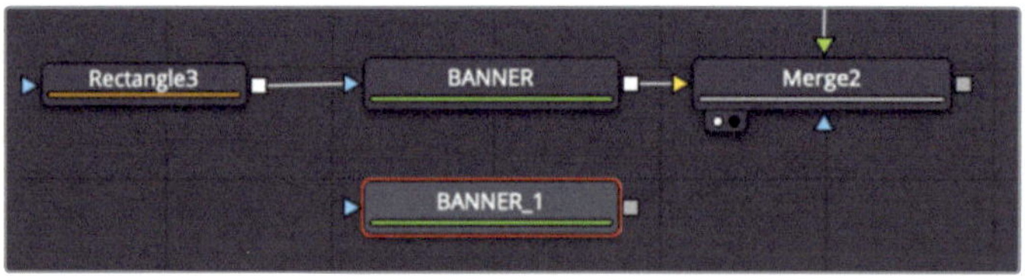

Now, we will make an instance for the rectangle.

3 Select the rectangle 3 node and press cmd-C (macOS) or Ctrl-C (Windows).

4 Click directly under the rectangle 3 node, and then right-click and choose paste instance.

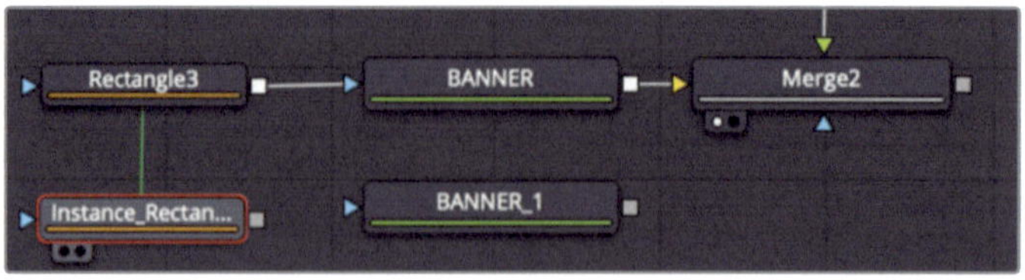

The instance has a thin green connection line that connects it to the original rectangle node. This instance line is the main way you can tell a node is an instance of another node.

5 Drag the output of the instanced node into the blue effect mask input of the duplicated banner node.

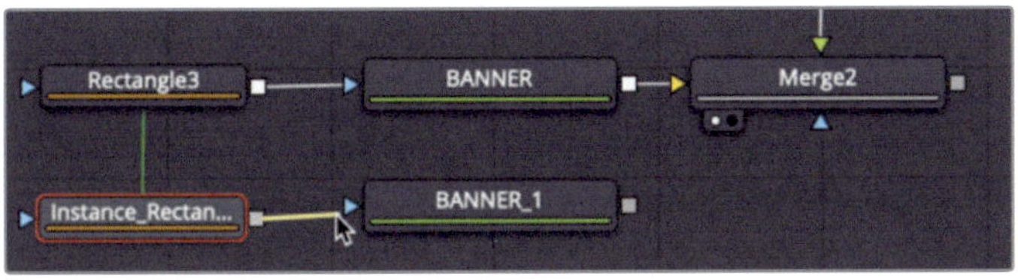

6 Select the banner 1 node, and then press 1 to see an exact copy of the banner.

If you change the shape or size of the rectangle, the instance changes as well. However, you can choose to deinstance any parameter to make small modifications while maintaining live links between the remaining parameters between the instance and the original.

7 Select the instance rectangle3 node.

8 To use the instance as an outline for our shape, in the Inspector right-click over the solid checkbox label and choose deinstance.

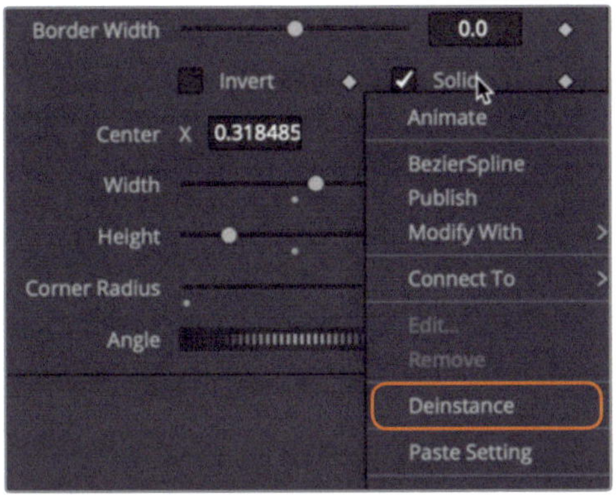

9 Right-click over border width and, again, choose deinstance.

You can now freely change these two settings without modifying the original banner node.

10 Click the solid checkbox to remove the fill color of the rectangle.

11 With the slider, drag the border to around 0.01 to create a thin grayish outline.

This has created a nice rim for the banner. Next, you'll change the color of the banner rim by changing the background gradient color.

Creating a Gradient Highlight

To create a nice highlight rim around that banner, you'll change the gradient color of the duplicated banner node to simulate a highlight.

1 Select the banner 1 node (the duplicate) and in the inspector, set the type to gradient.

  Unlike the vertical setting, which provides only two color swatches, the gradient setting uses the same gradient bar as the text. This gives you a bit more freedom when creating the gradient.

2 Set the first and last color stops to a very dark (almost black) blue and set a center color stop to white.

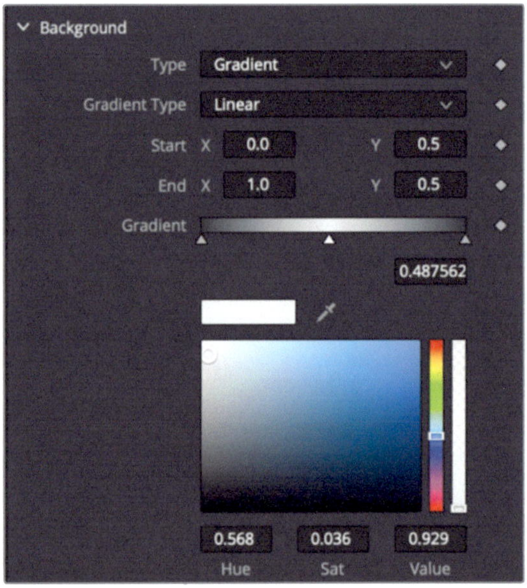

  In the viewer, a thin green line represents the length and direction the gradient color spreads over in the frame. You only need the gradient to spread as wide as the banner outline. You can change the spread by adjusting the line in the viewer.

3 Drag the red control handle on the left side of the line to position it on the left edge of the outline banner.

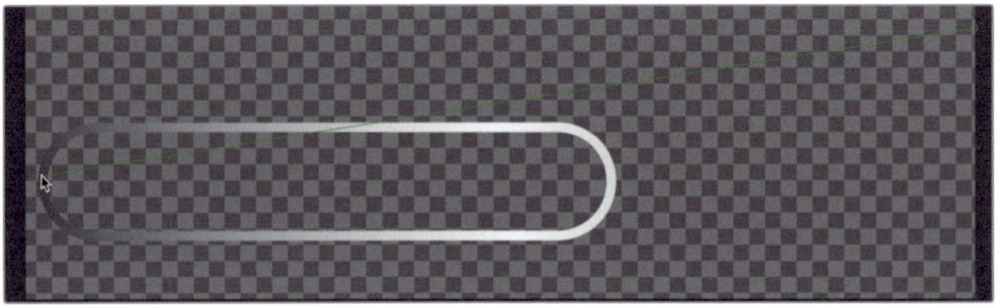

4 Do the same for the right control handle, positioning it on the right edge of the outline banner.

5 In the inspector, drag the first and last color stops in toward the center to simulate a highlight beam in the center of the banner outline.

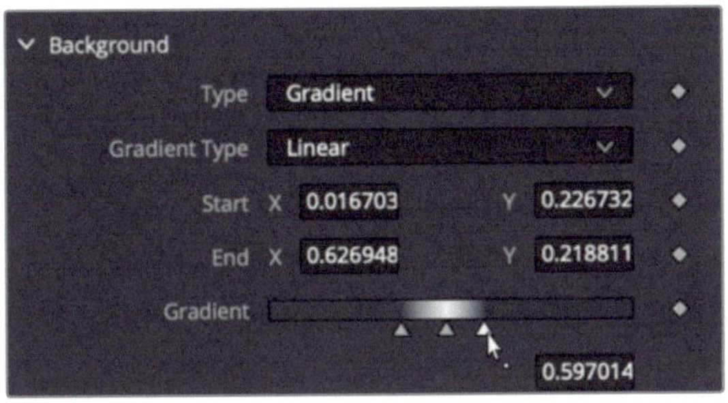

The gradient is now spread across the banner outline but no further, creating a white highlight in the center. The gradient controls in the inspector include an offset position that can be animated. Animating the offset simulates a subtle highlight that moves across our banner rim.

6 Move to the first frame in the render range.

7 In the inspector, click the keyframe button to the right of the offset parameter.

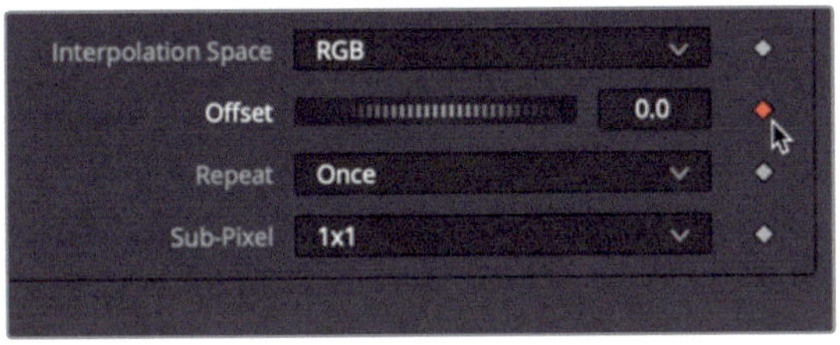

8 Drag the offset dial to the right until the white highlight is off the screen to the right. The parameter will be around 0.6.

9 Move to the last frame in the render range and drag the offset dial to the left until the white highlight is off the screen to the left. The parameter will be around - 0.6.

10 Move to the first frame in the render range, and then play the animation.

This created a nice, slow moving highlight that travels across the rim of our banner.

Animating with Instances

You can animate instanced nodes, using either the original or the instanced node. Deinstanced parameters can be animated independently while the remaining parameters will animate identically on the original. We'll do this animation with the banner rectangle and instance outline rectangle shapes. First, let's merge the banner and outline so we can see how the animation works on both shapes.

1 Since we want the outline to be the foreground over the banner fill shape, drag the output of the copied banner 1 node to the output of the merge 2 node.

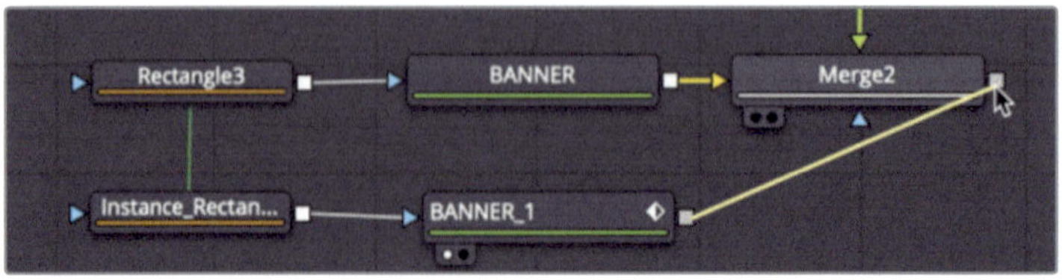

2 Select the merge 3 node and press 1 to see the combined banner and outline in the viewer.

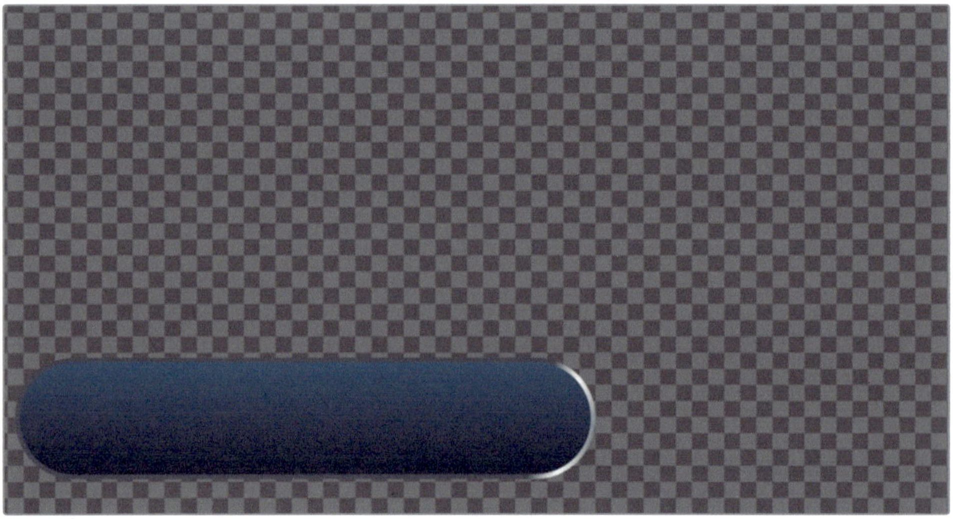

Since the two rectangular shapes are instanced, if we animate one of them the other will follow.

3 Move to the start of the render range.

4 Select the instance rectangle node in the node editor.

5 In the inspector, click the keyframe button to the right of the width slider.

6 Drag the width slider to 0.

Because the width slider for both the banner and the outline rectangle of the banner are instanced, setting the slider to 0 causes both elements to squeeze down and disappear.

7 Move to frame 10 and drag the width slider back to around 0.6, or until it completely encompasses the text.

8 Lastly, so the entire lower third is rendered back into the edit page timeline, connect the output of the merge 3 to the media out node.

9 Move to the start of the render range and click the play button to see the animation.

The animation happens on both nodes simultaneously. Instancing can help speed up the formatting and animation of similar elements.

Animating with the Follower

You can animate words using keyframes as you did to move the text in and out of the mattes, but you also have access to a special modifier designed specifically for animating text character by character. The *follower* is a sequential animation modifier that applies keyframe animation to your text with a custom delay between each character. For our animation, we will create a bright highlight that travels across our letters. To create the bright highlight, we need to add a solid outline around the text that contains the look of our highlight.

1 Select the BRIAN node, and in the inspector, click the shading tab.

The shading tab is for styling the appearance of the text. However, the text node has eight different shading options you can apply to the text. From adding shadows to inserting outlines, these additional shading options are fully customizable text elements. Fusion has given each of them a default name and look such as "shadow" or "red outline," but you can rename them to whatever you want.

2 To add a second shading element, select element 2, and click the enabled checkbox.

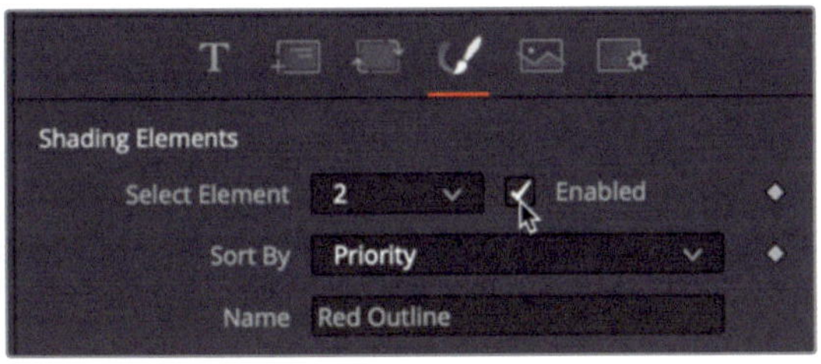

TIP After selecting a Shading number (1–8) from the pop-up menu, you will not see that chosen shading effect until you select the enable checkbox.

This default shading element is set as a red outline. We'll change the color to a bright yellow.

3 In the inspector, change the color to a pale yellow.

Next, you'll set the softness and glow amount you want for the highlight. To start, you apply the look to the entire text. Later, you will animate it to so it moves across the text character by character.

4 At the bottom of the inspector, open the Softness section and set the X and Y parameters to around 2.0

This adds a small amount of softness to the outline. Next you'll apply a glow.

5 Drag the glow slider about halfway to the right to add a glow to the soft outline.

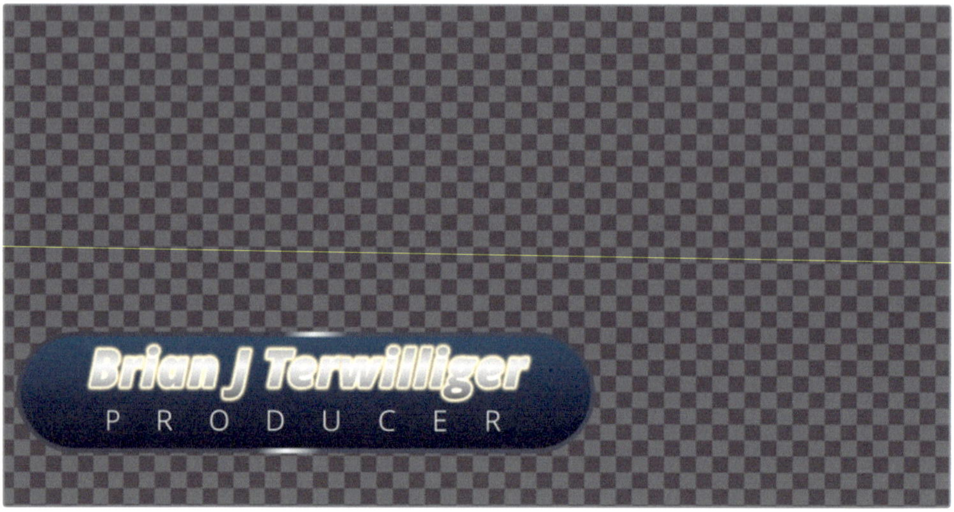

You can always return here and modify the color and softness later. For now, this is the highlight look we will use. Next, we'll use the follower modifier to animate this highlight.

6 Click the text tab, and in the inspector, right-click the styled text box, and choose follower.

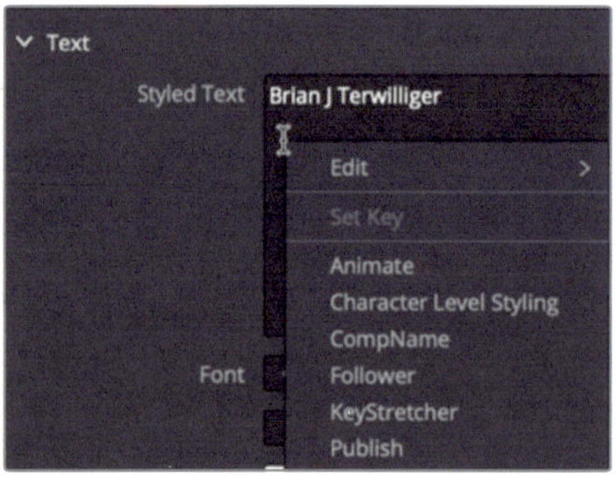

Like the Character Level Styling you used in the previous lesson, the follower is a modifier, and as with other modifiers, its controls appear in the modifiers tab.

7 At the top of the inspector, click the modifiers tab.

To animate the follower, you keyframe the text how you want. You can keyframe position, rotation, size, or even color. Then, once you have designed the animation, you can decide how much delay to place between characters so the animation ripples through your entire text string.

8 In the follower section of the inspector, click the Shading tab.

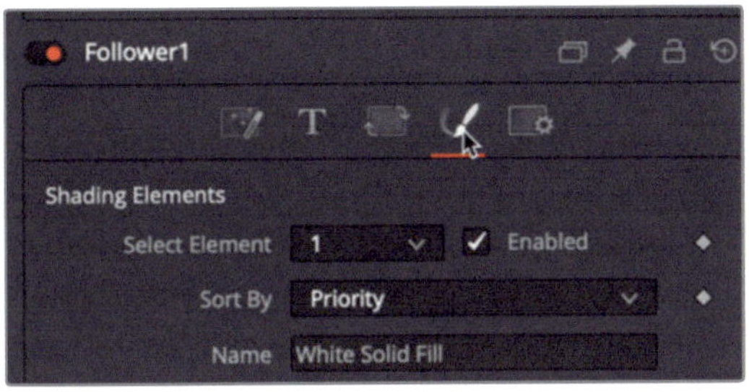

We want to create a glow, so we will set the soft edge and glow properties of the outline on shading element 2. Then, we can control the animation using the thickness size of the soft edge.

9 Move the playhead near frame 30, where the text is fully on-screen.

10 From the shading element pop-up, select number 2.

> **TIP** When using the follower, you must enable keyframing on a parameter before you will see any changes in the viewer.

You will animate the thickness so the soft edge is not visible at the start, then expands, and finally disappears again at the end of the animation.

11 Click the keyframe button to the right of the thickness parameter and drag it to 0.

The viewer now shows the text without the glowing outline. We'll make this a quick glow that lasts only 10 frames.

12 Move the playhead to frame 35 and drag the thickness back to its default value of 0.02.

Now, you'll lower the thickness back to 0 to fade the glow effect off again.

13 Move the playhead to frame 40 and drag the thickness back to 0.

You've created a 10-frame animation with a glow that fades on and off. Now, using the timing tab, you can delay when the fading of the glow occurs for each letter in the sentence.

14 Click the timing tab and drag the delay slider to 2. Play the animation.

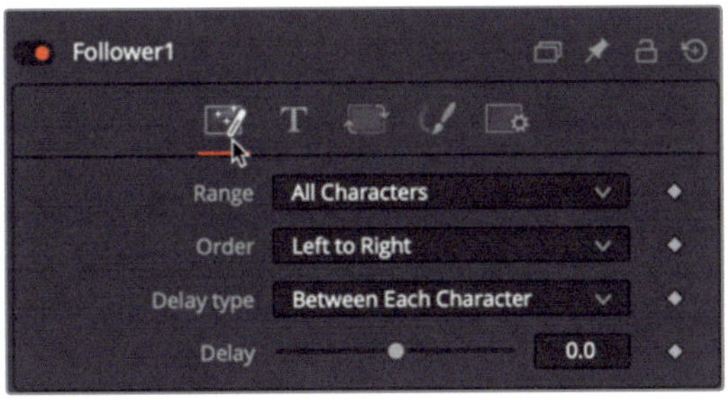

Because you set the delay to 2 frames, the first letter begins to glow, and 2 frames later, the second letter begins to glow. Each subsequent letter animates 2 frames after the one before it.

15 Stop playback. From the order menu, change the animation to right to left and play through the animation once to preview the changes.

The animation reverses direction now from the character on the left starting the sequence to the character on the right starting the sequence.

Almost all the controls in the text tools tab of the inspector can be animated using the follower, including rotation, size, shearing, and color. By playing around for only a few minutes, you can see why this modifier is a fantastic tool for motion graphics.

Adjusting Keyframe Timing

Up until now, most of our keyframing hasn't been any different than what you can do in the edit page. However, the Fusion page has more tools for refining animation, making it the best way to get the exact timing and motion you want. We'll start by looking at a way to adjust the timing of keyframes using the *keyframe editor* .

1 In the upper right of the Fusion window, click the keyframes button.

To open more room for the keyframes editor, you can temporarily hide the node editor.

2 In the upper-left of the Fusion window, click nodes to hide the node editor.

3 In the upper-right corner of the keyframe editor, click the zoom to fit button to fill the panel with the keyframe tracks.

TIP Dragging in the time ruler at the top of the keyframe editor will expand the tracks horizontally so you can zoom in to a specific area.

The keyframe editor shows tracks much like multilayered timelines on the edit page. However, the track stacking order in the keyframe editor has no bearing on the arrangement of images in the viewer. The tracks only adjust the timing of elements and keyframes.

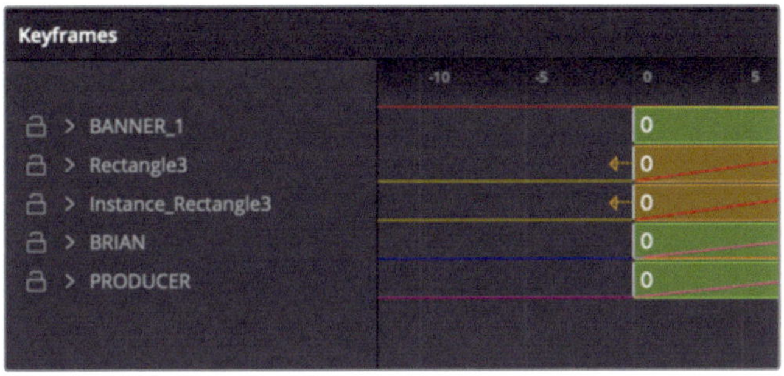

The nodes in the node editor are listed to the left in the header, and you can choose to view all nodes or only nodes with keyframes applied.

Thin vertical lines represent the keyframes on each node, but you can expand each track to show individual tracks for each parameter with keyframes.

4 In the header area to the left of the keyframe editor, click the disclosure arrow next to the BRIAN track to display the individual keyframe tracks.

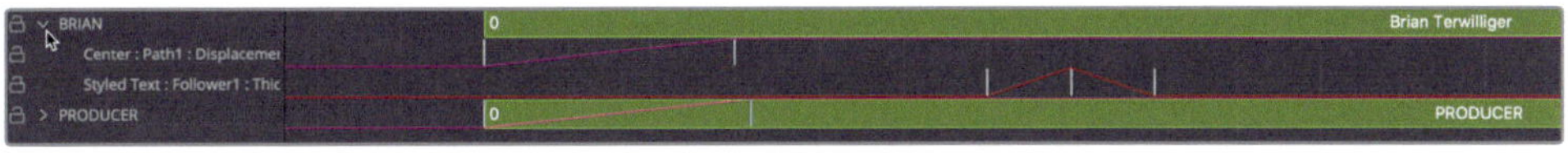

TIP Selecting a tool's name in the header also displays its controls in the inspector and selects its node in the node editor.

Let's delay the slide in of the BRIAN text until the banner is fully on-screen. To do this, you'll move the keyframes a few frames later in the timeline.

5 Drag a red playhead to frame 20, where the banner is fully on-screen.

6 In the revealed BRIAN track, drag a selection rectangle around the two keyframes on the BRIAN track Center: Path 1: Displacement.

TIP When a keyframe is selected in the keyframes editor's timeline, you can press cmd-delete (macOS) or Ctrl-backspace (Windows) to delete the keyframe.

These keyframes represent the text's Y animation. Dragging a selection rectangle around the keyframe selects them, as indicated by the yellow selection color.

7 With the keyframes selected, drag them so the first keyframe lines up with the red playhead.

As you drag, a tooltip in the lower-left corner of the window displays the current frame number.

TIP You can click in a gray area of the timeline to move the playhead by holding down cmd-option (macOS) or Ctrl-alt (Windows) and clicking where you want the playhead to move.

Now, you can do the same thing for the producer text.

8 In the header area, click the disclosure arrow next to the PRODUCER track and drag the two keyframes to line up the first one with the red playhead.

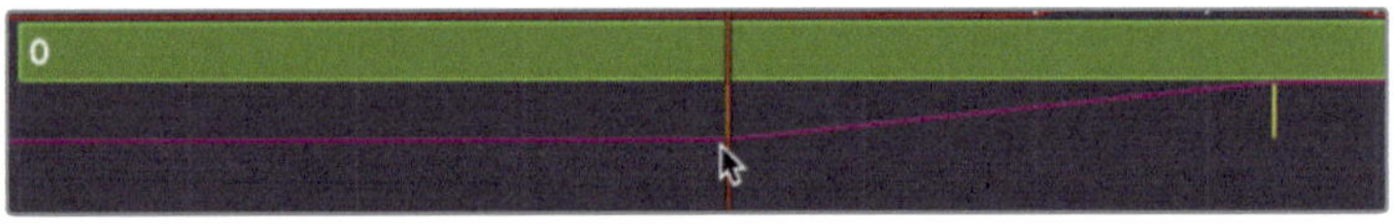

TIP When selecting media in nodes, you can use the keyframes editor's timeline tracks to trim and slide the start and end points, much as you would trim them in the edit page timeline.

Instead of just dragging the keyframe, you can enter an offset value to move a keyframe a specific number of frames. Let's move the BRIAN text highlight so it appears later in the composition by about a second.

9 Select the three keyframes on the BRIAN track Styled Text: Follower1.

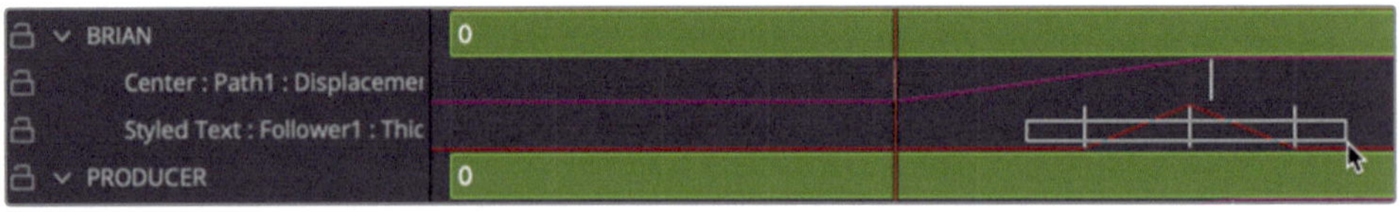

10 With the keyframe selected, in lower right of the keyframes editor, from the time editor pop-up menu, choose T offset.

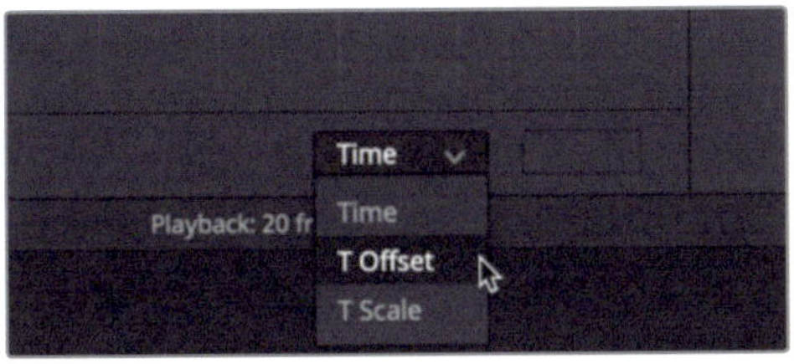

Entering a value in the T offset field will move the selected keyframes forward or backward by the number of frames entered.

11 In the T offset field, enter **24** to move this keyframe 24 frames forward.

12 Click the play button to view the results.

Fusion's keyframe editor contains a lot of additional functionality for adjusting the timing of animations. However, you cannot actually manipulate the interpolation between keyframes. To modify how keyframes accelerate and decelerate, you must switch to the spline editor.

Diving Deeper into the Spline Editor

The keyframes editor adjusts the interpolation between keyframes. On a very basic level, you can set how smoothly elements animate on and off the screen. You can control the position, value, and interpolation for any animated parameter in your project using a spline graph interface.

1 In the upper right of the Fusion window, click the spline button to open the spline editor, and then click the keyframes button to close the keyframe panel and give yourself more room for manipulating the splines.

In this exercise, you will smooth the banner width and the text animation in the spline editor.

2 In the spline editor header to the left of the panel, select the checkbox for width (this will select both the Rectangle and Instance Rectangle width).

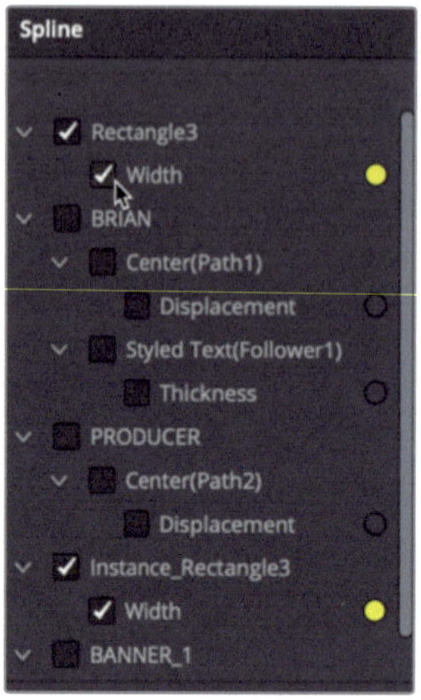

Each animated parameter has a color next to its name that corresponds to its spline color in the spline editor. This color coding makes it much easier to verify that you are adjusting the correct spline.

> **TIP** You can change a spline's color by right-clicking the parameter's name in the header and choosing change color.

Instead of looping the entire clip when you play the animation, you can temporarily confine the render range to a small work area.

3 In the render range out field to the left of the time ruler, enter **60** to place the out point of the render range at frame 60.

4 In the upper right of the spline editor, click the zoom to fit button to zoom in to the keyframes in the graph area.

To make this animation less rigid, we'll convert these keyframes from linear acceleration to a smoother ease in/ease out acceleration.

5 Drag a selection rectangle around the two keyframes displayed in the spline editor graph.

TIP Press Option-right or left arrow (macOS) or Alt-right or left arrow (Windows) to move to the next or previous keyframe.

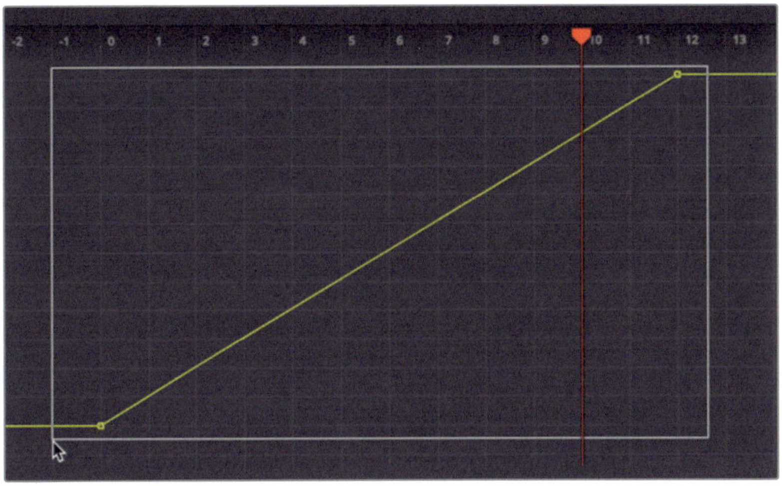

6 In the lower-left corner of the spline editor, click the smooth button or press Shift-S.

7 Click the play button to review the animation.

TIP You can stop the graph area from automatically scrolling past the area with keyframes as you play by right-clicking in an empty area of the spline graph and choosing scale > manual.

With only the few hints below, see if you can smooth the keyframes for both the BRIAN and PRODUCER displacement parameters.

8 Enable the displacement parameters in the spline editor header, select the keyframes and smooth them.

9 Click the play button to review the animation.

You now have a smoother animation for the text and banner.

Copying and Pasting Keyframes

Just as you animated the text on-screen, you'll want to animate the text off the screen as well. You can do this by copying and pasting the keyframes for each track, and then reversing them.

1 Click the checkboxes next to Rectangle 3 > Width and BRIAN > Displacement.

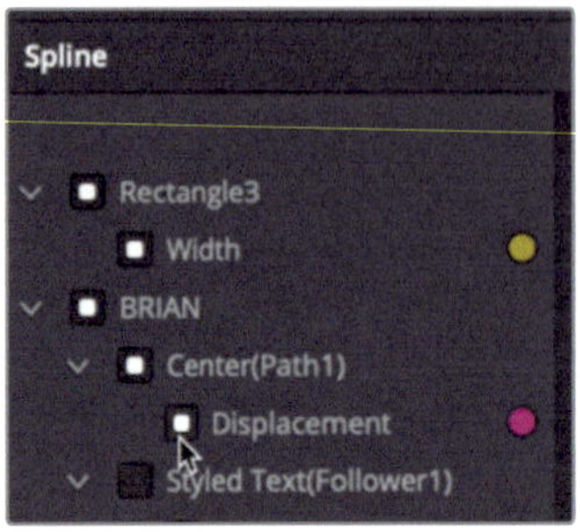

Clicking an active checkbox in the spline editor header once will cause the spline to remain visible but not be editable.

2 Click the checkboxes next to Rectangle 3 > Width and BRIAN > Displacement a second time.

Clicking an active checkbox a second time will hide the spline in the spline editor graph. The only spline still visible is the PRODUCER text animation.

3 Drag a selection rectangle around the two visible keyframes that represent the PRODUCER > Displacement animation.

4 Right-click over the selected keyframes and choose copy points or press cmd-C (macOS) or Ctrl-C (Windows).

With the keyframes now copied, you can move to a location in the time ruler where you want to paste them. You'll paste the keyframes where you want the PRODUCER text to animate off the screen.

5 In the time ruler, move the playhead to frame 90.

To make sure you paste on the same spline as the copied keyframes, select the spline using the name in the spline editor header.

6 In the spline editor header, click on the PRODUCER > Center (path) > Displacement text.

7 Click on the spline under the playhead to add a keyframe, and then choose paste points/values or press cmd-V (macOS) or Ctrl-V (Windows).

TIP If the spline editor is the active panel, pressing cmd-K (macOS) or Ctrl-K (Windows) will add a keyframe at the playhead position.

The keyframes are pasted precisely where the playhead is located. However, this is the same animation that brings the PRODUCER text onto the screen. We need to reverse this animation so the PRODUCER text animates off the screen.

8 Drag a selection rectangle around the two pasted keyframes.

9 In the toolbar below the spline graph, click the reverse button.

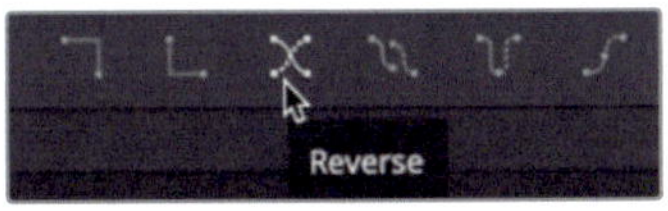

10 Move the playhead to the start of the animation and press the play button to review the animation.

You now need to do the same steps for the Brian > Displacement spline and the Width spline.

11 Move the playhead to frame 90.

12 Click the checkbox next to PRODUCER > Displacement twice to hide the spline and enable the spline for the BRIAN > Displacement.

13 Drag a selection rectangle around the two keyframes and copy them, and then click on the spline under the playhead to add a keyframe and paste the keyframes.

14 Drag a selection rectangle around the two pasted keyframes and click reverse in the toolbar at the bottom of the spline editor.

This takes care of the two lines of text; next, you'll make the banner shrink off-screen.

15 Click the checkbox next to BRIAN > Displacement twice to hide the spline and enable the spline for the Rectangle 3 > Width.

16 Drag a selection rectangle around the two keyframes and copy them.

17 This time, move the playhead to frame 100 and paste the keyframes.

18 Right-click in the time rule and choose Auto Render Range to reset the render range to the original length of the clip.

19 Move the playhead to the start of the animation and press the play button to review the animation.

Now your animation is almost complete. You have a nice-looking lower third that animates on and off the screen.

Squishing and Stretching Keyframes

Although the text animation is good, the highlight across the text is a bit too slow. It would be nicer to see a quicker movement. You could just drag the keyframes closer together, but doing so wouldn't retain the relative space between the keyframes. It's better to use the time stretch tool to stretch or squish the keyframes while keeping their relative timings.

1 Hide the keyframes for Width, and then enable the BRIAN > Styled Text(Follower) > Thickness.

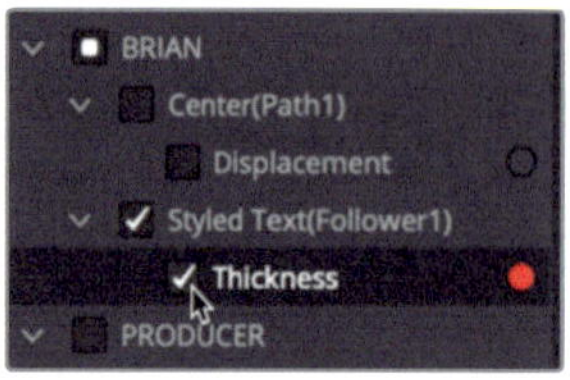

The spline editor now shows the three keyframes that make up the highlight animation. By selecting all three keyframes, you can stretch or squish them, thereby making the animation longer or shorter. Using the stretch tool, you are able to keep the relative timing of the keyframes as well!

2 Drag a selection rectangle around all three keyframes in the graph .

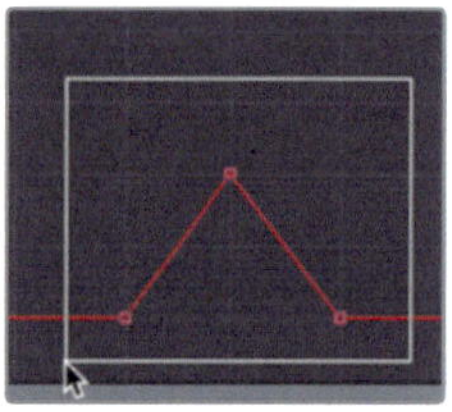

3 In the spline editor toolbar, click the time stretch tool.

Enabling the time stretch tool places two bars around the selected keyframes that can be used to squish and stretch the keyframes.

4 Drag the rightmost white time stretch bar in toward the middle until you have made the duration of the three keyframes about half the size.

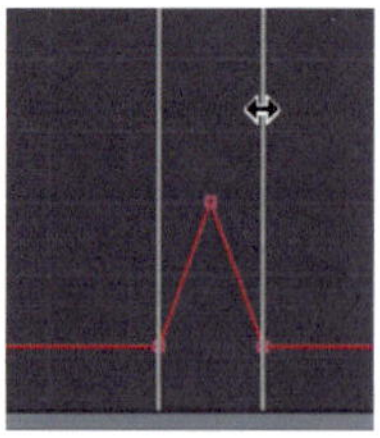

5 In the spline editor toolbar, click the time stretch tool to disable it.

6 Press the spacebar or click the play button to view the results.

The animation is now faster, but the relative distances between the keyframes is maintained.

Trying Out Versions

When designing the look of your title, you should pick colors with a purpose. Generally, you try to impose some hierarchy by focusing the audience on one dominating big element, and then diminish the visual emphasis for less important aspects of the design. You've started this process already by using a larger type size on the BRIAN text. Let's refine the visual emphasis by choosing different colors for this banner.

1 When choosing colors for your text, you don't need the spline editor. So, in the upper right of the user interface toolbar, click the spline button to close the panel; then, in the upper left, click the nodes button to open the node editor.

   Each node can have up to six saved states, called *versions*. Each of these saves a snapshot of the inspector that you can return to at any time during your project. It's a great way to try out different node settings without losing your previous work.

2 Move to the middle of the render range where all the elements are on-screen.

3 Select the banner node, and in the inspector header, click the versions button.

   The versions are numbered starting with the default selection, version 1. Clicking another number selects that version. Any changes you make from that point on will be saved for the selected version.

4 Click the version 2 button to select it.

5 In the top color swatch, choose a bolder red color.

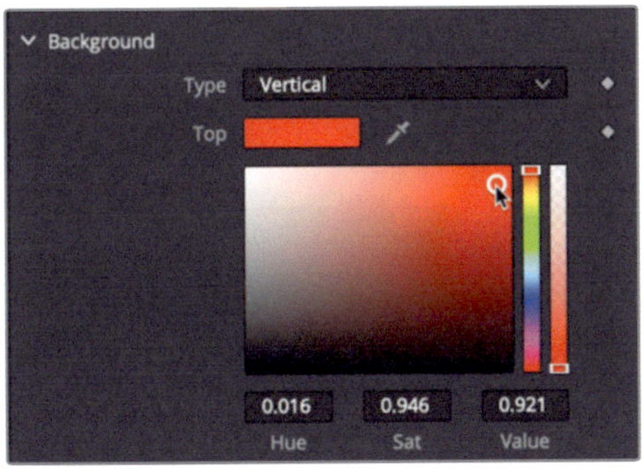

6 Set the bottom color swatch to a dark red color.

Now you will change the banner outline color as well.

7 Select the banner 1 node.

8 In the header, click the versions button and click version2.

For this gradient, make sure you change the color for both the starting and ending color stops.

9 Select the first color stop on the gradient bar and set it to a medium gray.

10 Select the last color stop on the gradient bar and set it to the same medium gray as the first color stop.

You now have two nodes with different versions. You can switch back to version 1 anytime by clicking on the version 1 button.

11 Select the version 1 button at the top of the inspector to see the initial outline gradient.

12 Click back on the version 2 button to return to the updated look.

TIP Versions are saved only for the current node in the project. Adding a second node of the same type will not include the saved versions. However, you can save the current configuration as the default setup for a node by right-clicking over it and choosing Settings > Save Default.

Any node can take advantage of versions as you work out your design or composite. It makes for quick design changes if you or your client would rather use an earlier look you created. It's just one click away.

Timeline Versions

If you intend to make a more fundamental change to the composition—such as changing the color and animation or rethinking the nodes used in the comp—you may want to make a *timeline version*. Timeline versions are entire compositions that are saved for each clip that enable you to make minor or major changes in a composition while retaining access to previous versions.

1 In the upper left of the user interface toolbar, click the clips button.

 The clips button displays a thumbnail timeline at the bottom of the node editor that shows every edit in the currently loaded timeline.

2 Right-click thumbnail number 2, which is the currently selected clip, and choose create new composition.

 The previous animation is now saved as composition, 1 and composition 2 is currently loaded. You can change the look of each element in this comp, and it will be saved as composition 2.

3 Select the banner 1 node, and in the effects library, choose tools > effect, and click the highlight tool.

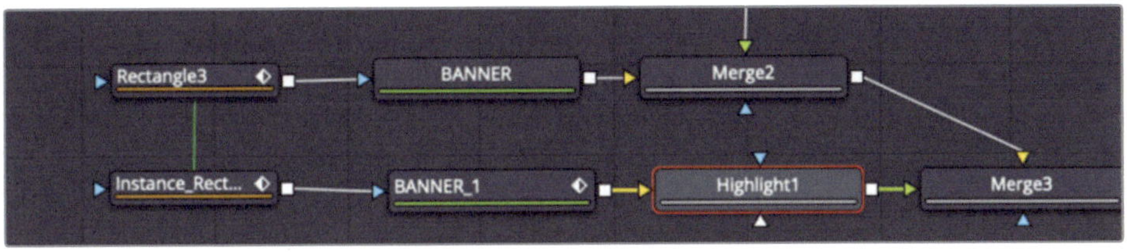

 The highlight node can be used to add shiny glints to objects.

4 In the inspector, drag the low threshold slider to around 0.7.

 Creating a lower threshold will cause the highlight to appear over the white part of the banner outline.

5 Click the play button to preview your changes.

Now you have two clip versions saved, one with the highlight and one without. You can switch between them at any time using the same menu you used to create the version.

Saving a Template

After going through the effort of creating this lower third design, you may want to reuse it with small tweaks to fit different projects. Rather than opening this entire comp over and over, you can save it as a single template in the edit page effects library. The first part of creating this template is to use Fusion's macro creation feature to collapse all the nodes into one single node that displays only the parameters you want to be able to tweak.

1 Click in an empty area of the node editor, and then select the BRIAN node.

2 Hold the cmd key (macOS) or the Ctrl key (Windows) and select the PRODUCER node.

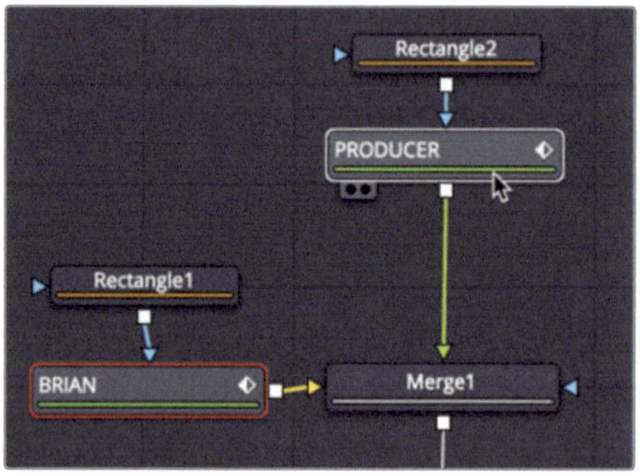

3 Now, press cmd-A (macOS) or Ctrl-A (Windows) to select all the remaining nodes.

The order you select the nodes is the order they appear in the macro list, making it easier to place the most important nodes at the top of the list.

4 Right-click over any of the selected nodes and choose macros > create macro.

The macro editor window shows every node and every parameter in the node tree. There is a field at the top to name the macro.

5 In the macro name field at the top of the window, name the macro 2D Airplanes Lower Third.

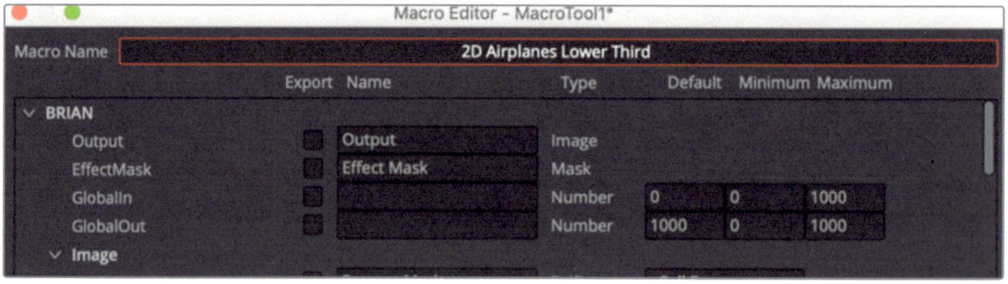

The list of nodes and parameters can be a bit daunting at first, but you are really only interested in the top two text nodes. We'll only want the template to provide controls for changing the text and size.

6 In the macro list, click the disclosure arrow next to Image to hide those parameters.

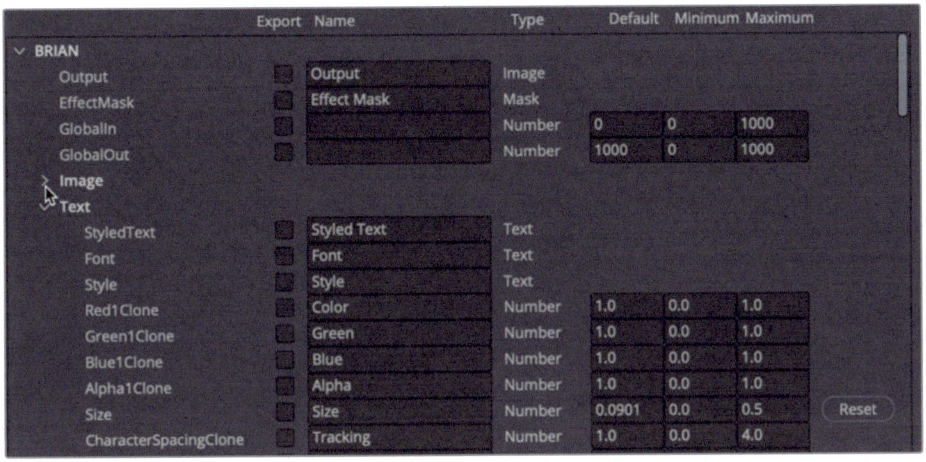

7 Click the checkbox to the right of the Styled Text field, and then a little further down in the list, click the checkbox to the right of the Size field.

By clicking the checkbox next to Styled text and size, you have selected these two fields to be displayed in the edit page of the inspector.

8 In the macro list, click the disclosure arrow next to BRIAN to hide all those parameters.

9 Click the disclosure arrow next to PRODUCER and enable the checkbox for Styled Text and Size.

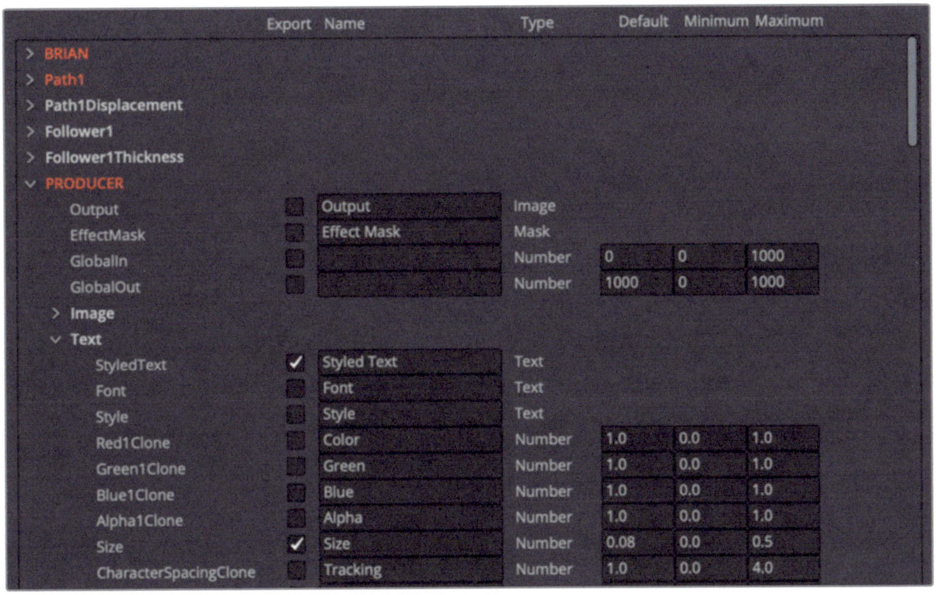

So this styled text and size is not confused with the previous settings related to the BRIAN text, let's rename these parameters.

10 Change the Styled Text name to Lower Text, and change the Size name to Lower Line Size.

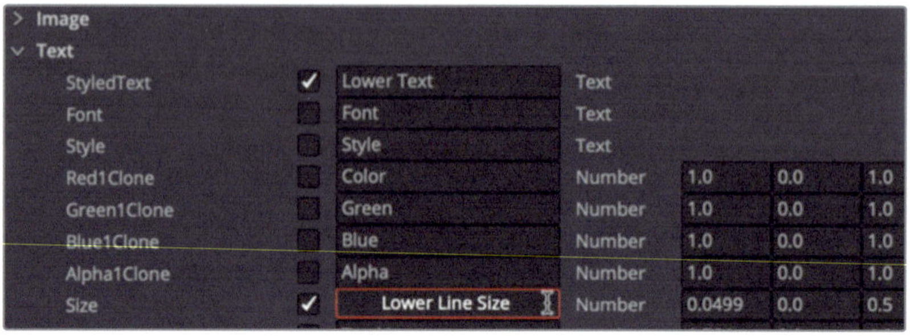

11 Click the close button and click Yes in the warning dialog to save the template.

A save window appears, allowing you to choose where you want to save the macro on your hard drive. The default location saves it for access through the Fusion page only. To get it to show up in the edit page, you'll need to change the location.

12 In the save window, select the templates folder, then select the edit > titles folder. (On Windows, the path is DaVinci Resolve > Support > Fusion > Templates > Edit > Titles.)

This location saves the template so it will show up in the edit page effects library.

13 Click save in the window, and then quit and relaunch DaVinci Resolve.

14 Open the Fusion lessons part 1 project and switch to the edit page.

15 In the effects library, select the titles category and scroll down to locate the 2D Airplane Lower Third template.

16 Drag the template to the timeline and replace the existing title.

17 Select it to view the controls in the inspector.

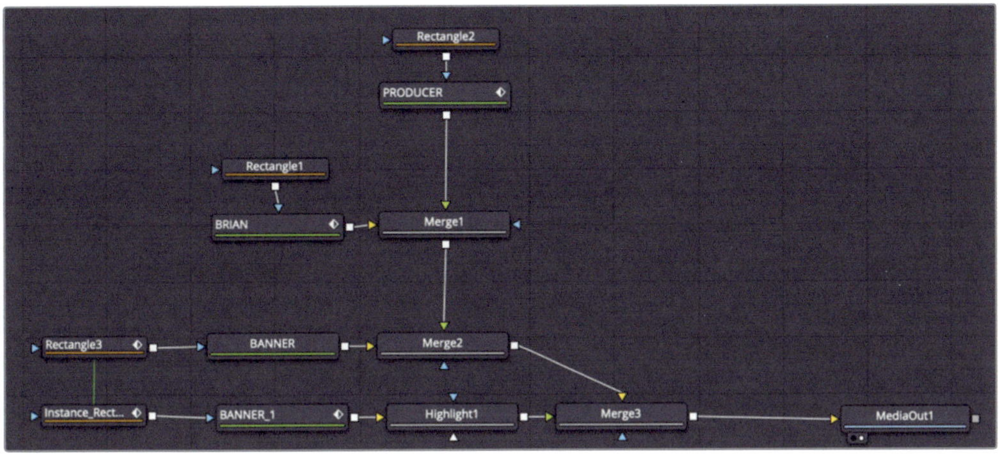

Completed node tree for Lesson 03.

You've completed this lesson on type and title animation by animating in multiple ways, modifying those animations, and designing a look. You started with the text+ tool in the edit page, then brought it into the Fusion page to build a more complex animation using keyframes and the follower. You used both node versions and timeline versions to save different looks and animations of your composition. Finally, you saved the title animation as a template that could be accessed in the edit page.

Lesson Review

1 True or false? Timeline versions only save the appearance of text.

2 True or false? You apply the follower to text by right-clicking in the styled text box where you type the text.

3 True or false? Only one word can be entered into a text+ node.

4 True or false? Adding a keyframe to a parameter in the inspector will cause that keyframe to appear in both the keyframe editor and the spline editor.

5 True or false? Clicking a version button in the inspector for a selected node will also save the settings for any node connected to the selected node.

Answers

1. False. Timeline versions save the entire composition in its current state.
2. True. You apply the follower to text by right-clicking in the styled text box where you type the text.
3. False. A single text+ node can include multiple lines of text.
4. True. The keyframe editor and the spline editor show all keyframes on all parameters.
5. False. Clicking a version button in the inspector saves only the settings for the selected node.

Lesson 4

Animating with Keyframes and Modifiers

In this lesson, you'll move beyond title animations into full motion graphic design. A motion graphics designer creates every animated logo, infographic, and design element in a commercial, TV show, or web video. When you take on the role of motion designer, you aim to convey a message by animating graphics. Although that often includes text, as we have explored in the previous two lessons, now you'll work on more general animation techniques. The motion design you'll create here borrows a lot from major Hollywood movies, reality tv shows, and promotional videos. You will create a vintage travel map look, complete with a plane and animated travel line. You'll use different animation techniques, including a mixture of keyframes, simple expressions, and modifiers as you sharpen your animation skills.

Time

This lesson takes approximately 50 minutes to complete.

Goals

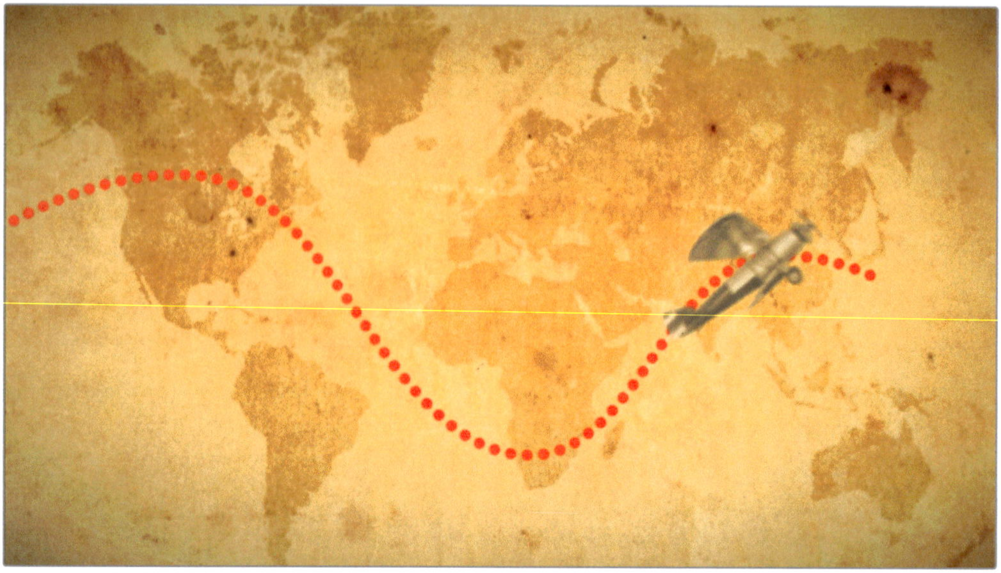

Completed composite for Lesson 4.

Identifying a Clip's Resolution

To start this Fusion composition, you'll use the same project and timeline you used in the previous two lessons. This project and timeline contain the vintage map, which you can use as the background for your plane and painted flight path line.

> **NOTE** The timelines bin includes a backups bin with timelines saved at various stages of the lesson and a completed projects bin with finished compositions. These bins are both available for reference and reverse-engineering the node trees.

1. From the edit page in the Fusion 16 Lessons part 1 project, load the Part 1-START timeline and move the playhead to the vintage map clip.

   This map is a simple 4K image in an HD timeline. You'll use it as the background behind your animated plane. Because a project's resolution in the edit page uses the master timeline settings, clips brought into a project are usually scaled automatically to fit the timeline's resolution.

2. Click the Fusion page tab or press Shift-5.
3. Select the media in 1 node, and press 1 to load it into viewer 1.

4 Hold the cmd key (macOS) or the Ctrl key (Windows), position the mouse pointer over viewer 2, and scroll the middle mouse wheel until you see the resolution in the upper-right corner of the frame.

The resolution displayed in the upper-right corner of the Fusion viewer shows a UHD resolution (3840 x 2160). Although your timeline resolution in the edit page was 1920 x 1080, the Fusion page sets its composition size based on the original clip's size. Therefore, you are always compositing with the highest resolution, direct from the source. On returning to the edit page, the media out node scales down the final result to fit the edit page timeline's resolution.

Keyframing a Motion Path

With your project set, you'll keyframe a plane across the map to create a motion path that you can then reuse for other elements in the project. To keyframe the plane's position, you'll first import the graphic and then apply a transform node. The plane graphic you can use in this composition is already in the media pool.

1 In the upper-left corner of the DaVinci Resolve window, click the media pool button to open the media pool.

2 From the travel map bin, drag the **Biplane with alpha.tif** file into an empty area of the node editor.

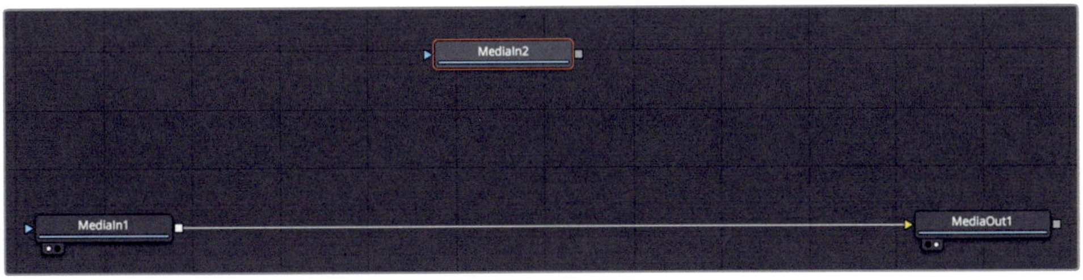

3 With the media in 2 node selected, press 1 to display it in viewer 1, then press F2, and rename the node **BIPLANE**.

4 Select the media in 1 node, and rename the node **MAP**.

For this lesson, you'll continue using viewer 1 in a single viewer layout to give youself a good amount of design space.

5 Drag the output of the biplane node to the output of the map node to create a merge node with the biplane as the foreground.

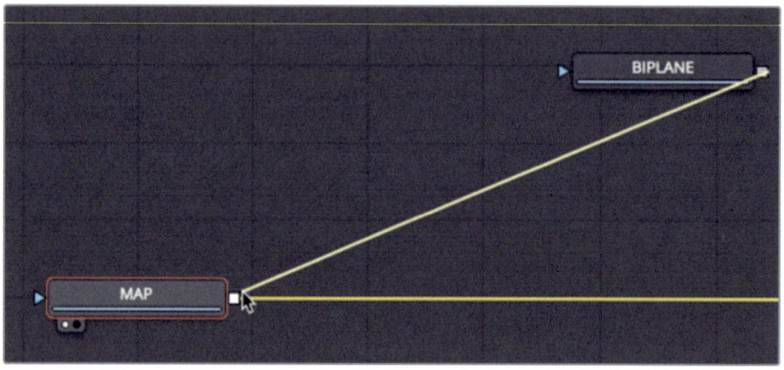

6 Select the merge node and press 1 to see the composite in the viewer.

The biplane is displayed over the background map, but it is clearly too large to fit within the vintage map background.

7 Select the biplane node, and in the toolbar, click the transform tool.

8 In the inspector, adjust the plane's size to 0.20.

The plane is now more proportional to your map. We can begin keyframing it across the map to simulate a flight path.

9 Drag the playhead to position it at the start of the render range.

10 Hold the cmd key (macOS) or the Ctrl key (Windows), position the mouse pointer over viewer 1, and scroll the middle mouse wheel until you have a bit of room around the frame. You'll use this space to position the plane out of the frame.

11 Using the plane's center control in the viewer, drag the plane off-screen left, just off the coast of Mexico.

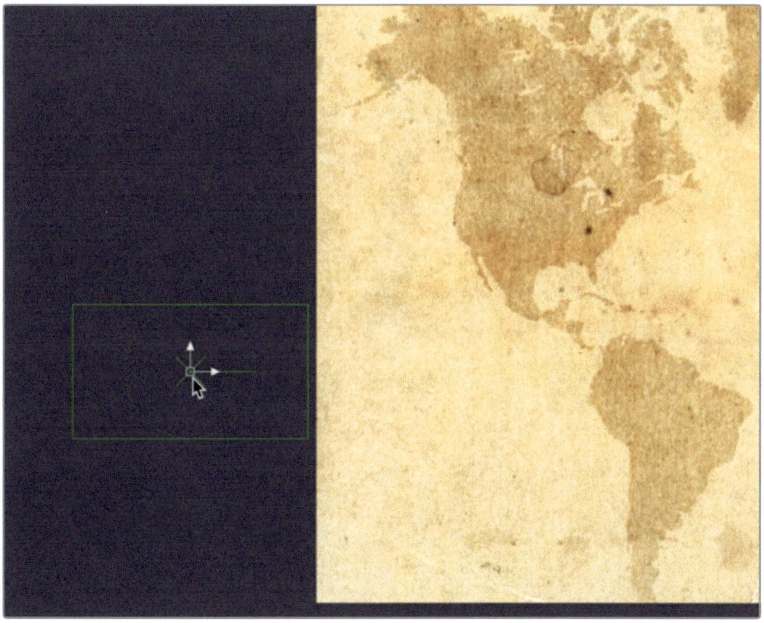

12 As you would in the edit page, click the keyframe button for the center X and Y controls in the inspector.

You have now set a keyframe for the first frame of this composition. As you move the playhead and drag the plane to a new location on the map, a keyframe will automatically be added and you will begin to create a motion path.

13 Move the playhead to frame 25, and then drag the plane in the viewer to the northernmost part of Canada to add a second keyframe.

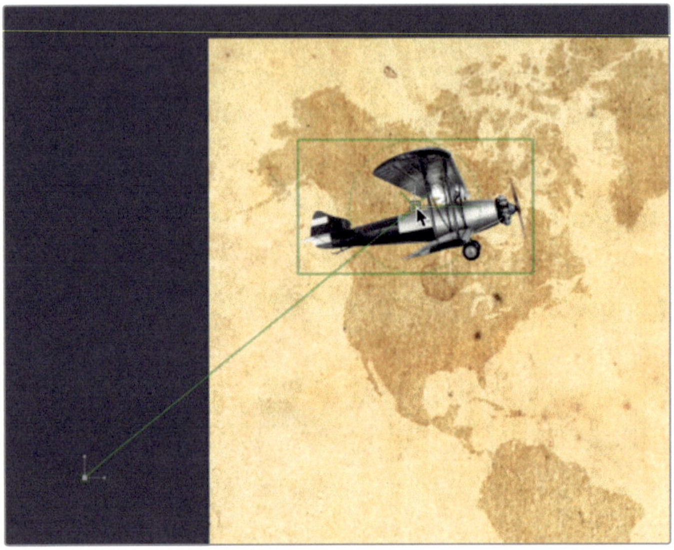

As you drag the plane, a line representing the motion path is drawn in the viewer.

14 Move the playhead to frame 50, and drag the plane to the southern tip of Africa.

15 On frame 75, move the plane over China.

16 Finally, move to frame 100 and drag the plane off-screen, somewhere near New Zealand.

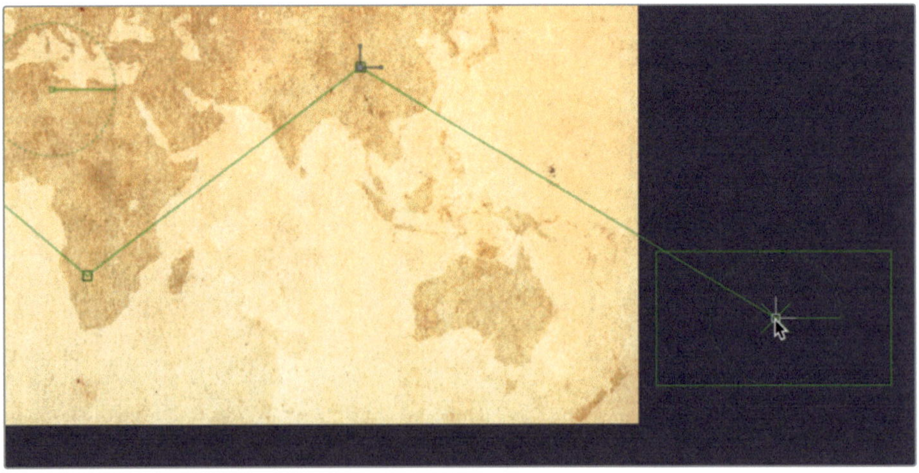

TIP Each control point along the path represents a keyframe. cmd-click (macOS) or Alt-Click (Windows) on the path in the viewer to add a control point that does not correspond to a keyframe. This allows you to change the spatial shape of the path without having to manage temporal adjustments.

With the basic flight path of the plane created, you'll quickly smooth the path, so it is not so linear. Above the viewer is a viewer toolbar, which is used to modify the spline path.

17 In the viewer toolbar, click the select all points button.

18 Press Shift-S to smooth the plane's path, or click the smooth button in the viewer toolbar.

The path for the plane now has smooth curves as it travels across the map.

19 Press the play button to review the animation.

NOTE Feel free to rearrange the nodes within the node editor as you continue to add nodes in this lesson.

Like the character level styling and the follower you used in previous lessons, the spline shape created by keyframing includes a modifier that converts the spline into a path. You can use that path modifier to drive the animation of other elements.

Auto-Orienting Objects

This animation would be better if the nose of the plane followed the direction of the plane's path. Instead of keyframing the angle of rotation for the plane, you can connect the angle to the path modifier and have Fusion dynamically animate the orientation of the plane. So, as the plane moves along the path, the plane's angle also changes accordingly, making it "turn" at each bend of the path. In the future, if you adjust the path in the viewer, the plane's angle also adjusts according to the path curvature.

1 Select the transform node in the node editor.

2 In the inspector, right-click the angle parameter and choose connect to > path > heading.

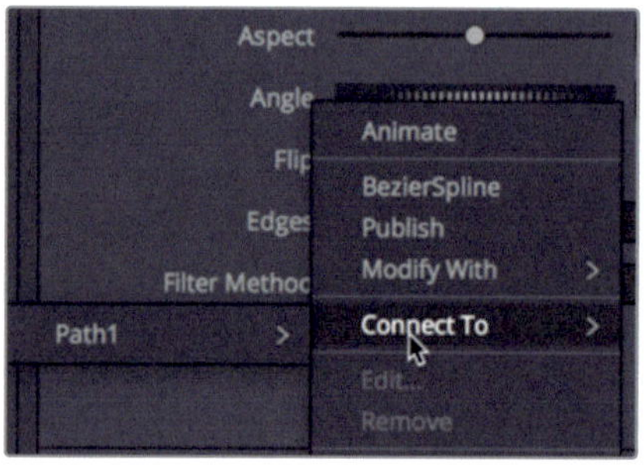

The connect to menu is used to link a parameter to an existing modifier. In this case, the modifier is the path you created by keyframing the plane. Connecting the angle parameter to the path modifier auto-orients the parameter.

3 Return to the start of the render range and play the animation.

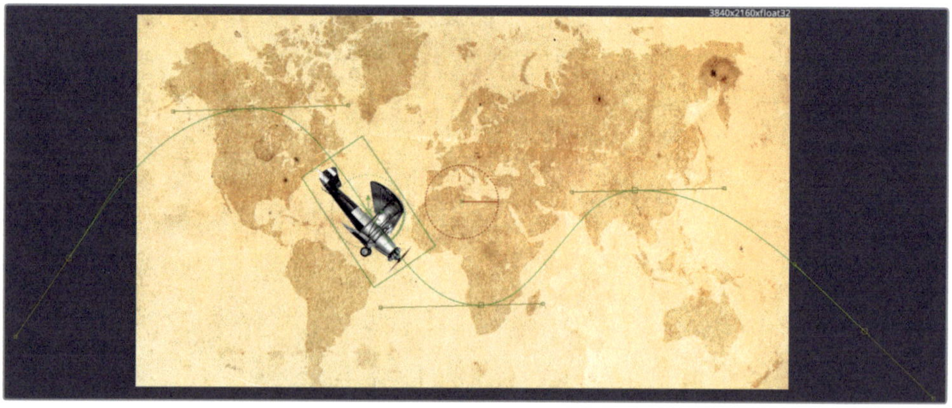

The path modifier changes the angle of the plane dynamically, based on its movement along the path.

Straightening Out Alpha Channels

To better integrate the plane with the map, let's give the plane a more weathered vintage look.

1 Select the transform node, and in the toolbar, click the color correction tool.

2 In the inspector, use the color wheel to tint the plane to better match the background, and increase the lift to about 0.25 to give the plane a faded, sepia-tone appearance.

As you adjust the lift on the plane, the background map also brightens. This is a common problem that occurs when you perform color correction on an image with a premultiplied alpha channel. To correct this problem, you can enable the pre-divide/post-multiply checkbox in the color corrector.

TIP If you increase the gamma instead of the lift, you are more likely to see a white halo around the image instead of a brightening background. However, the root cause of this behavior and its solution are the same.

About Straight and Premultiplied Alpha Channels

To display anti-aliased edges in an alpha channel, Fusion performs a multiplication operation on the edge pixels. At that time, some of the original color values shift slightly darker to achieve the anti-aliased look.

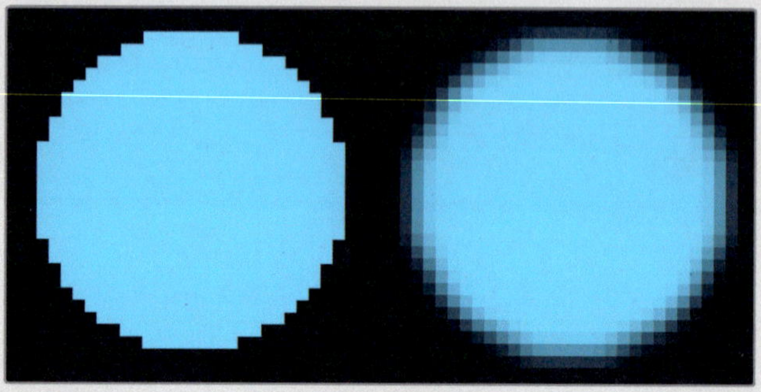

Straight alpha channel. Premultiplied alpha channel.

In Fusion, this color shifting is not noticeable because the merge tool expects to receive a premultiplied alpha channel. It becomes noticeable only when the black level or gamma of the graphic changes through some tonal value adjustment such as color correction. The result is either a brighter or darker outline around the edges of the graphic.

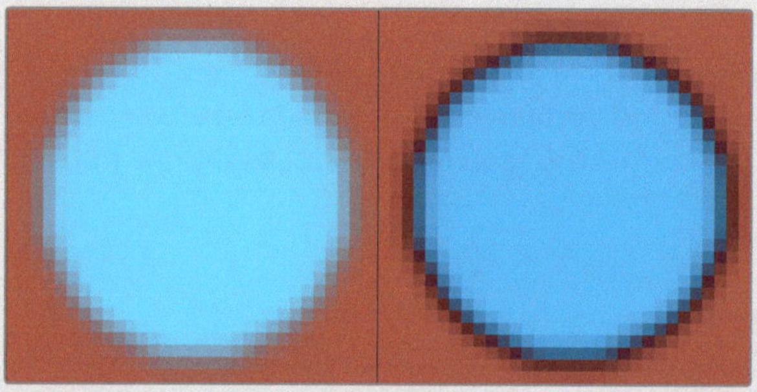

Composited premultiplied alpha channel. Tonal adjustment on premultiplied alpha channel.

Luckily, you have a relatively easy solution available. You perform an operation that divides the multiplied alpha channel, then make your color correction, and finally re-multiply the edges again. Most of the color correction tools in Fusion include a pre-divide/post-multiply checkbox that does the work for you. All you have to do is be aware of the issue and when it might need to be corrected.

3 In the inspector, click the options tab, and then select the pre-divide/post-multiply checkbox.

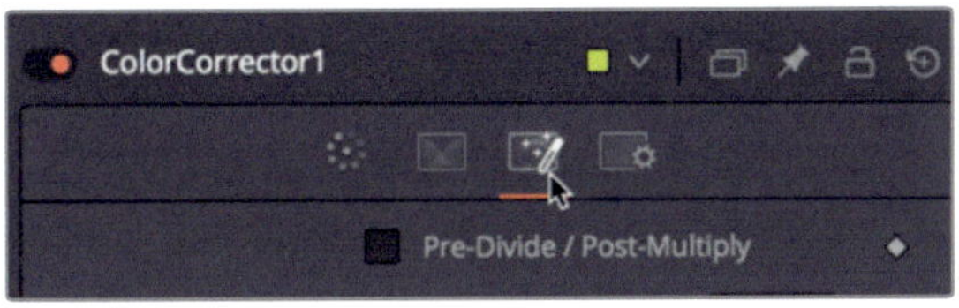

TIP Alternatively, you could deselect the checkbox, and insert an alpha divide node before the color correction and an alpha multiply after the color correction.

With the airplane looking good, you can enhance this project by having a classic red-dotted line animate along with the plane, highlighting its flight path.

Painting a Motion Path

In Fusion, you can use paint strokes for visual effects and animated motion graphics. No matter what the task, painting always starts with a paint tool. Unlike using multiple brush and shape tools to achieve the painting style you want, Fusion includes a single versatile paint tool that incorporates many brush types and paint styles. To paint the path across our map, you'll add a paint node directly after the map node.

1 Select the map node, and in the toolbar, click the paint tool.

The paint tool is added directly after the map node. At the top of the viewer is a toolbar for the selected paint node. The toolbar offers several strokes and paint styles suitable for motion graphics or retouching shots. The paint stroke style comparable to drawing a spline path is the polyline stroke brush.

2 In the viewer toolbar, click the polyline stroke.

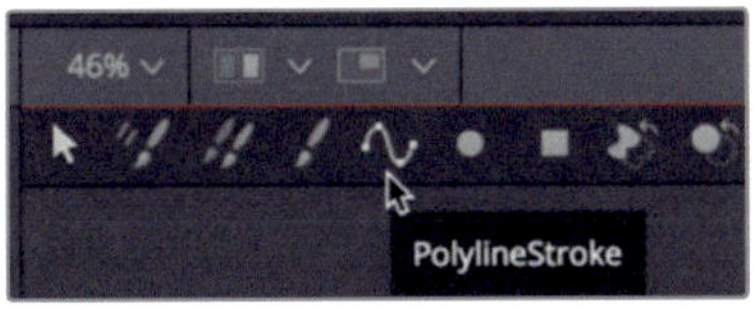

Unlike other paint stroke styles with which you could *paint* out the path using one continuous brushstroke, the polyline stroke functions similarly to a Bezier-style drawing tool: You click to add control points that create a *painted* stroke.

For your plane path, you'll only need to add two points to create a line of any length, anywhere on the screen. Once you have a polyline stroke, you can modify it to use the form of the path.

3 Click over Brazil, and then click again over Australia.

You now have a polyline paint stroke that we can modify with the path modifier. However, the path modifier currently only exists on the transform node where you created the keyframes. You'll need to make it available to all other tools.

4 Select the transform tool and click the modifiers tab.

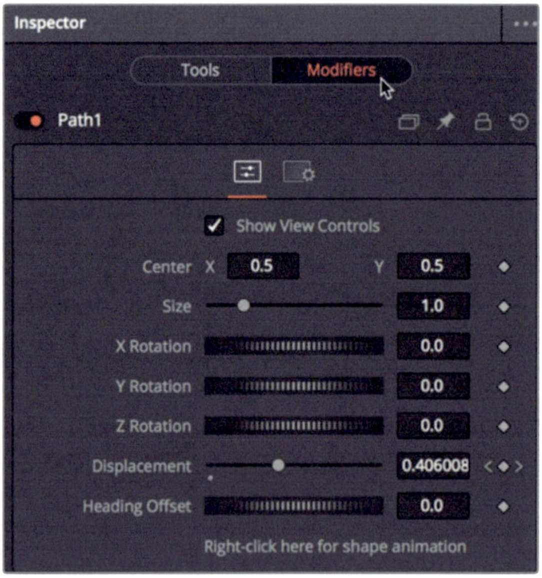

The modifiers tab shows the Path 1 that you created when you keyframed the plane. Modifiers are just optional extensions to a tool's main feature set. For instance, you used the transform tool to create a spline through position keyframes. Creating a spline using position keyframes automatically creates a path modifier. That modifier can then be *published* and made available to any object that can make use of a path. So the first step is to publish the path so the paint stroke can use it.

5 At the bottom of the modifiers tab, right-click over the label "Right-click here for shape animation" and choose publish.

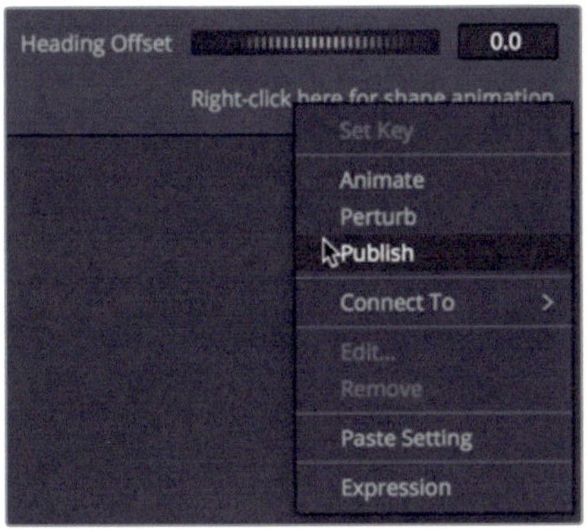

At the top of the inspector, the modifiers tab shows the path: polyline that is now published and available for other tools.

6 Select the paint node and select the modifiers tab, if it isn't already selected.

7 At the bottom of the inspector, open the stroke controls.

From here, you can connect to the published path polyline.

8 At the bottom of the stroke controls, right-click over the label "Right-click here for shape animation" and choose Connect to > Path 1: Polyline > Value.

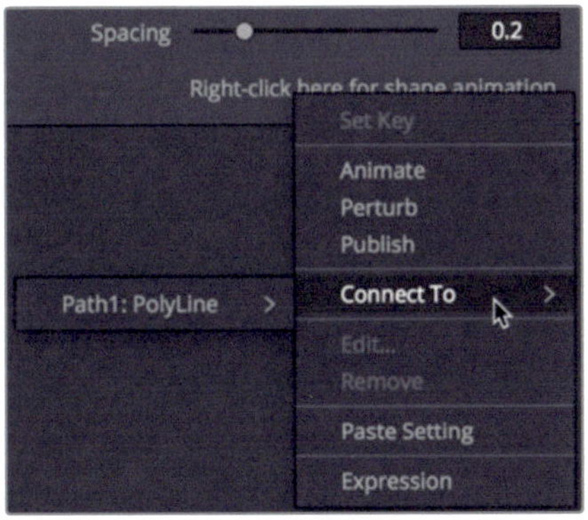

The simple polyline paint stroke you created now takes the shape of the motion path. Once you have a paint stroke, any changes are made in the modifiers tab.

9 At the top of the modifiers tab in the inspector, open the brush controls. Change the brush type from soft to circular, and the size to about 0.01.

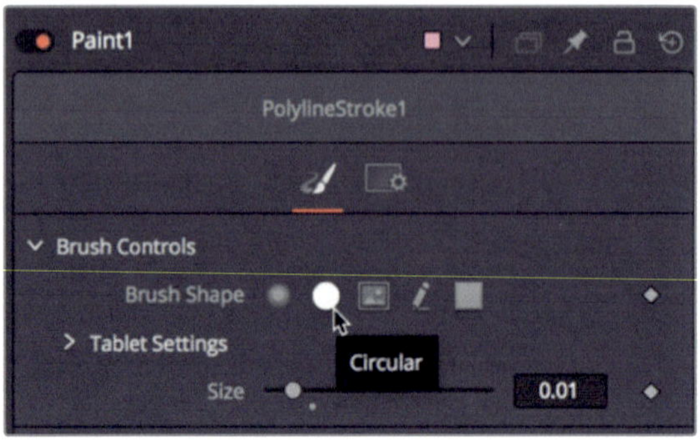

10 In the apply controls, change the color to a vibrant red.

11 In the stroke controls, drag the spacing slider all the way to the right to create a dotted line.

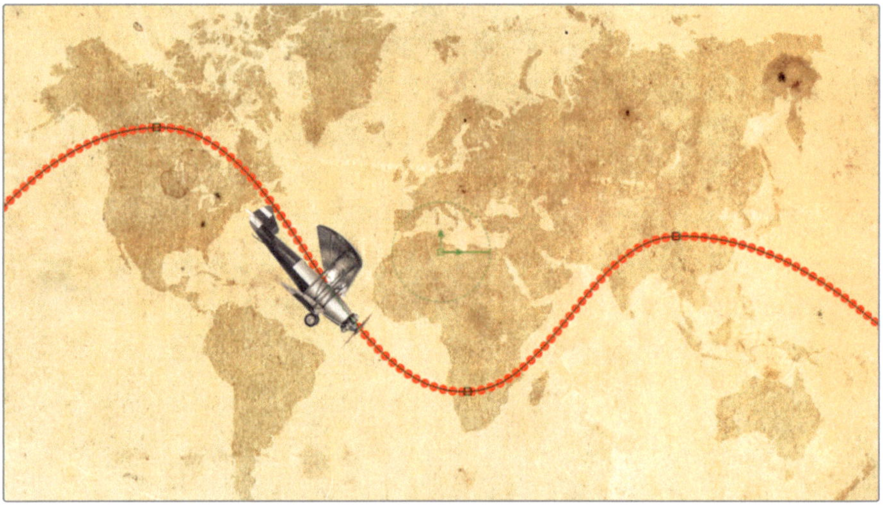

Sliders in the Fusion page can only go so far, but that does not mean the parameter's value is restricted. You can enter values larger than the limit of the slider. Doing so will scale the slider's range to allow for the new larger value.

12 In the spacing numeric field enter 1.5 and press enter.

The dotted line in the viewer now has more spacing between dots, but more importantly, the slider range has grown so you can drag to 1.5 and even a bit beyond. Almost every slider in the Fusion page includes this extended range capability.

Linking Parameters

To add more visual interest, you'll animate the dotted line so it follows the plane. The hard way to do this is just to keyframe the line, but if you change the plane's speed, you'll then need to go back and adjust the paint stroke's animation as well. Fusion allows you to link the animation of one parameter to any other parameter, even if the parameters are completely different.

1 In the modifiers tab, drag the write on end slider back and forth to see the results in the viewer.

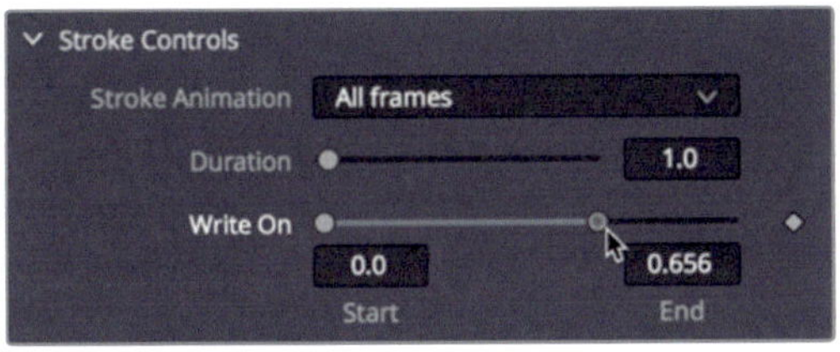

The write on control has start and end values that you can change by dragging the sliders. Dragging the end slider changes the end of the paint stroke, creating a paint-on effect. You can animate this control by connecting it to the path's displacement property so it writes on at the same rate as the plane's position along the path.

2 Right-click over the write on's end handle and choose Connect to > Path1 > Displacement.

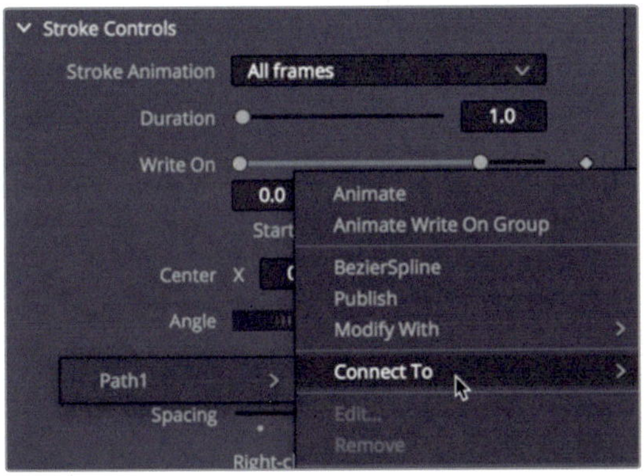

3 Drag the playhead through the render range to watch the path write on, following the plane.

The Connect To menu command is a very easy way to connect two parameters, but it doesn't allow much flexiblity. For instance, what if we wanted the paint stroke to lead the plane by just a few frames instead of following the exact same position as the plane? The Connect To menu has no way to do that, but you can link parameters in other ways. These alternatives may require a bit more work but give you much greater flexibility. First, you'll remove the connected displacement from the end handle of the write on.

4 Right-click over the write on's end handle and choose Remove Path1 Displacement.

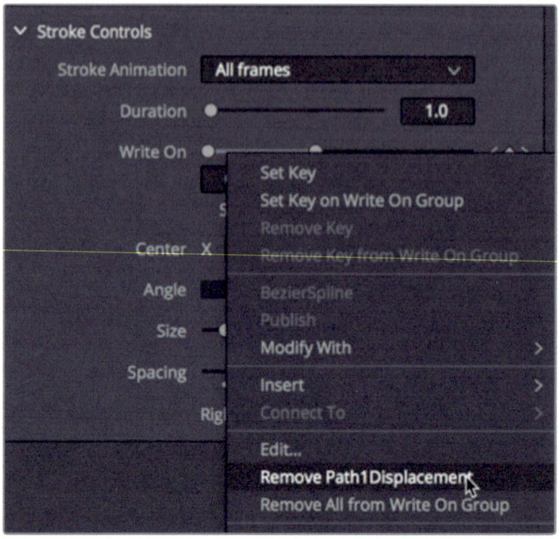

Now, the displacement parameter you want to link to is located on the modifiers tab of the transform node, since this is where the path was generated.

5 In the node editor, select the transform node and click the modifiers tab.

The transform's path modifier includes the displacement and heading parameters that we connected to earlier.

6 Drag the playhead back and forth to see how the displacement parameter animates based on the time in the composition.

As you drag through the render range, the displacement parameter is animated to move the plane along the path. This is the control to which you want the end slider of the paint write on to link.

7 With the transform node selected, from the upper-right corner of the path1 inspector's section, click the pin button to prevent the path1 from closing when you select another node.

By pinning the path1 inspector, you can select another node and see the two nodes' controls at the same time. This setup makes it possible to link the two parameters from different nodes.

8 In the node editor, select the paint node, and then select the modifiers tab in the inspector.

9 In the write on end field type = (equals sign) and press return or enter.

Typing an equals sign in any numeric field opens a simple expression field. This expression field gives you the option of linking two parameters by typing the name of the parameter you want to link, or by linking the two parameters using a *pick whip*. A pick whip is a graphical line you draw between two parameters as a shortcut method of creating an expression.

10 In the inspector, drag from the plus icon on the left of the expression field up to the displacement label in the path1 controls.

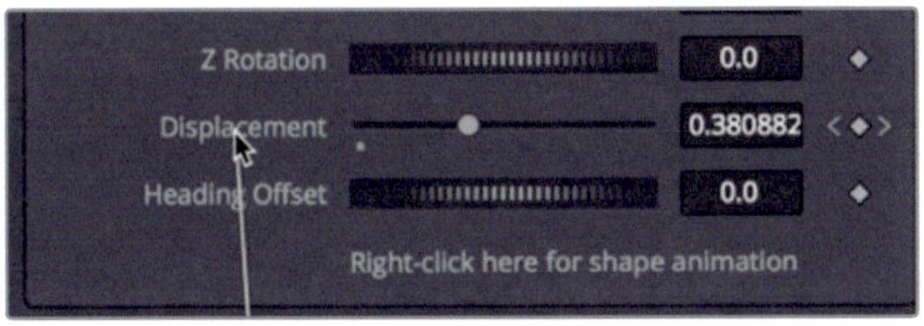

11 Play the composition to see the linked animated paint stroke.

These animated elements look good so far, but as we said, this method offers some flexibility over the Connect To menu. You can customize the speed or position of the dotted path by modifying the expression that you added with the pick whip.

12 In the expression field, insert **+.1** at the end of the expression. By adding +.1 to the expression, you are offsetting the displacement by 10%, moving it ahead of the plane's displacement.

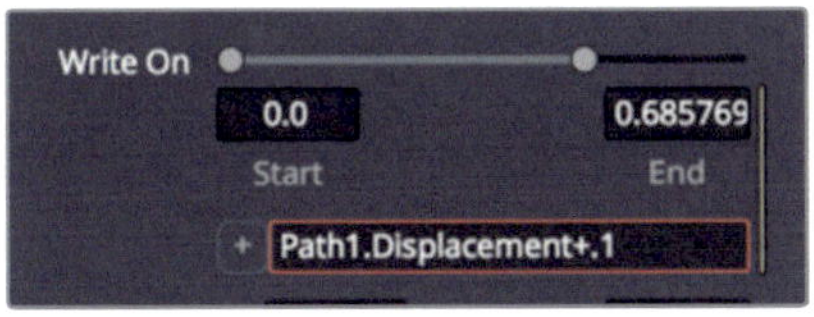

13 Play the composition to see the offset paint stroke.

Since you won't be linking any more parameters right now, you can unpin the path 1 from the inspector.

14 In the node editor, select the transform node.

15 In the upper-right corner of the inspector, click the pin button to release the path 1 controls in the inspector. The path 1 controls will close when you select another node.

This simple pick whip example animates one parameter based on the value of another. If you modify the speed or acceleration of the plane, the paint stroke inherits the same adjustment.

Making Acceleration Adjustments

Unless you are trying to create a very robotic animation, you'll seldom use linear motion as we have here. Controlling how quickly and smoothly elements move from one state to another is a crucial step in motion graphics. By default, the Fusion page applies a linear interpolation between keyframes, so animations move at a constant rate. However, we can return to the spline editor and have the plane slowly accelerate as the path dips into southern Africa, and slow it down again when it reaches the top of the path on the right. As we make the change, the linked paint stroke follows along, inheriting the same acceleration changes.

1. In the upper right of the Fusion window, click the spline button to open the spline editor.

   The spline editor header on the left side of the panel shows all the animated parameters.

2. Select the checkbox next to the displacement parameter.
3. In the upper-right corner of the spline editor, click the zoom to fit button to fill the graph area with the selected displacement spline.

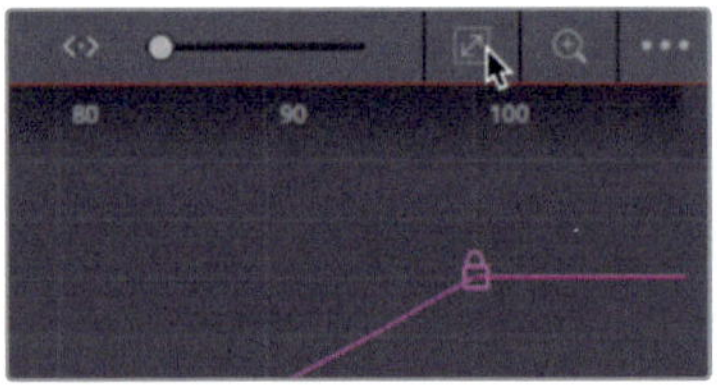

   Changing the acceleration of the plane along the path consists of smoothing the spline between keyframes and then adjusting the spline handles to increase or decrease the rate of acceleration.

4. Drag a selection rectangle around all the keyframes in the spline editor.

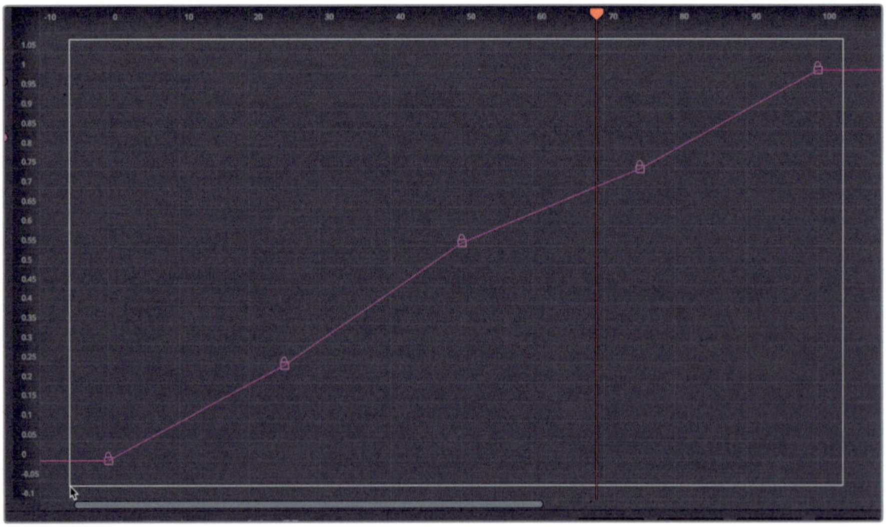

5 In the lower-left corner of the spline editor, click the smooth button or press Shift-S to smooth all the keyframes.

6 Select the next-to-last keyframe in the spline editor, and then adjust the handles to make them horizontally flat. This will slow the animation just before and just after the keyframe.

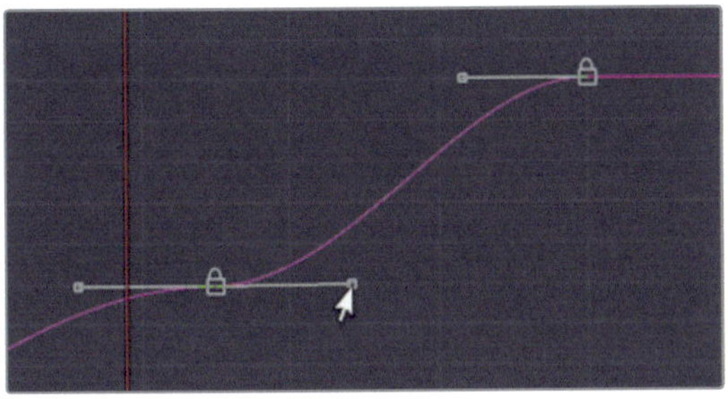

7 Select the second keyframe in the spline editor, and again adjust the handles to make them horizontally flat.

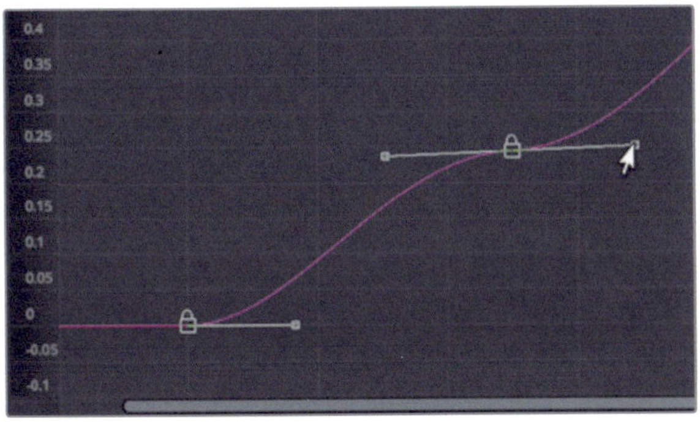

As the plane reaches the top of Canada, it will slow down and then pick up speed as it heads down to southern Africa.

8 Play the animation to see the results.

You produced a more realistic animation just by smoothing the interpolation and flattening the rate of change on specific keyframes.

Applying Random Animation Modifiers

Modifiers can do so many amazing things for your animations, from converting splines to paths, creating rippling animations for text, and even animating any parameter using randomly generated values. Random animation can come in very handy when trying to create a wiggling position animation or flickering lights using brightness controls. To complement this vintage animation, you'll use one of DaVinci Resolve's built-in filters to create a flickering old-style film look.

1 Close the spline editor and select the merge 1 node in the node editor.

  You'll add a filter effect to the entire composition by placing it directly after the merge 1 node.

2 In the upper left of the interface, click the effects library, and from the OpenFX category, open the ResolveFX texture subcategory of filters.

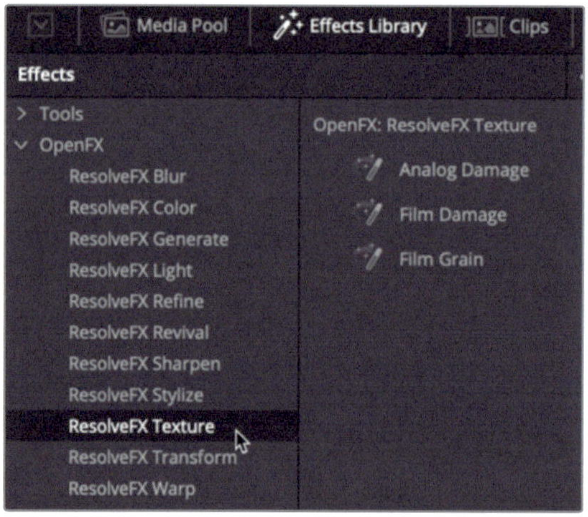

3 Click the film damage tool to add it to the node editor, and then press 1 to see the results in the viewer.

4 Play the composition to review the newly added filter.

The film damage filter adds some film scratch lines, a slight vignette, and a sepia tint to the composition. With all the parameters in the inspector, you have a great deal of control to customize the effect. One addition that would make it more realistic would be a bit of flickering light simulating an old film projector bulb on its last bit of energy. Instead of animating this effect, we can use the perturb modifier to automatically generate random animation.

5 In the inspector, right-click over the focal factor label.

The focal factor parameter adjusts the strength of the vignette. If you apply a random animation to this slider, it will give the appearance of a weakening light bulb.

6 Choose modify with > perturb from the contextual menu.

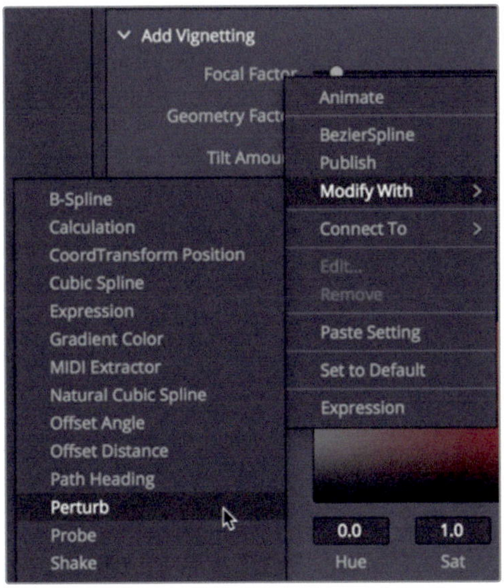

TIP The perturb and shake modifiers are both used to generate random animation. The differences in the results are subtle, but the controls in the perturb modifier are a bit more precise.

7 Play the composition and let the animation cache to RAM to review the flickering animation.

The effect is less of a flickering projector bulb and more of a slow pulse. Like many modifiers, you can control a number of properties to get the look you want.

8 Click the modifiers tab at the top of the inspector.

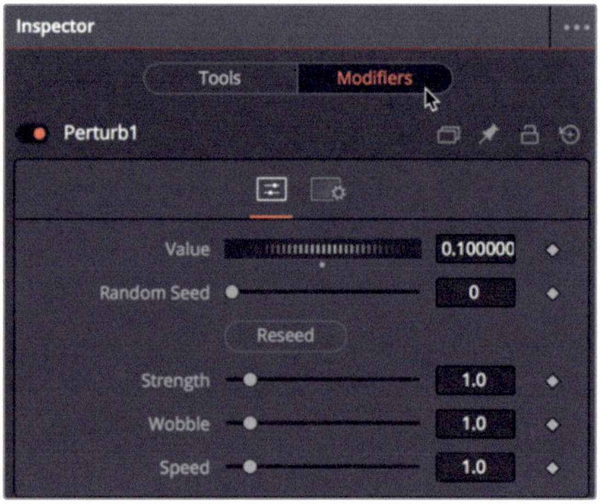

At the top of the modifier tab is the value setting. The perturb value setting uses the current focal factor setting as a starting point. Dragging the value slider is similar to dragging the focal factor slider.

Other controls—such as strength, wobble, and speed—control how far the focal factor parameter varies from the initial value setting, how erratic the motion is, and the speed of the motion. Keeping strength at a lower value keeps the flickering from getting too dark or too bright. Keeping the wobble and speed higher create a faster, more chaotic energy to the animation.

9 Lower strength to 0.5 to make the change in brightness less drastic.

10 Increase both wobble and speed to 10 and play back the animation.

The perturb modifier is extremely flexible; you can add it to polylines, grid meshes, and even color gradients to wiggle or randomize almost any parameter you want to animate.

Customizing Motion Blur

As we did with the text in the previous lesson, the last refinement you'll make to this animation is to increase the photo-realism with some motion blur.

1 Move the playhead in the middle of the timeline where the plane is on-screen.

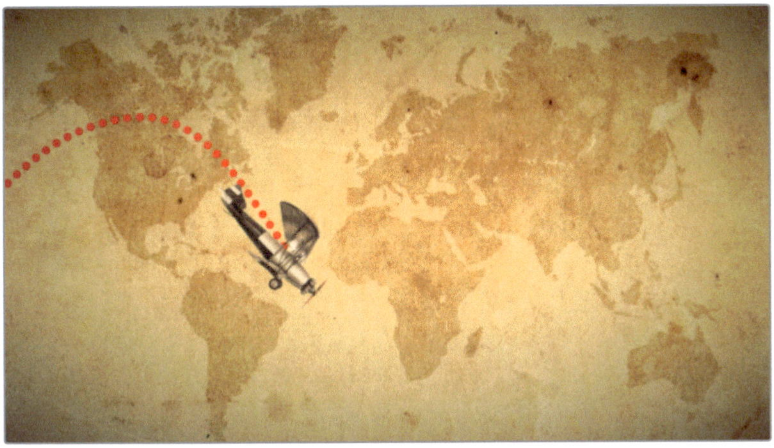

2 Unlike the previous lesson where the text+ node created the motion, here the transform 1 node is creating the motion. So start by selecting the transform 1 node connected to the biplane.

3 In the inspector, click the settings tab, and select the motion blur checkbox.

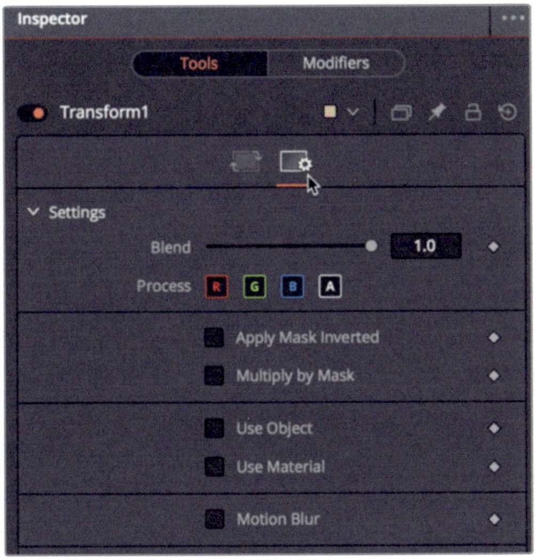

The quality and shutter angle of the motion blur are the two most commonly used controls for improving the look and spread of the blur; however, increasing these two controls also increases rendering time.

4 Increase the quality to 6, and the shutter angle to 200.

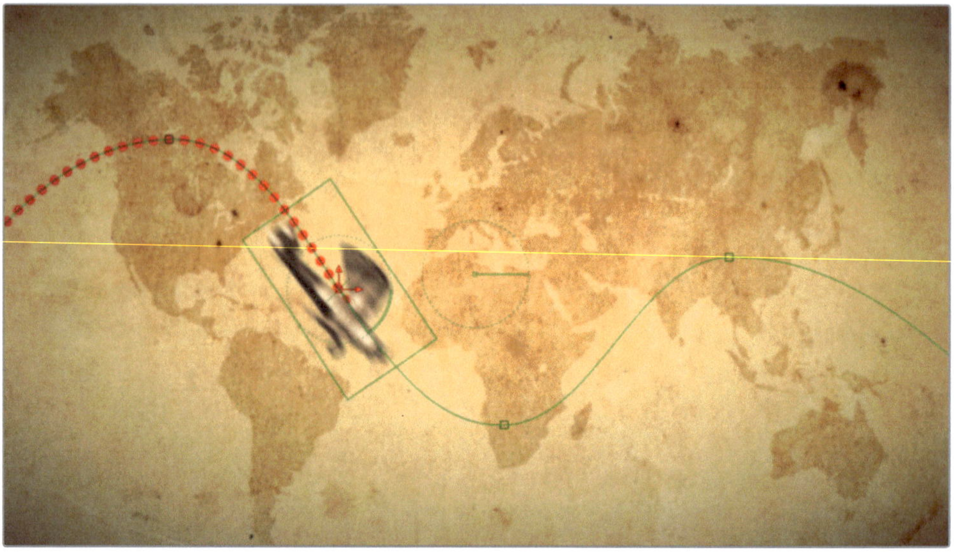

The quality parameter increases the number of times the image is replicated to create blur. The shutter angle simulates the shutter angle in a camera. Higher numbers create a smoother "smear" between samples. Setting this parameter to 360 is similar to having a shutter in a camera open for an exposure of one whole frame.

TIP Right-clicking to the right or left of the transport controls allows you to disable motion blur for the entire composition.

Now, you can return to the edit page and use DaVinci Resolve's smart cache to render and view your animation.

5 Click the edit page, and then choose playback >render cache > smart. Once the cache is complete, play back the animation.

You have completed your vintage biplane animation and are ready to show your clients your design.

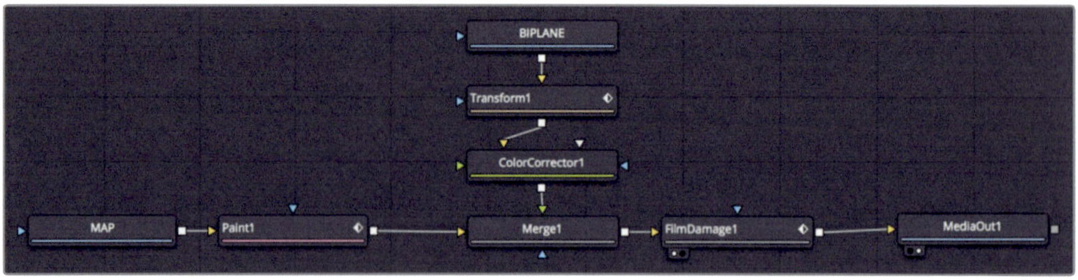

Completed node tree for Lesson 4.

Practice Exercise

To reinforce the concepts you learn, there will occasionally be optional exercises at the end of lessons. These exercises will be very similar to what you will have just completed. However, they will have brief directions for guidance instead of detailed step-by-step instructions. Time for a practice exercise.

Located in the travel map bin in the media pool, you'll find **Dr Jones with alpha.tif**. Try to add this silhouette to your plane comp and have it follow an inverse path of the plane's path (When the plane is at the top of the map, Dr Jones is at the bottom, etc.).

1. In the media pool travel bin, drag the Dr Jones tiff to your comp in the Fusion page, and resize him to appropriate proportions.
2. Use a transform node to keyframe Dr Jones so he follows the same path as the biplane, but inverted.
3. Create a blue painted path that follows Dr Jones.

   Here's a final image of how your composite might look.

Lesson Review

1. True or false: You cannot connect the output of a media in node directly to the input of a paint node.
2. True or false: To attach a paint stroke to a path, you must publish the path.
3. True or false: When using an image with a premultiplied alpha channel, you must divide the alpha prior to color correcting it. Then it must be multiplied again before being connected to a merge node.
4. True or false: To auto-orient an object along a motion path, you right-click over the center X and Y parameter and choose connect to > path > heading.
5. True or false: The displacement parameter controls an object's position along a motion path.

Answers

1 False. You can connect the output of a media in node directly to the input of a paint node.

2 True. To attach a paint stroke to a path, you must publish the path.

3 True. When using an image with a premultiplied alpha channel, you must divide the alpha prior to color correcting it. Then it must be multiplied again before being connected to a merge node.

4 False. To auto-orient an object along a motion path, you right-click over the angle parameter and choose connect to > path > heading.

5 True. The displacement parameter controls an object's position along a motion path.

Part II
Visual Effects Creation

Lesson 5

Compositing Split Screens

The Fusion page is not limited to motion graphics. It excels at creating photo-realistic visual effects composites. The next four lessons in this book focus on visual effects compositing. Whereas motion graphics center on text-heavy animated graphic design, visual effects compositing is primarily about combining multiple, distinct images to make a believable new whole.

One of the most critical aspects when compositing is to have all the different elements used in a composite follow the same camera motion. These are often called *match moves* because you analyze the movement of a background and apply it to a foreground or vice versa. The Fusion page provides three primary techniques for extracting the camera's motion: point tracking, planar tracking, and 3D camera tracking. In the course of this book, you will learn all three. We'll begin with the standard point tracker in this lesson.

Time

This lesson takes approximately 45 minutes to complete.

Goals

Although not the most sophisticated tracking tool, the point tracker is a real workhorse because it works on the broadest range of shots.

This first visual effects lesson will analyze the motion in two shots to create a split-screen effect. Now, when we say spit screen, we are not talking about a simple picture-in-picture effect. A split screen in visual effects terms is a technique where different takes of a shot are combined to make one better overall take. For instance, let's say we have two actors in a shot. One actor may deliver an excellent performance in take 1 while the other actor delivers their best performance in take 2. The director may want to combine the two performances and make it appear as a single take.

Completed split screen for Lesson 5.

Using Layers from the Edit Page

Part II of this book uses a new project for its four lessons. You will start by restoring a new archive that includes all the media, bins, and timelines you'll need.

1. Open DaVinci Resolve 16, and in the project manager, right-click, and choose restore project archive.
2. In the navigation window, go to the R16 Fusion lessons folder and open **Fusion 16 Lessons part 2 dra**.
3. Click the edit button, or press Shift-4 to return to the edit page if you are still on the Fusion page.

> **NOTE** The timelines bin includes a backups bin with timelines saved at various stages of the lesson and a completed projects bin with finished compositions. These bins are both available for reference and reverse-engineering the node trees.

Let's look at the first edit in the timeline, which includes two video layers, and see how you can bring both elements into the Fusion page.

4 From the timelines bin, double-click Part 2 -START.

5 Move the playhead over the first red marker in the timeline.

This first edit consists of a clip of a cowboy in a darkened barn with a dog playing outside. It's a rather somber shot. Video track 1 has a different shot from inside the barn without the cowboy or the dog playing in the yard.

6 In the timeline, select the clip on video 2, and press D to disable it and view the yard without the dog.

7 Press D a second time to enable the foreground clip again.

Your job as the visual effect artist is to perform a split-screen effect. The goal is to combine the inside of the barn with the cowboy and the outside without the dog.

To start the split-screen effect, you must bring both takes into the Fusion page. To take more than one clip from the edit page into the Fusion page, you must create a Fusion clip.

8 Select both clips in the timeline. Right-click the selected clips, and choose New Fusion clip at the top of the menu.

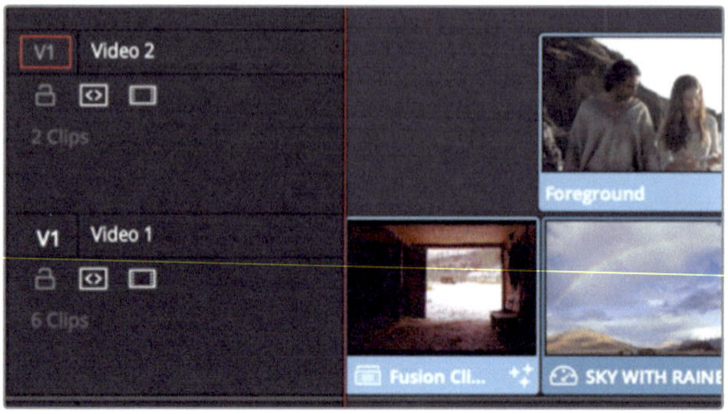

A Fusion clip is created in the timeline and added to the selected bin. The two layers are collapsed into the container.

TIP You can display all the layers of a Fusion clip in the edit page timeline by right-clicking the Fusion clip and choosing open in timeline.

9 With the playhead still positioned over the clip, click the Fusion page button, or press Shift-5.

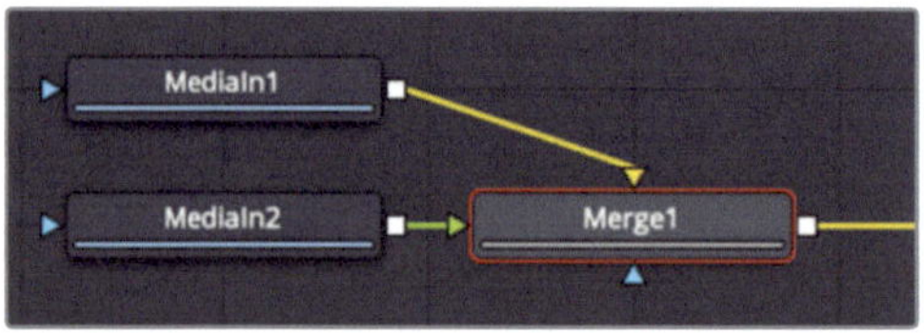

The two layers are brought into the node editor and combined in a merge node.

TIP Changing a clip's size, rotation, or position in the edit page will carry over into the Fusion page when a Fusion clip is created.

The media in 1 node represents the clip on video track 1, and the media in 2 node represents the clip on video track 2. The clip from video track 1 connects to the background in the merge node, and video track 2 connects to the foreground input.

To quickly identify each media in node and what they represent, you can change their default names to be more descriptive of the contents.

Image Processing across DaVinci Resolve Pages

The layout of the pages at the bottom of the DaVinci Resolve interface represents the project workflow. Media is ingested in the media page, assembled into a timeline edit or cut page, undergoes compositing in the Fusion page, and is then graded in the color page before rendering out through the deliver page.

This order of operations also corresponds to the image-processing pipeline, with some exceptions. The video signal of the source media in the edit page (or cut page) flows to the Fusion page, where it is processed and output to the color page. However, if an effect (ResolveFX or OFX plug-in) is applied in the edit page, it will *not* appear in the Fusion page but will instead appear on the color page. Edit page effects are applied to the image data signal after it has left the Fusion page. The only exception to this is resizing and repositioning in the edit page. To force visibility of an edit page effect in the Fusion page, you can convert the clip into a compound clip on the edit page timeline.

Image processing pipeline

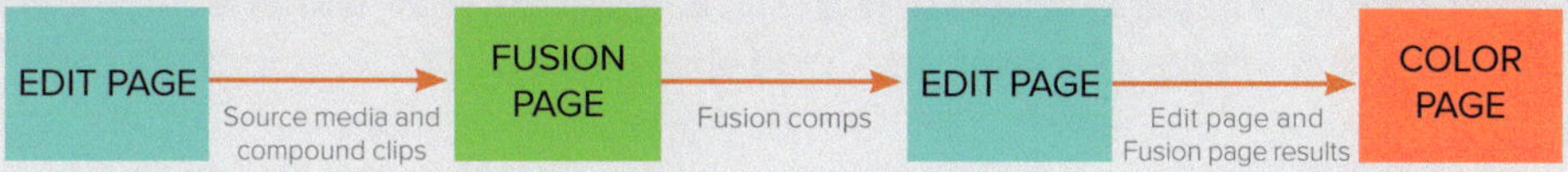

The viewers on all pages represent different stages a clip may go through. The edit page (or cut page) has two viewers. The source viewer shows the unaltered source media, and the timeline viewer shows the media after it has been processed by the edit, Fusion, and color pages (essentially, this is your final output). The Fusion page shows the unaltered source media or compound clip. The color page shows all edit and Fusion page effects as well as any color grading.

Viewer results

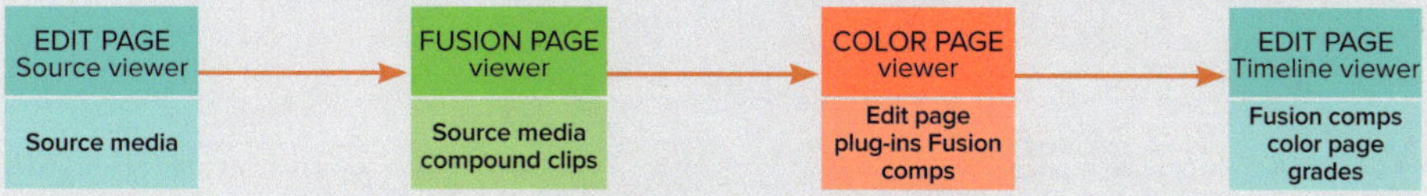

10 Rename media in 1 to **OUTSIDE** and media in 2 to **INSIDE**.

> **TIP** Node names cannot contain spaces, but you can use underscores to separate words.

Now, you'll continue with this split-screen composite by tracking the two clips and using that tracking data to stabilize them.

> **NOTE** This lesson uses dual viewers. If your workspace is still using a single viewer from the previous lesson, click the dual viewer button in the upper-right corner of the viewer.

Tracking in the Fusion Page

The first step in creating this split screen is to remove all camera motion from one or both clips. To remove camera motion, or stabilize a clip, you need to track the clip using a tracker node in Fusion. After at least one of the two clips is stabilized, it will be easier to line them up and make them appear as a single clip. We'll stabilize both clips to make them even easier to align.

1 Select the INSIDE node, and then open the effects library.

2 In the effects library, go to tools > tracking, and click the tracker tool to add it after the INSIDE node.

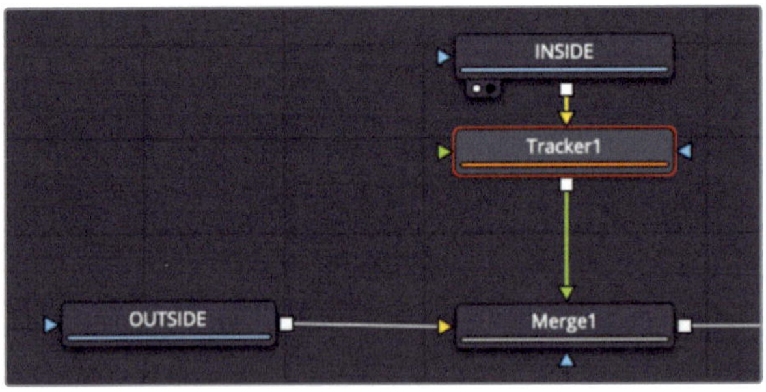

3 Select the tracker node and press 1 to see it in the viewer.

The first step in setting up the tracker is to locate a high-contrast pattern in the frame that includes the motion you want to track and position the tracker over it. Because you want to remove the camera motion, select objects in the frame that move only because of the camera moving.

4 Move the playhead to the start of the render range.

In the viewer, the tracker consists of two boxes. The inner box identifies the high-contrast pattern that the tracker should follow, and the outer box is the search area. You can position the inner pattern box by dragging it from the upper-left corner handle.

> **TIP** The tracker automatically chooses the clip connected to its background input as the clip to track.

5 In viewer 1, drag the handle in the upper-left corner of the pattern box to position the tracker over the upper-left corner of the doorway.

The pattern box expands and displays a magnified view of the area you are over so you can be precise with your selection. The upper-left corner of the door is a high-contrast, well defined area that stays in the shot the entire time. This makes it a good tracking point.

6 When the upper-left corner of the door is centered in the magnified view, release the mouse button.

The outer box of the tracker is the search box. As the tracker moves frame by frame through the clip, it looks for the pattern you have identified in the pattern box. The larger the search area is, the longer the tracking analysis takes. On slow-moving objects, the pattern probably won't move far from one frame to the next, so you can usually create a relatively small region for the search box. When you have a fast-moving object, you may need to increase the search box size.

Because the camera does not move very quickly in this shot, the tracked objects will not move very far from one frame to the next. So, we can leave the search box at its current size.

When stabilizing a shot, a single point only stabilizes the translation of that pattern in the frame. The pattern can still scale and rotate. You need at a minimum three points to prevent an image from moving, scaling, and rotating. To be on the safe side, we'll pick four points in the frame.

7 In the inspector, click the add button three times to add three more trackers to the tracker list.

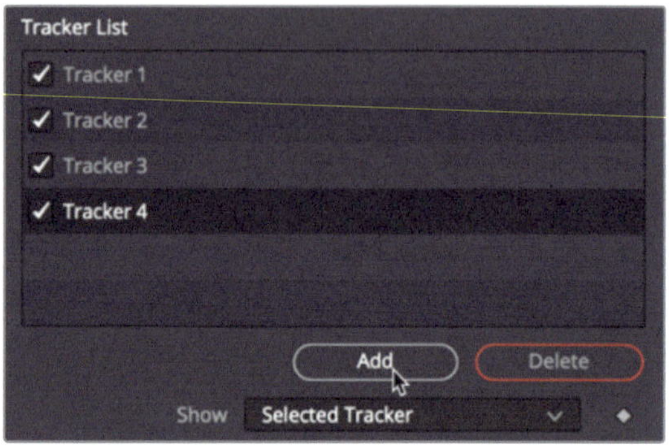

TIP You can double-click the tracker in the tracker list and rename it for organizational clarity.

8 In viewer 1, drag tracker 3 over the lower-left corner of the door.

The doorway corners are high-contrast areas. They are also *nailed to the set*, meaning they do not move except for the fact that the camera moves. The dog and the cowboy are not nailed to the set, so they would make a poor choice for tracking in this shot.

9 Drag tracker 2 over the upper-right corner of the door, and tracker 4 to the lower-right corner where the bench is located.

You can click the tracking analysis buttons along the top of the inspector to track forward or backward.

10 Click the track from first frame button to begin the tracking process.

Viewer 1 displays the progress of the trackers until the analysis is complete. A dialog box is displayed showing some information about the tracking process.

11 Click OK in the dialog to confirm the completion.

You now have four reliable trackers that follow the camera movement. However, tracking is not the effect; it is only the means to an end. The next step is to put that tracking data to use in stabilizing this clip.

Using a Tracker for Stabilization

After the tracking analysis completes, you typically change the tracker's operation mode from tracking to utilizing the tracking data.

1 In the tracker's inspector, click the operation tab.

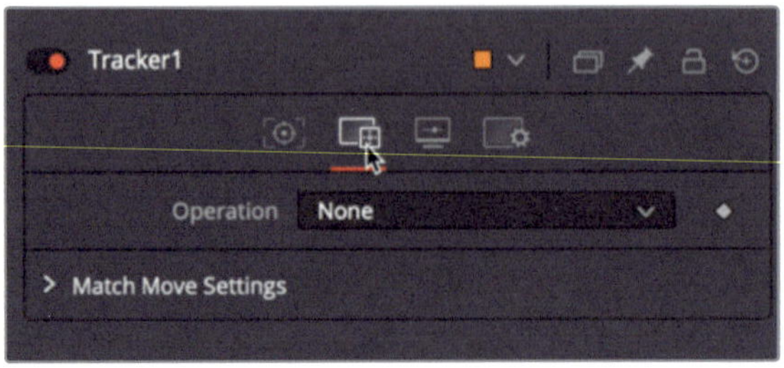

The operation tab is where you determine how the tracking data gets used. The operation menu at the top of the inspector contains all available options.

2 From the operation menu in the inspector, choose match move.

The operation menu options are dependent on the inputs connected to the tracker node. To stabilize a clip, you must have the clip connected to the background input as we have here. Then you set the match move operation to apply to the background.

3 From the merge menu, choose BG only.

4 Play the clip to view the stabilized clip.

As the clip plays in viewer 1, notice that it reveals some of the checkerboard background on the lower-left corner of the frame. Stabilization works by taking the camera motion from the tracker, reversing that motion, and applying it to the same clip. For instance, if the camera moves down, the tracker moves the entire frame up to offset the move. However, this exposes the background because the clip is the same resolution as the output. You'll have to fix this at some point, but for now, if you are only looking at the barn door, you have a very steady clip. You now must repeat the process for the OUTSIDE clip.

5 Select the OUTSIDE node, add the tracker tool to it, and press 2 to view it in viewer 2.

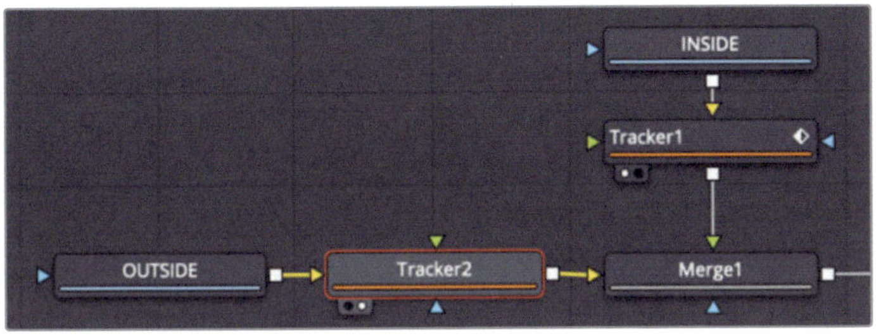

6 In the inspector, click the add button three times to display four tracking boxes in the viewer.

Now, you'll locate four high-contrast patterns in the frame. Since you want to remove the camera motion on the area outside the barn, select patterns from the yard. Selecting patterns from the yard ensures parallax motion between inside the barn and outside, and does not cause a problem for stabilization.

7 Move the playhead to the start of the render range at frame 0.

8 In viewer 2, drag tracker 1 over the wheels of the trailer, and then position tracker 2 over the middle fence post.

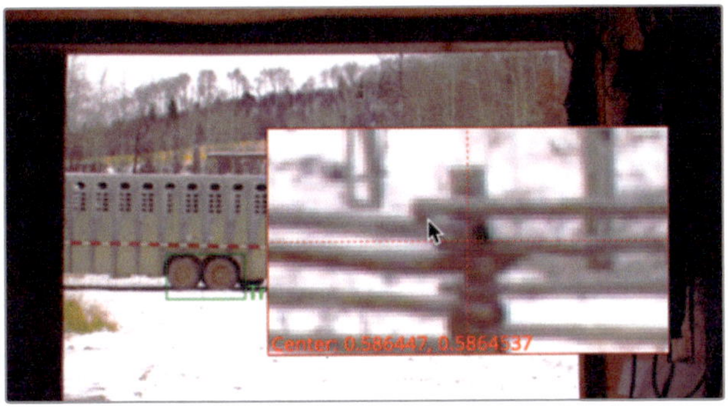

9 Drag tracker 4 over the large isolated speck of dirt on the lower right, and then place tracker 3 over the gray square on the far left.

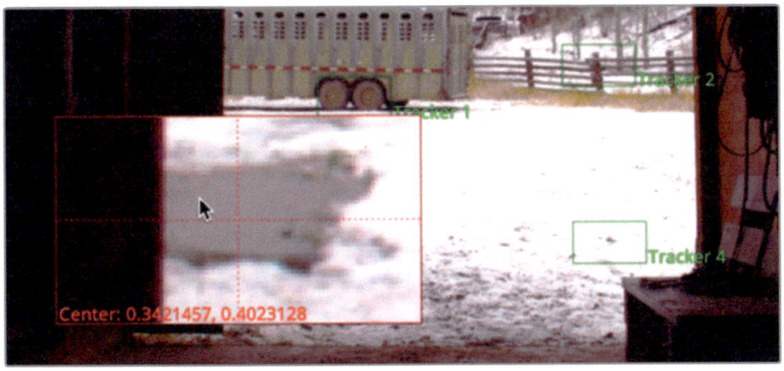

10 Click the track forward button in the inspector, and then click OK in the completion dialog.

   To switch from tracking to stabilization, you'll use the operations menu again.

11 In the tracker's inspector, click the operations tab.

12 From the operations menu in the inspector, choose match move.

13 From the merge menu, choose BG only.

14 Play the clip to view the second stabilized clip.

As you play the shot, the OUTSIDE barn is stabilized. In the next exercise, you'll introduce a matte that will combine the two stabilized clips, using only the areas from each shot that we want to be exposed.

Cutting a Hole with a Polygon Matte

A considerable part of compositing has to do with drawing mattes, sometimes called *rotoscoping*. Mattes help you isolate specific areas of your image so that effects are applied only where they are needed.

To make this composite work, we must cut a hole in the INSIDE clip to see through to the OUTSIDE.

Similar to the text animation you created in Lesson 3, you'll use a spline shape to draw a matte around the barn door of the INSIDE node. You will then use that matte to cut a hole that makes the doorway of the INSIDE node transparent.

1 Move to the first frame in the render range.

   It is essential to add the polygon mask shape while you are positioned at the start or end of a render range because mask shapes auto-animate. Meaning, if you change the shape on any other frame, the point of the shape interpolates. You are not changing the shape for the entire clip, just on the specific frame. This makes it very quick to rotoscope a moving object but confusing if you are unaware of the behavior.

2 From the toolbar, drag the polygon shape tool into an empty area in the node editor below the merge node.

   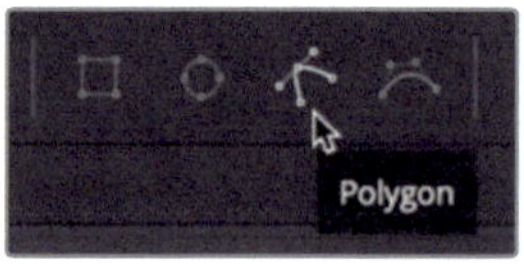

   When a polygon mask tool is selected, drawing tools appear above the viewer, and you can begin to draw a shape even before connecting the node into your composite.

3 To increase the size of the viewer and see more detail around the area you are rotoscoping, position the mouse pointer on the line between the transport controls and the toolbar, and then drag down to increase the viewer's size.

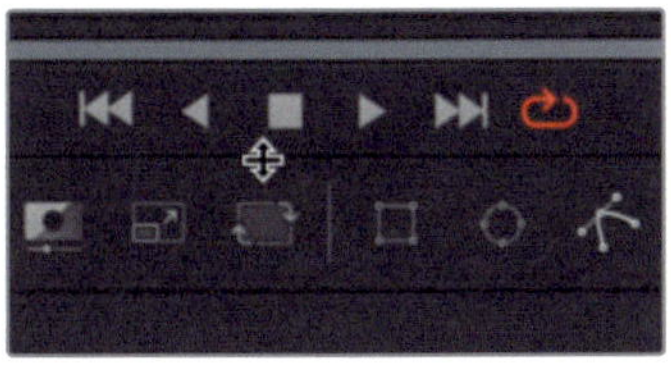

We won't need two viewers for this task, so we can gain more working space by just viewing a single viewer for the INSIDE node.

4 Select the INSIDE node and press 1, and then In the upper-right corner of viewer 1, click the single viewer button.

Now, viewer 1 takes up the entire top half of the screen to give you a good canvas on which to draw your matte.

NOTE The terms *matte* and *mask* are often used interchangeably. In this book, matte refers to a grayscale image that identifies transparent and opaque pixels. A mask is the application of a matte. That is, you use a matte to mask off part of an image.

5 To zoom in on the barn door area in the frame, move the mouse pointer over viewer 1. Hold down the cmd (macOS) or Ctrl key (Windows), and scroll the middle mouse wheel to zoom in until the entire viewer fills with the barn door opening.

To draw a polygon shape around this doorway, you'll need no more than 25 to 30 control points along the inside edge of the barn door. You want to use as few points as possible but as many as needed to cover the door's shape correctly.

TIP You can always add and subtract points later, but the more points you add, the more you'll have to manage.

6 Select the polygon node, and then in the viewer, click in the upper-left corner of the doorway to add a control point. Then, move your pointer down to the lower corner of the door and click to create a second control point on the corners.

7 Move the mouse pointer across the entryway until you reach the bench, and then click to add points outlining the bench.

TIP Holding the middle mouse button and dragging in the viewer will pan the viewer, making it easier to see various parts of the frame while you are zoomed in drawing your matte.

You do not need to add many points along the straight edges of the doorway, since they are more or less straight. However, you will want to be more detailed as you continue the shape up the right side of the doorway

8 Move your pointer up along the right side of the doorway clicking to add new control points along the small uneven details.

9 To complete the matte, move the mouse pointer from the top right of the doorway and hover the mouse pointer over the first control point in the upper left.

10 When the mouse pointer displays a circular icon, click the first point you added to close the shape, or press Shift-O.

TIP Many problems occur when you assume you have closed a shape, but it is still open. Making sure you close the shape either by using the keyboard shortcut (Shift-O) or clicking directly on the first control point, will save you a considerable amount of troubleshooting time later.

Now you have a rough shape that will be used to cut through the INSIDE barn door, making it transparent. However, when drawing mattes, it is rare to have all sharp corners on objects in a frame, as you have drawn here. You'll need to create some curves to ensure the polygon matte follows the shape perfectly.

Creating Curved Edges

Some areas of the barn door along the right side are not truly sharp corners, as you may have drawn. They have a curvature to them. You'll need to follow the curve of the doorway with your matte shape until they match perfectly. You can do this by smoothing the control points where needed.

However, before making changes to a shape, let's set the polygon shape to a modify-only mode.

1 Above the viewer, click the modify button in the viewer toolbar.

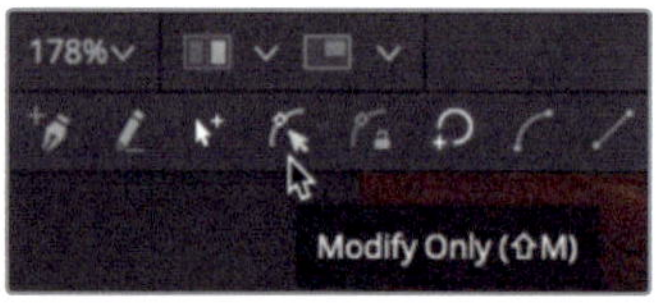

The modify only mode allows you to manipulate the existing control points on a polygon shape without worrying about adding additional controls points.

TIP Accidently selecting a handle on a spline instead of a control point can change your shape in unwanted ways. You can choose to show or hide control handles by clicking the show/hide handles button in the viewer toolbar, or by pressing Shift-H.

2 Select the control point along the right side of the protruding rope, and then in the viewer toolbar, click the smooth button or press Shift-S.

3 Use the handles that appear on the control point to form the curve to the shape of the rope.

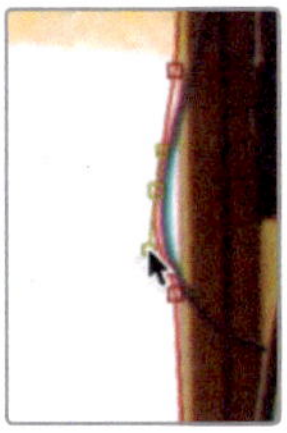

4 Smooth any other control points that require a curvature rather than a hard corner.

TIP To remove a keyframe, position the playhead over it in the time ruler, and in the inspector, right-click over the “Right-click here for shape animation” label and choose remove key.

The beauty of stabilizing the INSIDE clip is you will not have to keyframe the polygon shape mask at all. Since the clip does not move, the shape only has to be drawn once, and you are done.

Aligning and Blending Elements

With all three pieces of the puzzle in place (a stabilized background, stabilized foreground, and a mask shape) it’s time to create the split-screen composite. The next step is to connect the polygon mask into the composite, so it cuts a hole for the inside barn boor. To embed the matte into the INSIDE image, we’ll use a new tool called the matte control node. The matte control can be used to combine a foreground alpha channel into a background image.

1 Select the tracker 1 node connected to the INSIDE node.

2 From the toobar, select the matte control and press 1 to see it in the viewer.

3 Drag the output of the polygon node to the green foreground input on the matte control.

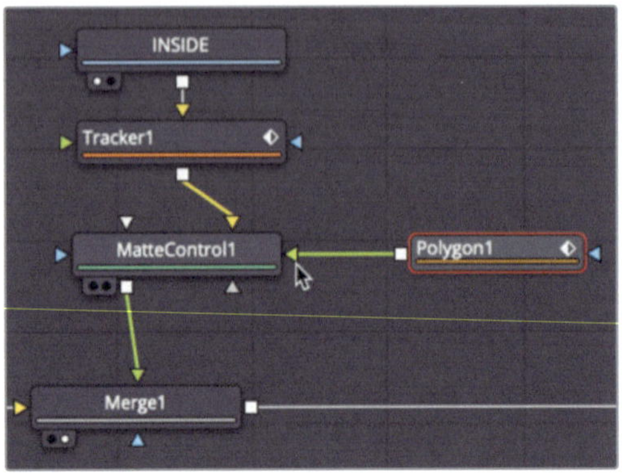

4 Select the matte control node, and in the Inspector, set the combine menu to combine alpha.

This causes the foreground alpha from the mask to be copied into the alpha channel of the background image. Instead of copying the matte, you want to subtract the matte from the image.

5 From the combine op menu in the inspector, select subtract.

When combining a matte with an image, you will typically need to premultiply the alpha. You can perform that operation directly in the matte control's inspector.

6 At the bottom of the inspector, click the post multiply image checkbox.

The effect mask subtracts the yard with the dog from the image, leaving a transparent hole that you can now fill with the empty yard.

7 Select the merge and press 1 to see the combination of the two clips using the matte.

Now we have the correct parts of the two clips showing in the viewer. You will need to align the OUTSIDE node to fit realistically in the hole cut by the matte. To do that, you'll add a transform node after the OUTSIDE's tracker and reposition the stabilized clip into the barn door.

8 Select the tracker 2 node in the node editor, and then click the transform tool in the toolbar.

The barn door opening in the OUTSIDE node is smaller than the opening created by the polygon mask. It is also offset to the left. You can use the transform node to scale and reposition the OUTSIDE node into place.

9 In the inspector, use the transform's size slider to scale the OUTSIDE node to around 1.3.

10 Use the on-screen controls to move the OUTSIDE barn door up until the top of the door is aligned with the INSIDE node's door, and then position it to the right until the left side of the barn doors in both shots line up.

11 Adjust the angle dial slightly, so the top of the barn doors are level.

12 Play the composite to see the two clips combined in a split screen.

You can now see the basic split-screen effect, but there are many things you can do to enhance this effect. One thing you'll want to do in most cases is to add the original camera's motion back into the clip. The only reason to stabilize the INSIDE node was to make rotoscoping the barn door easier. With that complete, restoring the motion will enhance the realism.

Restoring Motion

The key to adding the camera motion back into the shot is to reuse the tracker that removed it. Remember, the tracker node can perform many different operations once the tracking analysis is complete. The same node that tracked the clip for stabilization can be duplicated and reused to apply motion, sometimes called a "match move."

1 Select the tracker 1 node in the node editor, and then choose edit > copy or press cmd-C (macOS) or Ctrl-C (Windows) to copy the node.

2 Select the merge node and press cmd-V (macOS) or Cltr-V (Windows) to paste a copy of the tracker 1 node after the merge node.

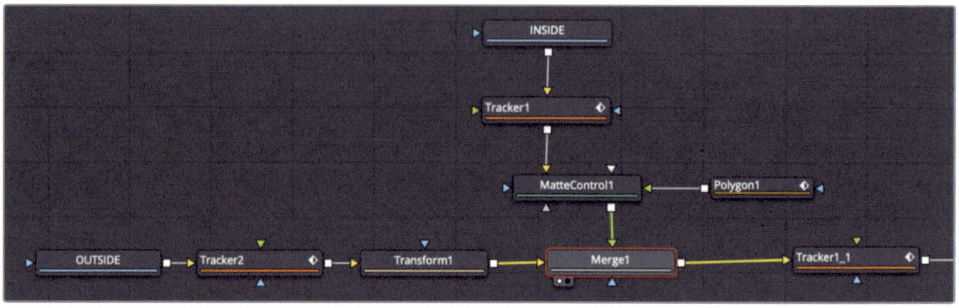

3 Select the newly pasted tracker 1.1 node and press 1 to see it in viewer 1.

We mentioned earlier in this chapter that the tracker node's operation is dependent on the connections to the node. To stabilize, you need a background connected. To add motion, the tracker must have a foreground clip connected. Right now, we have only the merge connected to the yellow background input.

4 Drag a second output from the merge node and connect it to the green foreground input of the tracker 1.1 node.

5 In the inspector, click the operation tab and select FG over BG from the merge.

6 Play the composite to view the split screen.

The foreground (FG) over background (BG) setting basically uses the tracker node as a merge node. The major difference is that in a tracker, the foreground image not only gets composited over the background, but it also inherits the motion captured by the tracker node.

Finishing the Shot

Your shot is almost complete. One or two final touches can make this shot perfect. First, you will feather the bottom of the mask to make the transition from the outside ground to the inside more gradual. Second, you will always need to do some color correction to match the two shots.

1 Move to the start of the render range.

  Earlier, we mentioned that the polygon mask is auto-animated. If you change any of the points on the spline, it creates a keyframe. By positioning the playhead on the first frame, we are changing the shape only at the beginning of the render range. Whatever we do will carry on through the entire composite.

2 Select the polygon node, and then, above the viewer, click the make double poly button in the viewer toolbar.

  Instead of blurring the entire shape to create a gradual transition, the double polyline option allows you to create a second outer spline shape anywhere along the spline for feathering the edge.

3 Press the tab key to select the outer polyline, or right-click over a control point in the view and choose controls > select > polygon1 outer polygon.

  Pressing the tab key with the double polyline option enabled toggles the selection of the spline shape from the inner shape to the outer shape.

  The outer shape appears as a green dashed line instead of the solid line appearance of the inner shape.

  By default, the points on the two shapes are identical so that any changes made to the inner shape will also be made to the outer shape. However, this relationship is one way, meaning adjustments to the outer shape can be made without affecting the inner shape.

4 In the lower-left corner of the polyline, drag the control point down to create a feathered edge, and then drag the lower-right control point down to create the feathering along the entire bottom of the shape.

TIP Holding the cmd or Ctrl key and dragging on the control point will also select and modify the outer polyline.

Since you modified the outer polyline on the first frame in the composition and there are no other keyframes on this polygon mask, the feathering will remain static for the entire shape. Now you can perform a small amount of color correction on the OUTSIDE node to better match the two clips.

5 Select the transform 1 node, and then in the toolbar, click the brightness contrast node.

6 In the brightness contrast node, drag the lift up to around 0.2 and the gamma up to around 1.2.

Your split-screen composite now has a better overall look with a soft edge matte and a bit of color correction. You've taken these two shots from the edit page timeline into a completed composite, using the tracker node, a simple polygon mask, and the merge node.

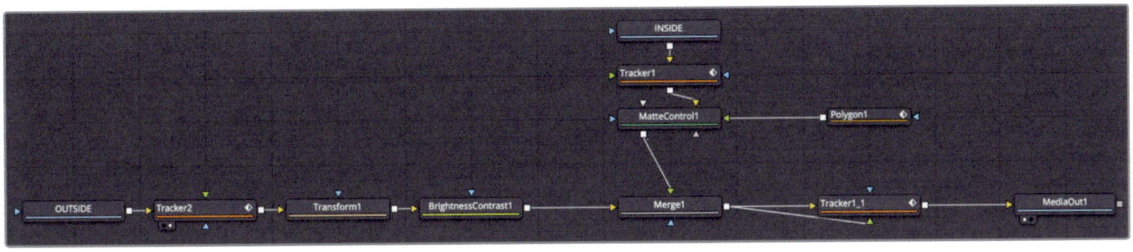

Completed node tree for Lesson 5.

Lesson Review

1 True or false: A Fusion clip is used to take multiple timeline layers from the edit page into the fusion page.

2 True or false: Green is the color of the effect mask input on a matte control node.

3 True or false: You must click the keyframe button in the inspector to animate a polygon matte.

4 True or false: The tracker is used to stabilize clips in the Fusion page.

5 True or false: In the tracker's operation tab merge menu, choosing BG only will stabilize the clip that is connected to the yellow background input of the tracker node.

Answers

1 True. A Fusion clip is used to take multiple timeline layers from the edit page into the Fusion page.

2 False. The green input on any node is the foreground input.

3 False. No button needs to be enabled. Polygon mattes auto-animate by default.

4 True. The tracker is used to stabilize clips in the Fusion page.

5 True. In the tracker's operation tab merge menu, choosing BG only will stabilize the clip that is connected to the yellow background input of the tracker node.

Lesson 6

Replacing a Sky

One of the most common problems with shooting outdoors is the sky. Either the location or time of day you are shooting lacks the drama your scene calls for, or your camera lacks the dynamic range to expose the foreground without overexposing the sky. The good news is, as far as visual effects go, sky replacement isn't all that difficult once you know the basic node structure.

In this lesson, you will learn an indispensable node structure for a sky replacement. During the process of learning how to perform a sky replacement, you'll also take a closer look at compositing images of different resolutions.

Time

This lesson takes approximately 50 minutes to complete.

Goals

Completed sky replacement for Lesson 6.

Understanding the Resolution of a Fusion Clip

This lesson will continue with the same project you restored in the previous lesson. We'll open the same timeline and begin with the clips located at the second red marker.

1. Open DaVinci Resolve 16, and from the project manager, open the Fusion 16 Lessons part 2 project.
2. In the timelines bin, double-click the **Part 2 - START** timeline.
3. Move the playhead over the second edit and the second red marker.

DaVinci Resolve in general, and the Fusion page specifically, is resolution independent. You can efficiently work with any number of elements at different resolutions. But knowing what resolution independence means and dealing with it in your node tree are two very different things. When constructing composites with mixed resolutions, you must be aware of how images are handled, not only between the edit and Fusion pages but also within the Fusion page itself.

Like the clips you worked with in Lesson 5, this second edit in the timeline also includes two layers.

4 Select the foreground clip on video track 2 and press D to disable this clip in the viewer and reveal the sky with rainbow clip on video track 1.

In this two-layer composite, the sky with rainbow is a still photo with a resolution of 3888 x 2187, and the foreground clip on video track 2 is 1920 x 1080. The timeline is also set to 1920 x 1080 HD.

As you can see, the sky with rainbow has some buildings along the bottom, which we do not want to see in the sky replacement composite. Since this is a high-resolution photo, we can scale it up without losing any quality.

5 Select the sky with rainbow clip on video track 1 in the timeline, and then in the inspector use the zoom controls to scale the clip up to about 1.5.

Because Fusion is fully integrated into DaVinci Resolve, the layering and transforms you perform on the edit page timeline carry over to the Fusion page.

6 Select the foreground clip on video track 2 and press D to enable this clip in the viewer.

7 Select both clips in the timeline, then right-click and choose new Fusion clip from the menu.

As you saw in Lesson 5, a new Fusion clip is created in the timeline and added to the selected bin. The two layers are collapsed into a container and can be brought into the Fusion page.

8 Position the playhead over the new Fusion clip, then click the Fusion page button, or press Shift-5.

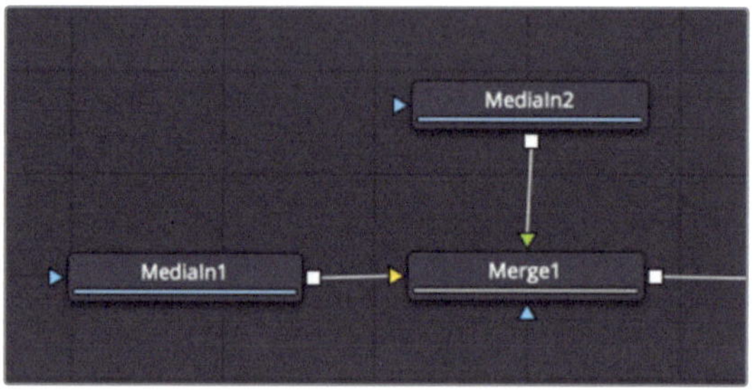

NOTE This lesson uses dual viewers. If your workspace is still using a single viewer, click the dual viewer button in the upper-right corner of the viewer.

The two layers are brought into the node editor and combined via a merge node. The media in 1 node represents the sky clip from video track 1, and the foreground clip from video track 2 is represented as the media in 2 node.

9 Hover your mouse pointer over viewer 2, hold the cmd key (macOS) or the Ctrl key (Windows) and scroll the middle mouse until you can see the resolution in the top-right corner of the frame.

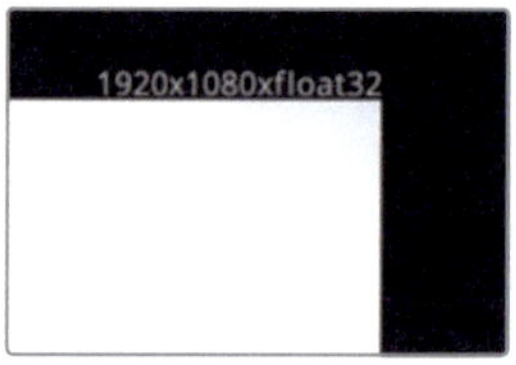

When using a Fusion clip, the resolutions of the nested clips are set to timeline resolution. This is the working resolution of the composite.

10 Select the media in 1 node, and then click the transform tool in the toolbar. Press 1 to see the transform node in the viewer.

The transform node is added to the sky clip. Since this is a large-resolution image in a 1920 x 1080 frame, we should be able to scale it back a considerable amount and see the remaining image that falls outside the frame boundary.

11 Select the transform node, and in the inspector, drag the size slider down just a small amount.

As you scale the image back, it reveals transparency around the edges as if the background clip has a resolution of 1920 x 1080. The beauty of a Fusion clip is that it allows you to use the intuitive tools of the edit page to layer, trim, and align your clips before bringing them into the Fusion page. However, the new Fusion clip is created at the timeline resolution. If your source clips and timelines have the same resolution, as you had in Lesson 5, then the Fusion clip is the most efficient way to set up your composite. However, if you are dealing with clips larger than the timeline resolution, then a Fusion clip will resize all the sources to fit. This is not the best setup for dealing with a large photo of a sky where we only want to use a small area. Let's look at a different technique that will allow us to work with clips at mixed resolutions.

Maintaining Resolution Independence

Unlike a Fusion clip, bringing a single clip from the edit page to the Fusion page maintains the clip's original resolution, despite how the timeline resolution is set. You experienced this using the map in Lesson 4. This means you are always using the highest quality compositing in the Fusion page when dealing with a single layer from the edit page.

1 Return to the edit page and choose Edit > Undo to undo the Fusion clip in the timeline.

With the clips returned to individual layers in the edit page, we can disable video track 2 and bring the sky image into the Fusion page.

2 Select the foreground clip on video track 2 and press D to disable the clip.

3 Select the Sky image on video track 2 and reset the scale you applied in the inspector.

4 Switch to the Fusion page.

If a clip or track is disabled, clicking the Fusion page button takes the first enabled clip under the playhead. In this case, it switches to the Fusion page with only the sky image from video track 1.

The resolution seen above the frame is now defined by the original source image resolution (3888 x 2187). Now you are compositing with the highest-quality sky image, but you still need to bring in the foreground clip.

5 Open the media pool, and from the sky replace bin, click on the metadata badge in the lower-right corner the foreground clip.

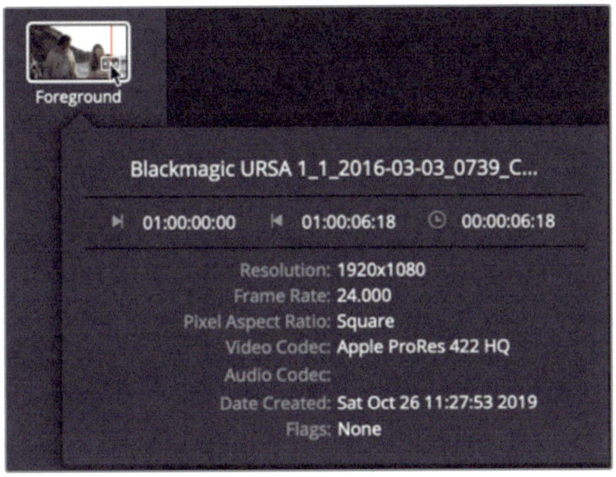

The metadata badge reveals a pop-up with some basic metadata about the clip. One of the pieces of the metadata is the resolution, which is 1920 x 1080. Dragging this clip from the media pool into the node editor retains the clip's native resolution.

6 Drag the foreground clip to an empty place in the node editor, and then press 1 to see it in viewer 1.

Above the frame, the clip displays its resolution as 1920 x 1080.

Just like single elements brought in from the edit page, elements from the media pool always retain their original resolution.

To keep the media organized, let's rename the nodes, so the names are more descriptive.

7 Rename the media in 1 node to **SKY** and the media in 2 node to **ACTORS**.

With all of the media at its highest resolution and our nodes named appropriately, we can begin creating the sky replacement.

8 Select the sky node and from the toolbar, and click the merge tool.

9 Drag the output of the actors node into the green foreground of the merge node.

Viewer 2 now shows a composition that uses the resolution of the larger sky image with a small 1920 x 1080 HD foreground in the middle.

A fundamental concept in Fusion is that the background input of a merge node determines the resolution of the merge node's output. This is how you can control the resolution when mixing footage of different sizes. In our case, the background input is the large sky image, not the full HD resolution of the foreground clip and timeline. To correct the resolution of our composite, we need to change the resolution of the sky image to be 1920 x 1080. There are a couple of ways to control the resolution of a composite, but we will use a background tool.

TIP The resize and crop tools also modify the resolution of a clip.

10 Drag the background tool from the toolbar to an empty place in the node editor.

11 In the inspector, click the image tab.

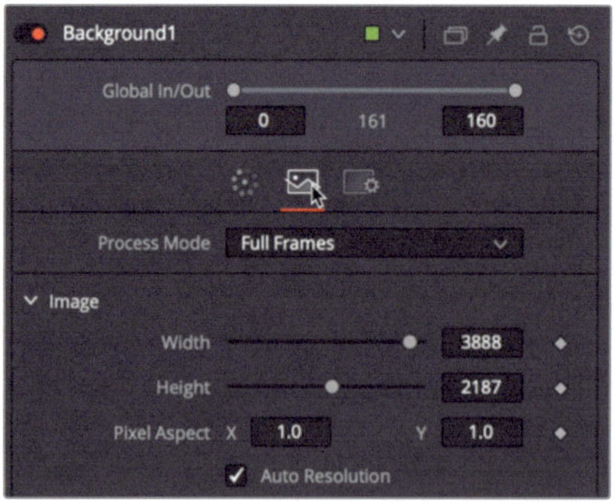

The background tool's image tab in the inspector has controls for the background image resolution.

12 Click the auto resolution button to disable the automatic resolution setting and enter **1920** as the width value and **1080** as the height.

The background input of the merge tool determines the resolution of the merge tool's output. So, by connecting a 1920 x 1080 HD image to the merge's background input, we can set the composite's resolution. The easiest technique to do this is to use a background node.

13 Select the background node, and from the toolbar, click to add a merge node.

14 Select the merge node and press 1 to see the black solid color of the background node in the viewer, then hover your mouse pointer over viewer 2, hold the cmd key (macOS) or the Ctrl key (Windows) and scroll the middle mouse until you can see the resolution in the top-right corner of the frame.

The merge node now uses a resolution of 1920 x 1080. If we connect the larger sky node to the foreground, it will be cropped (not resized) by the merge's resolution.

15 Click the connection line near the input of merge 1 to disconnect the sky node.

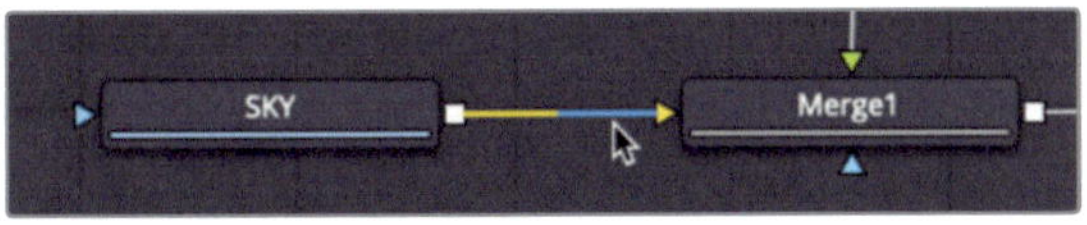

16 Drag the output of the sky node to the green foreground input of the merge 2 node.

As soon as the connection is made, viewer 1 shows the media in 1 node at its full resolution but cropped by the 1920 x 1080 merge node.

17 Drag the output of the merge 2 to the background input of the merge 1 node.

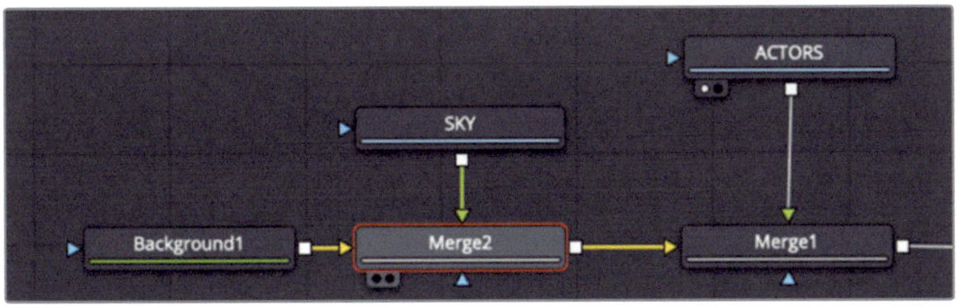

The merge loaded in viewer 2 shows exactly what we had on the edit page. One layer lies over another without any transparency. That's because the merge places the SKY foreground over the background, and since there is no mask or alpha channel in the foreground, there is no transparency.

Combining Tools to Create a Matte

Compositing typically requires an alpha channel or matte, as we used in Lesson 4 with the plane. Mattes or alpha channels are used to tell the software which areas of the foreground it needs to cut away to reveal the background. In our case, we have two elements, the live-action scene of the actors and our sky picture, neither of which have an alpha channel. When starting a sky replacement, some people might be inclined to reach for a luma keyer immediately. A luma keyer is a tool that creates a matte based on the luminance in an image, like an overexposed sky. However, it is usually not the right tool for sky replacement. At least, it's not the right tool to handle the edges of a sky replacement. A much better choice is to start with an apply mode located in the merge you already have.

1 Select merge 1 and, in the inspector, set the apply mode to darken (sometimes called minimum).

The darken apply mode shows the darkest pixel wherever the foreground and background overlap. Since the sky image is darker than the overexposed sky in the actors' clip, most of the sunset sky displays. However, there are lots of issues we need to deal with before getting a good looking sky replacement.

TIP The Channel Booleans node also contains a minimum mode and can be used in place of a merge.

2 Select the Sky in the node editor, and in the toolbar, click the transform node.

The larger sky image is cropped (not resized). You still have access to the entire sky image. You are just looking at it through a 1920 x 1080 window. The transform node will allow you to reposition the larger sky frame within the crop window.

3 In viewer 1, drag the center on-screen control to the left edge of the frame to include more interesting cloud formations.

Although the edges are taken care of with the darken apply mode, we'll still make use of a luma keyer to handle the transparency we see over the actors and hillside. Since the actors' clip has no alpha channel built into it, a luma keyer can create one for us.

4 In the effects library, choose the tools > matte category.

5 Drag the luma keyer to an empty area of the node editor, near the actors node.

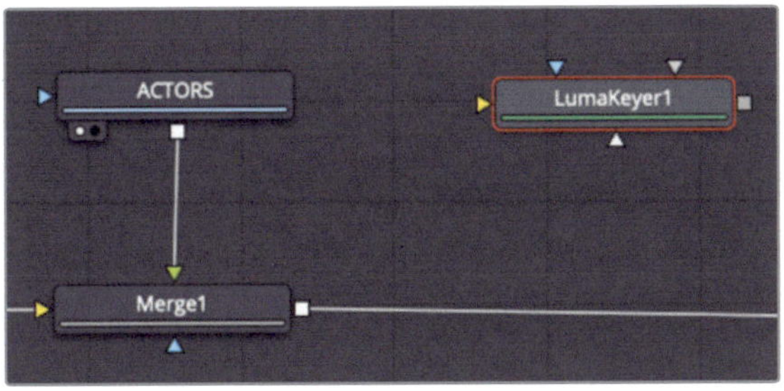

6 Drag a second output from the actors node and connect it to the yellow input of the luma keyer.

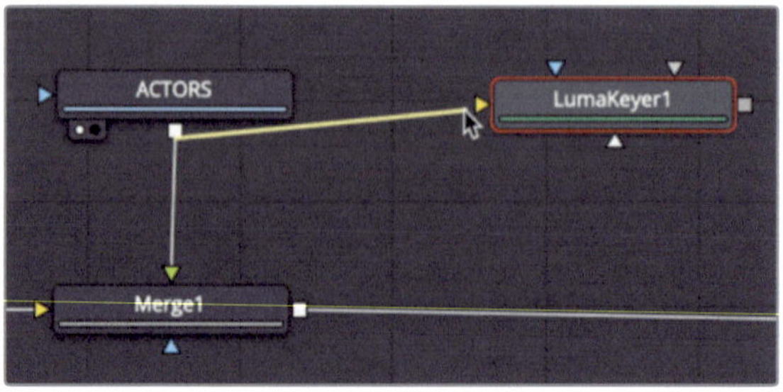

Nodes can have multiple outputs without causing any reduction in quality.

7 Press 1 to see the luma key in viewer 1.

Right away, we can see a semitransparent checkerboard pattern that represents transparency from the luma keyer settings. Because you will create a matte, it can be helpful to view the alpha channel in the viewer instead of the RGB image that is currently visible.

8 In viewer 1's toolbar, click the color controls button to view the alpha channel.

The viewer displays a grayscale image that represents the transparency based on the luma key. Areas that are pure white will be transparent, and areas of pure black will be opaque. Areas of semi-transparency will be some shade of gray.

9 In the inspector, drag the high threshold slider to the left until the majority of the sky appears solid white. The high threshold should end up around 0.8.

An important thing to understand about a matte is that they need to contain pure black and pure white. Gray areas, as mentioned above, will be a semi-transparent mix of foreground and background. That is OK for glass objects, but most of the time, areas will either be all foreground or all background. Right now, the sky is completely white, but the foreground of the actors isn't pure black.

10 Drag the low threshold slider slightly to the right to darken as much of the actors and rock as you can without darkening the sky. The low threshold should end up around 0.75.

11 To smooth the harsh edges, drag the blur slider between 2.0 and 2.5.

We now have areas of mostly pure black and pure white. Unfortunately for us, there are many highlights on the actors and some darker areas in the upper left of the sky that will cause problems in our matte.

Fixing Holes in a Key

The remaining problems we have are the areas where white still shows through due to the very bright highlights over the actors, the rocks, and the water. There is also a problem in the upper-left area of the sky. Let's tackle the rocks and actors first. Instead of laboriously drawing a matte or painting out the holes, we will use an easier method. Enter the erode dilate node. An erode dilate node expands or shrinks the edges of a matte.

1 In the node editor, select the luma keyer and from the filter category in the effects library, and click the erode dilate node to add it to the node editor.

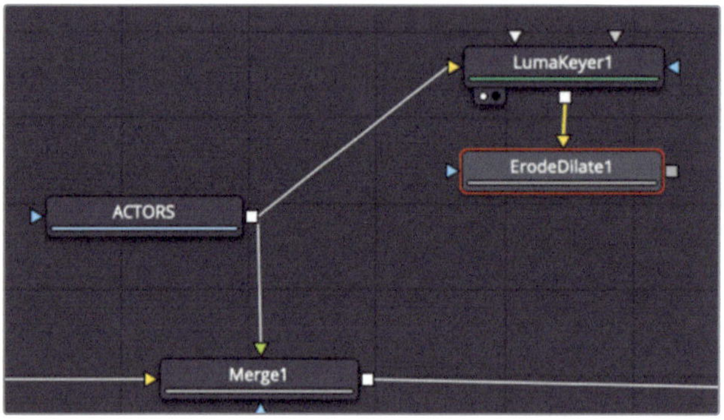

2 Press 1 to see it in the viewer.

3 In the inspector, in the amount slider, type **-0.005**.

Moving the amount slider left (negative) expands the edge of the matte covering up the holes. Unfortunately, it also expands the border of the rocks. How do we fix that? We'll add another erode dilate and set the amount in the opposite direction.

4 In the effects library, click the erode dilate to add a second node directly after the first erode dilate.

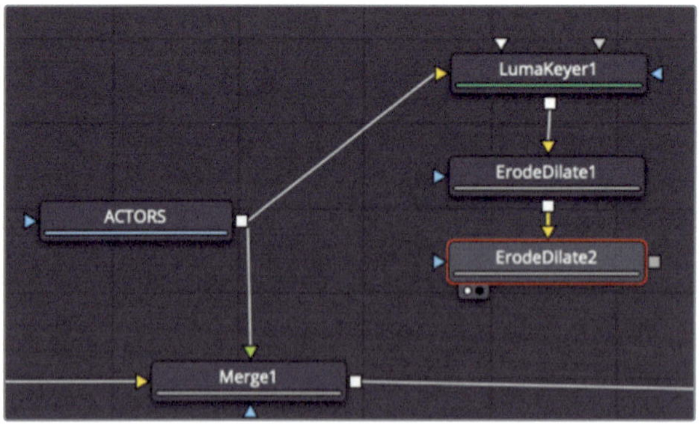

5 Press 1 to view the second erode dilate in the viewer.

6 In the inspector, in the amount slider, type **0.009**.

TIP If a slider in the inspector does not reach the value you want, you can often type in the number to extend the slider's range.

The erode dilate has brought us closer to having a clean black and white matte, but visual effects is often a game of whack-a-mole, where you fix two problems and reveal another. The erode dilate has not removed a few white spots on the actors that we will have to fix. We also need to combine this newly generated matte with the actual RGB image of the actors to get an idea of our finished shot.

Embedding Alpha into an Image

As you did in Lesson 5, you'll again use a matte control node to copy the alpha channel from the foreground and background image.

1 From the toolbar, drag the matte control tool to the node editor near the erode dilate 2 node.

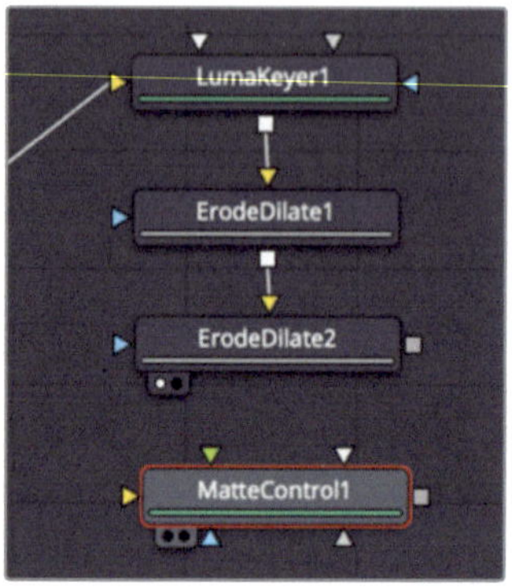

2 Drag a third output from the actors node to the yellow background of the matte control.

The background input on a matte control is used to connect the image that will have the alpha embedded.

3 Press 1 to view the matte control in the viewer and set viewer 1 to view the RGB channels instead of the alpha channel.

4 Drag the erode dilate 2 node output to the green foreground input of the matte control.

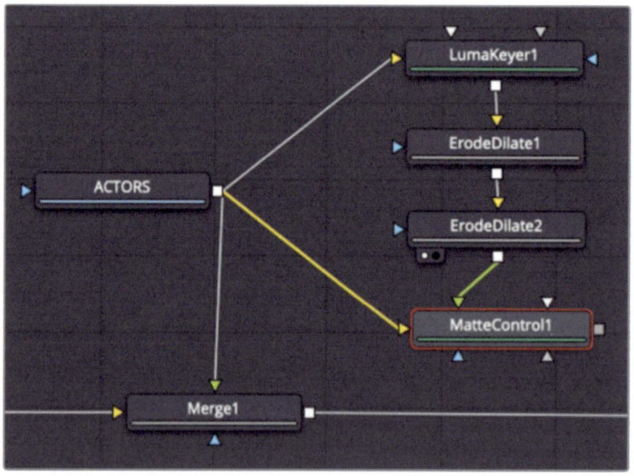

5 To embed the alpha channel into the actors image, in the Inspector, set the combine menu to combine alpha.

You now have the actors with an embedded alpha channel.

Typically in a matte, the white areas are opaque, and black areas are transparent. That is the opposite of what we have. So, we'll invert the matte.

6 In the matte control's inspector, click the invert matte checkbox.

To make a clean matte for compositing in the merge node, alphas must be premultiplied.

7 Click the post multiply button in the inspector.

Before going any further, it helps to view the composite of the foreground with the sky. The matte control output forms our new foreground with alpha, while the output of merge 1 contains the nice clean edges for our composite. So, now you will merge the two together.

8 Select the merge 1 node and click the merge tool in the toolbar to add a third merge.

9 Drag the output of matte control to the green foreground input on the merge 3 node.

10 Click the play button to review the composite.

The results are displayed in viewer 2. The image appears a lot closer to what we want, but there are still a few nagging semi-transparent areas to fill in over the actors before finishing off our sky replacement.

Tracking the Sky into Position

As you play the composite, it looks fine when the image is still, but as soon as it plays and the camera moves, the illusion falls apart. For this sky replacement to work, the sky needs to move as the camera moves. Instead of using the tracker node as you did in the previous lesson, you'll apply the tracker as a modifier.

1 Select the merge 2 node that is connected to sky node in the node editor.

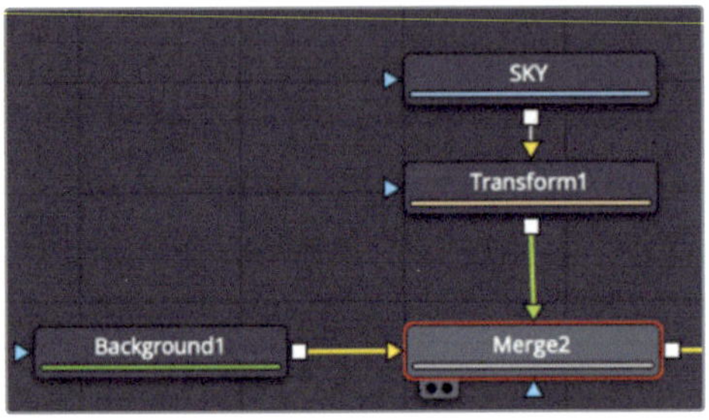

You'll need to apply the tracker modifier to a node that contains position controls. The merge tool includes center X and Y controls that can be used to reposition the clip connected to the foreground input.

2 In the viewer, right-click over the merge's center on-screen control, and from the menu, choose merge 2 center > modify with > tracker position.

TIP Alternatively, you can right-click over the center control in the inspector to attach the tracker.

The tracker modifier is attached to the merge's center X and Y controls. Although the process of tracking is the same as the tracker node, the tracker modifier is a single-point tracker that tracks up and down and side to side movement, so it is more limited than the multi-point tracker you used in the split screen. Still, it is perfect for quick, simple tracks, as we have here.

Adding a tracker modifier displays the tracker pattern and search rectangles over the center position in the viewer. Since it is a modifier, the tracker controls are displayed in the merge 2's modifier tab.

3 At the top of the inspector, click the modifier tab to display the tracker controls.

The tracker modifier assumes you want to track the background of the merge node. In our case, we want to track the actors' node.

4 From the node editor, drag the actors node to the tracker source field at the top of the inspector.

NOTE As you drag a node from the node editor into an inspector source field, the node will appear to move at first, but it'll snap back into its original location once the mouse pointer leaves the node editor.

5 Move the playhead to the start of the render range.

Just as you did in the previous lesson, you need to position the tracker over a high-contrast pattern that moves precisely how you want the sky to move. For this shot, the ridge in the background has a few rocks that stick out. These would make good high-contrast tracking points that we can use for the tracker.

6 In viewer 1, drag the handle in the upper-left corner of the pattern box to position the tracker over the very small, sharp rock sticking up from the ridge, between the two actors.

Unlike when viewing the tracker node, the pattern box does not expand to show a magnified view of the area you are over. However, the inspector does contain a small preview of the selected area.

Since this camera does not move very fast, we can leave the search rectangle at its default size.

7 In the inspector, click the track from first frame button to begin the tracking process.

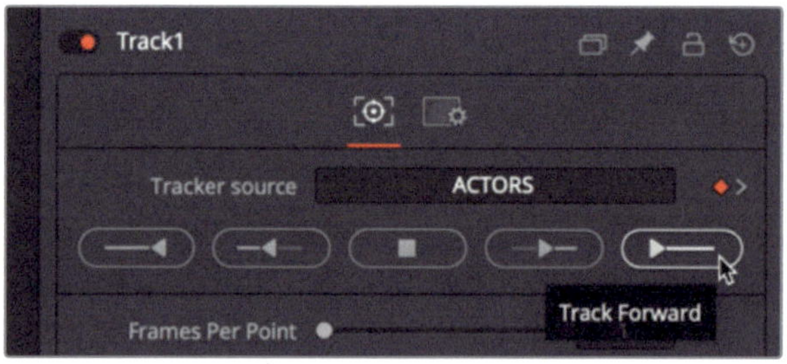

Viewer 1 displays the progress of the track until the analysis is interrupted around frame 90 when the actor's head obscures the rock. You'll need to find a way to continue the tracking for the remainder of this clip. Although the rock you initially selected is fine for the first half of the clip, it won't be the solution for the second half.

TIP If the tracker is interrupted prior to frame 90, try repositioning the pattern box better or making the pattern box slightly smaller and track the clip again.

Fixing Interrupted Trackers

When the tracking process is completed, a dialog appears to display information about the number of frames tracked and the time it took.

1. In the dialog, click OK to continue with the composite.

   You now have a solid track for the first 90 frames or so, but you'll need another solution for the remaining 70 frames. Instead of giving up hope, you can keep the tracking data of the initial rock you selected for the first half of the shot, and now identify a new tracking point for the remainder of the shot.

2. In the render range, drag the playhead to frame 90.

   This is the last good frame of tracking data from the first rock you tracked. At this point, you need to locate a new high-contrast object to track.

3. In the inspector, set the path center menu to track center (append).

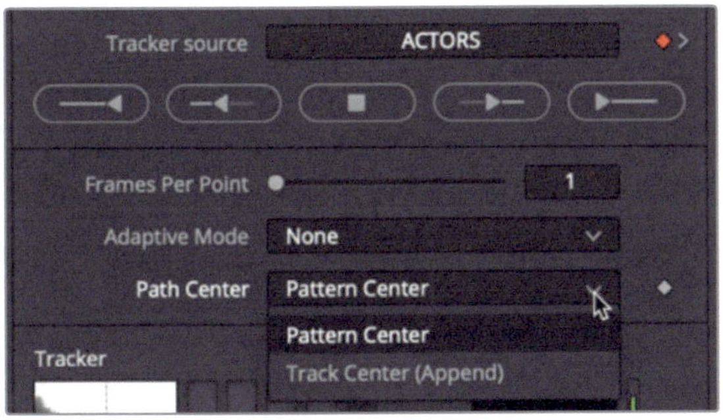

   The track center (append) setting allows you to move the pattern box over a new object and continue the tracking based on this new point.

4 In viewer 1, drag tracker 2's handle in the upper-left corner of the pattern box to place the tracker over the sharp rocks on the right side of the actress.

5 Click the track forward from current frame button to pick up the tracking process from frame 90.

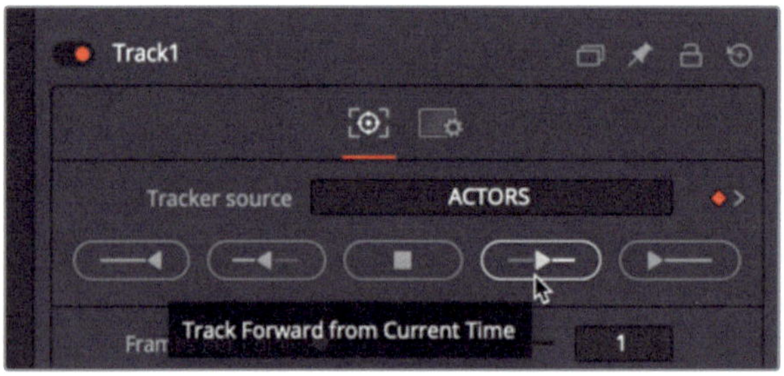

6 When tracking is completed, in the dialog, click done to continue.

Fusion automatically blends the two analyses for tracker 1 to create one seamless motion path. You now have a steady track that follows the position of the camera movement. Since you applied the tracker as a modifier to the merge center x and y, the motion of the sky automatically follows the tracker.

Blending In the Original

If you leave the sky replacement as it appears now, it just looks like a pasted-on sky. One of the main tricks when performing a sky replacement is to always mix back in a bit of the original sky. That's the final touch we need to apply to our sky.

1 Select the background node and press delete to remove it from the composition.

The only reason for the background node was to format the resolution of the merge correctly. If you have an image that uses the desired resolution, you can just use that image instead of adding a background node. For instance, in our composite, the actors node is the resolution we want for the final output. So we'll replace the background by taking yet another output of the actors node.

2 Drag a fourth output from the actors node to the background input of the merge 2 node.

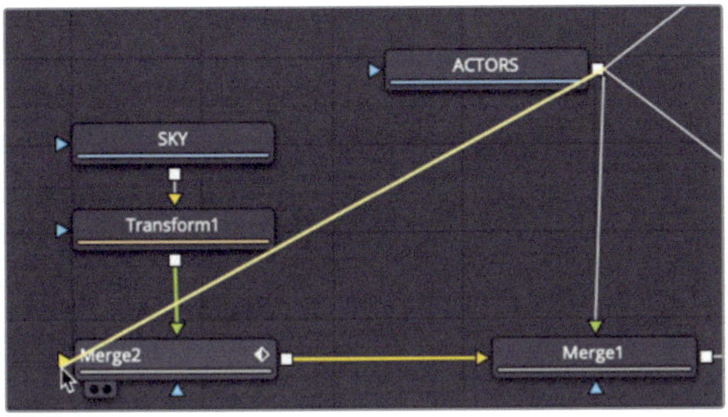

The composite returns to the viewers and looks no different than it did with the background node. However, a significant benefit of using the actors node is that we can blend back some of that original sky with the sky connected in the merge.

3 Select merge 2, and in the inspector, drag the blend slider down to around 0.7.

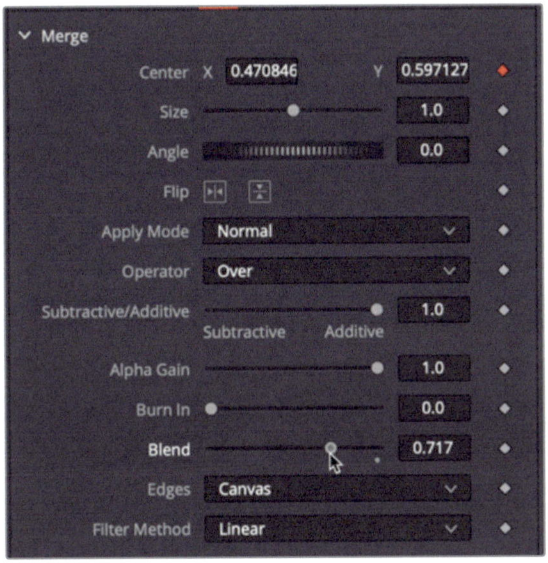

That one change has turned your pasted sky into a realistic replacement.

Some sky replacements require a few mattes, and some require none. Some might require a bit of color correction to get the foreground and background to match better. The goal of this lesson was to give you a basic structure you can build on for any sky replacement you may come across.

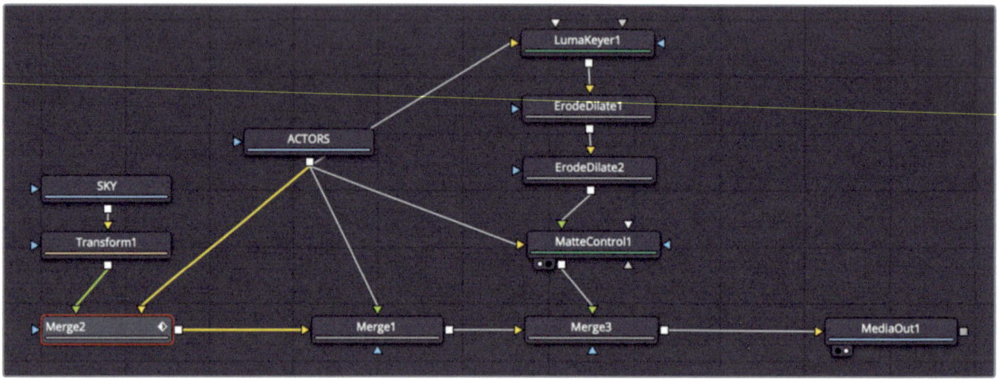

Completed node tree for Lesson 6.

Practice Exercise

In this practice exercise, you'll need to use the masking and tracking skills you've learned from these past two lessons. Using the third edit in the timeline, you'll simulate depth of field by creating a matte around the robot in the front, tracking the matte, blurring the remaining robots, and adding a separate stronger blur to the background.

1. From the edit page timeline, go to the third edit in the timeline (first blue marker) and switch to the Fusion page.
2. Bring both layers into the Fusion page.
3. Create a polygon matte for the front-most robot on the left.
4. Track the robot and apply the tracking data to the polygon matte.
5. Add a blur to the robots node and use the matte, so the blur applies only to the robots in the back.
6. Add a blur to the background room, so it is blurred more than the robots to simulate greater depth of field.
7. Color correct the robots to be slightly darker and with less saturation.

Here's a final image of how your composite should look.

Lesson Review

1. True or false: Clips from the media pool use the timeline resolution when added to the Fusion node editor.
2. True or false: The merge node contains apply modes like screen, multiply, and darken.
3. True or false: To offset a tracking point when a pattern becomes obscured, you must create a new tracking point.
4. True or false: Tracker modifiers have all the same controls and functionality as the tracker node.
5. True or false: A luma keyer is used to create a mask from the brightness in an image.

Answers

1. False. Adding a clip from the media pool to the node editor uses the clip's native resolution.
2. True. The merge node contains apply modes like screen, multiply, and darken.
3. False. To offset a tracking point you must set the path center menu to track center append.
4. False. Tracker modifiers track only a single tracking patterns, while a tracker node can track multiple patterns.
5. True. A luma keyer is used to create a mask from the brightness in an image.

Lesson 7

Replacing Signs and Screens

You've learned how to match move using the standard tracker; now it's time to learn a more sophisticated method that is ideal for replacing tablet screens, billboards, or signage on the side of vehicles. This type of match move is most efficiently done using a planar tracker. Planar tracking, as the name suggests, relies on there being a flat, planar surface for it to work.

In this lesson, you'll add a new sign to the side of a moving van. The process of adding a new sign to a moving vehicle (or screen to a tablet) is made up of three parts. First, you'll need to track the flat surface as it moves. Then, using Fusion's paint tool, you'll remove any tracking markers to create a clean surface. Once that is complete, you can composite a new logo using the tracking data.

Time

This lesson takes approximately 40 minutes to complete.

Goals

Completed composite for Lesson 7.

Tracking Planar Surfaces

Having Fusion built in to DaVinci Resolve comes with a number of benefits. One of those benefits is that it changes visual effects and motion graphics from a shot-based, isolated part of the workflow to an integrated, multi-shot workflow enabling you to change shots in an instant without importing, exporting, or timing the clips to the few frames you need.

1. Open DaVinci Resolve 16, and from the project manager, open the Fusion 16 Lessons part 2 project into the edit page.
2. In the timelines bin, double-click the **Part 2 START** timeline.
3. Move the playhead over the fourth edit and the third red marker, and press Shift-5 to switch to the Fusion page.

4 Press the spacebar to view the clip.

This shot of the van will ultimately need a sign added as it drives off. However, the first part of the composite is to put those black markers to use and track the side of the van.

The single point tracker is the simplest tracker in the Fusion page. Though it works well on many shots, it is not the most optimal tracker in some specific cases. For instance, what happens if the camera changes perspective? No matter how many point trackers you use, they are still independent 2D points that can only process limited camera motion. A planar tracker tracks multiple points across a defined flat surface, such as a billboard, TV screen, or in this case, the side of a van. In doing so, it calculates a more accurate 2.5D track that can contain more complex motion.

On this shot, you'll track the van's movement, so when you ultimately add the new logo, it will move across the shot as the van moves.

5 In the upper left of the user interface toolbar, click the effects library button to reopen the panel.

6 Select the tools > tracking category, and insert the planar tracker tool between the media in 1 and the media out 1 nodes.

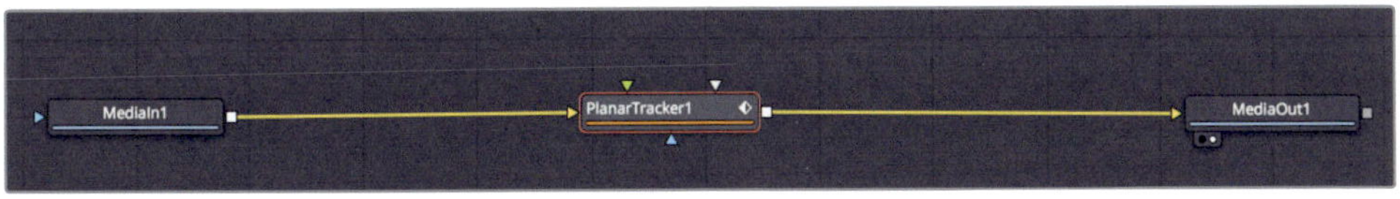

7 Press 1 to see the planar tracker in the viewer.

It's important to start planar tracking on a frame where the area you are tracking is clear and large in the frame. In this case, frame 65 is ideal because the side of the van is large in the frame but moving slowly, effectively eliminating motion blur.

8 Move to frame 65 in the comp.

Unlike the tracker that you used in the previous exercise, the planar tracker does not use one or two tracking patterns. Instead, it tracks the motion, scaling, and perspective distortions of an entire planar surface in a background clip. So, the next step is to identify the planar surface you want to track.

When the planar tracker is added, the polygon shape toolbar appears above the viewer. The same functionality you used previously to create polygon shapes applies here.

TIP If a clip has any significant lens distortion, it should be removed using a lens distort node before performing a planar track.

Selecting a Good Area for Planar Tracking

When drawing a shape around the area for planar tracking, keep the following guidelines in mind:

- Select as large an area as possible.
- Select an area that stays in frame as much as possible.
- Select an area that is clear of obstruction from moving foreground objects.
- Start your track when the area you are tracking is at its maximum size. You want as much detail in the area as possible when you begin the track.
- Start your track on a frame in which the area is the least distorted.

9 In viewer 1, draw a very simple shape that fits around the side of the van. Do not go outside the van, and do not include the wheels.

The area inside your shape is the pattern that will be tracked over time.

TIP When defining the area to track, include only pixels inside the shape that belong to the plane being tracked (in this case, the side of the van). Do not include any of the background area.

10 At the top of the inspector, click the set button.

The current frame is set as the reference frame for the rest of the track.

TIP If you stop before the tracking is completed, you must re-click the set button before resuming the tracking process.

11 At the bottom of the inspector, click the track to start button.

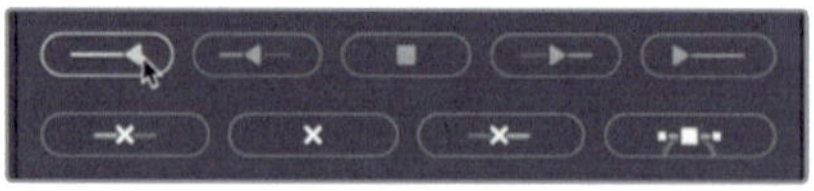

The viewer shows the tracking as it progresses backward to the start of the shot. When it is done, a series of dots appear in the render range to indicate that the track is completed.

12 When the first half of the track is done, return to the inspector and click the go button to move the playhead back to frame 65. Then click the track to end button.

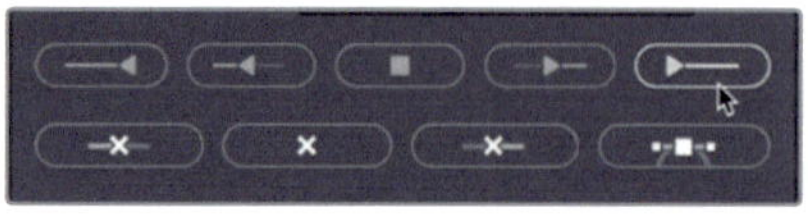

Despite the majority of the van disappearing off-screen, the planar tracker continues to track right to the end of the clip.

13 To test the track, in the inspector, change the operation mode pop-up menu to steady, and play the clip. (Make sure you have the planar tracker node and not the media in node displayed in a viewer).

The side of the van should stay locked in place throughout the timeline playback.

14 Reset the operation mode pop-up menu to track.

The steady setting is obviously not what you're trying to accomplish for the shot, but it's a good way to evaluate a track and ensure that there's no drift or bumps in the tracking data. If you do see errors, readjust your tracking shape at frame 65, and try again.

Painting with the Clone Tool

With the tracking done, we can now focus on removing the tracking markers. You will create a clean side of the van using the paint tool to clone white areas of the van over each marker. You just need to freeze a single frame to paint over, then cut out the clean side of the van and composite it over the live full-motion shot. Let's start by freezing the frame we want to paint over.

1 In the effects library, navigate to the tools > miscellaneous category, and drag the time stretcher tool into an empty part of the node editor.

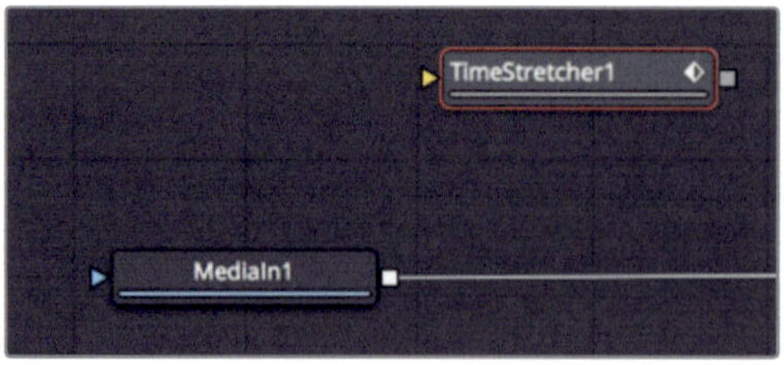

The first part of this job is to freeze the van on a clear frame. Freeze frames from the edit page do not transfer into the Fusion page, but it is easy enough to create a freeze frame using Fusion's time stretcher node.

2 Drag from the media in 1 output to the yellow input on the time stretcher 1 node.

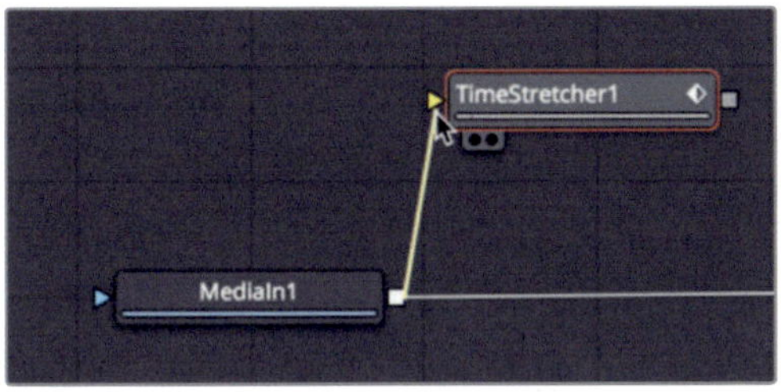

Dragging a second output from the media in node is similar to duplicating a clip in a timeline.

3 Press 1 to load the time stretcher node into the viewer.

The time stretcher automatically adds a keyframe when you connect a source to it. Doing so makes it easy to speed ramp a clip. However, you want to create a freeze frame, so you'll remove this keyframe and enter the frame number you want to hold for the entire shot.

4 In the inspector, click the red keyframe button to the right of the source time slider to disable it.

The frame you want to hold for the duration of the clip is frame 65, because it displays a clear, undistorted image of the side of the van.

> **TIP** The time stretcher frame numbers are based on the entire clip length, not the duration of the clip in the timeline.

5 In the source time field, enter **65** to hold frame 65 for the entire length of the shot.

If you do not keyframe this value, the time stretcher will hold the frame throughout the clip's duration. That's all it takes to create a freeze frame.

Now, you'll paint out the markers on this freeze frame.

6 In the node editor, select the time stretcher. In the toolbar, click the paint tool to connect it after the time stretcher.

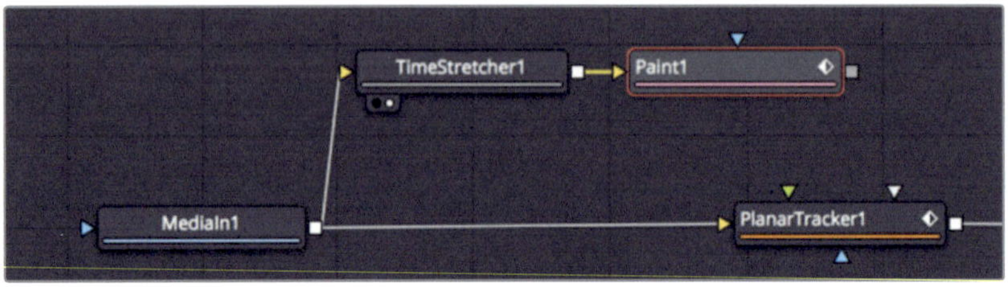

7 Press 1 to see the paint tool in the viewer.

With the paint tool selected, the inspector changes to show various paint controls, and a viewer toolbar is displayed above the viewer with the various paint stroke types.

The paint tool offers several stroke and paint styles suitable for motion graphics or retouching shots. In this exercise, you'll use a simple stroke brush in clone mode to copy white areas of the van and paint over the markers.

8 In the viewer toolbar, click the stroke tool.

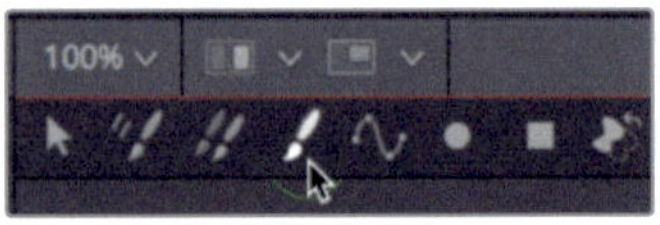

The stroke tool is the most versatile of paint tools, and the one you will use for most paint tasks.

9 In the inspector, click the clone apply control button to switch from painting with color to painting with a clone brush.

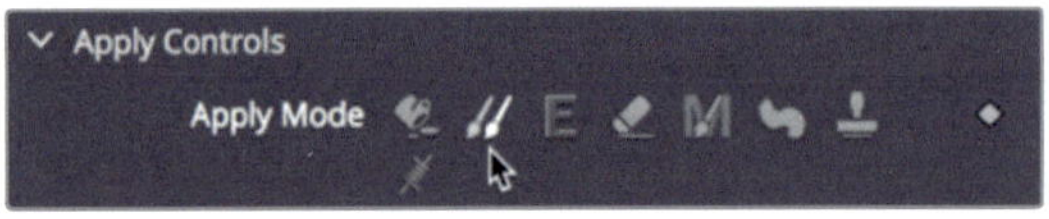

The clone brush works by selecting a source area from a frame and a destination area. The source area is the area in the frame you want to duplicate as you paint. The destination area is the area you will paint over using the content of the source area.

10 Option-click (macOS) or Alt-click (Windows) to the left of the first black marker in the upper left of the van to select the source area offset for the clone brush.

11 Once you have selected the source area offset for the clone brush, paint over the first marker.

As you paint, the source area offsets by the same amount. For instance, when you paint upward, the source area selection also moves up. It is often good to keep the source selection near the area you will paint over because doing so keeps the texture and color of the two areas as similar as possible.

TIP You can hold down the middle mouse button and drag in the viewer to pan.

12 If necessary, Option- or Alt-click on a new offset for the next marker and paint over it. Continue choosing a new offset and painting over each marker, until all the markers are removed.

TIP For cloning dozens or hundreds of small objects in a frame, a potentially faster alternative to using the stroke tool is the multi-frame clone tool. However, you must set the duration of a multi-frame clone tool before you begin painting.

You now have a clean van, perfect for compositing a new logo. In the next exercise, you'll import a logo and learn how to blend it onto the side of the van.

Using Photoshop PSD Layers

DaVinci Resolve can use various still image formats for graphics including TIFF, JPEG, and PNG. It can also use layered Photoshop files, allowing you some flexibility when it comes to selecting the layer you want to use from the PSD graphic.

1 Ensure nothing is selected in the node editor by clicking in an empty graph area.

 You'll add a graphic file to this graphics bin.

2 From the Fusion menu, choose Import > PSD. Then, navigate to the R16 Fusion Lessons folder > Fusion Files and import the EatatJoes.psd file.

3 Once imported into node editor, you will get a node representing each layer in the file and a merge node, labeled NORMAL, that combines the layers together.

4 With the NORMAL node selected, press 1 to see it in the viewer.

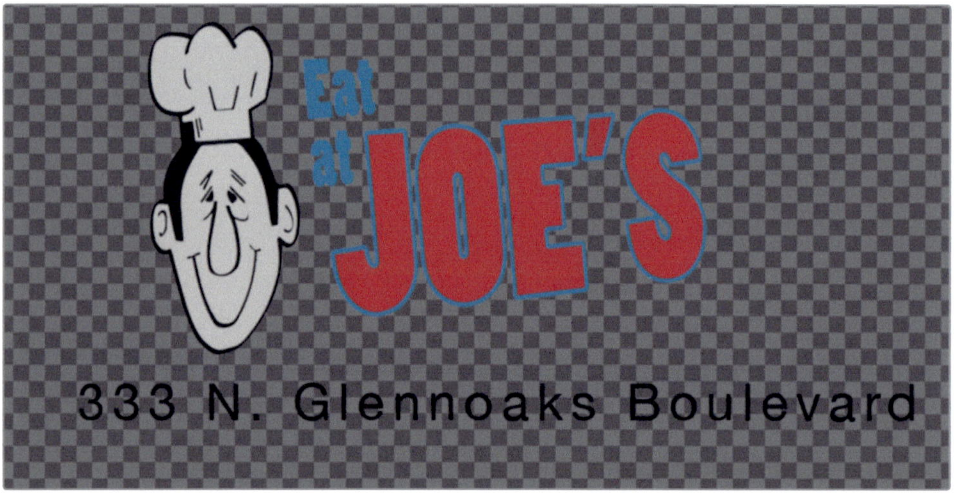

 This photoshop image is 1920 x 1080 resolution and includes two layers. One layer is the main logo, while the other layer is the address of the company. You can delete the NORMAL and ADDRESS nodes to use only the LOGO layer.

5 Select both the ADDRESS and NORMAL nodes and press delete or backspace on the keyboard.

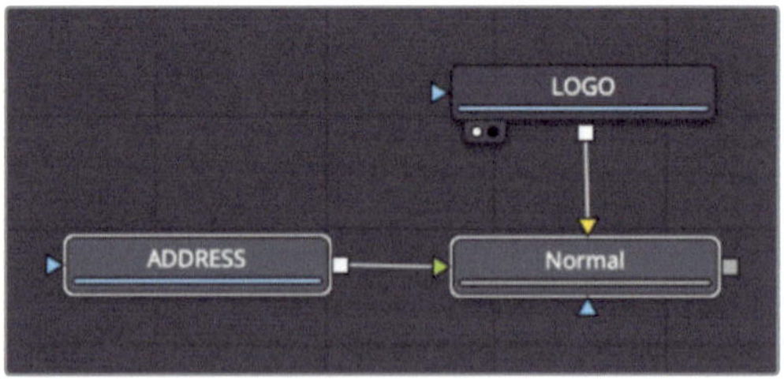

 With only the LOGO node remaining, you can composite it over the van.

6 Insert a merge node after the paint node, and press 1.

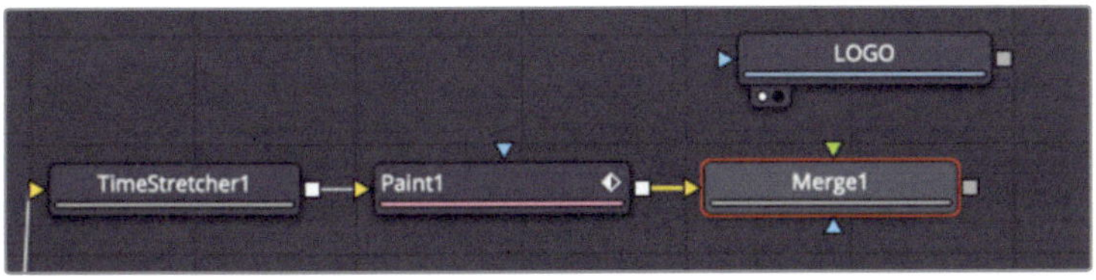

7 Connect the output of the LOGO node into the green input of the merge node.

That looks just like what it is: A photoshop document placed on top of the van clip. However, with some blending, color correction, and positioning, you can make it appear much more realistic.

8 Select the merge node, and in the inspector, set the apply mode to soft light.

The soft light apply mode is a good choice because it softly lightens the logo based on the white van color, giving the logo a nice diffused look, However, since the crease lines in the van are darker than much of the logo, the logo is darkened in those areas. The end result is good but maybe a bit too light and diffused. We can improve it a bit using some simple color correction.

9 From the toolbar, insert a brightness contrast node between the LOGO node and the merge 1 node.

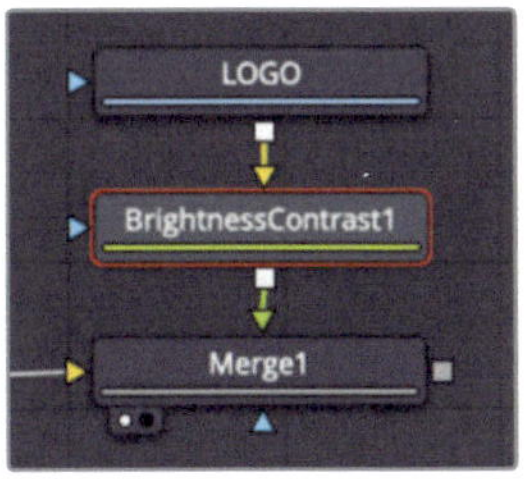

10 Adjust the lift down to darken the black part of the logo and increase the contrast slider until the logo appears less diffused.

Although looking much better, the edges of the graphic have become a bit ragged.

This is due to the fact that we have color corrected a graphic with a premultiplied alpha channel. So, just as we did in Lesson 4, we must pre-divide before the color correction and post-multiply after it. We can do both with just one checkbox in the brightness contrast node.

11 With the brightness contrast node selected, click the pre-divide/post-multiply check box.

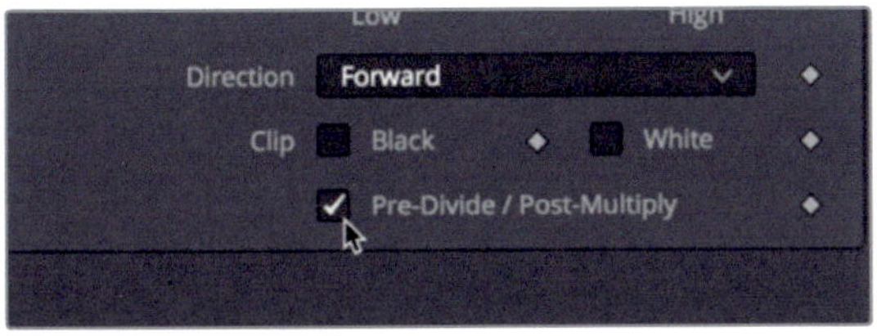

You now have a great looking logo that just needs a bit of positioning to appear like it is actually painted on the side of the van.

Corner Pinning an Image

Positioning this logo could be done using a transform node, but that wouldn't allow you to replicate the perspective distortion from the camera's viewpoint. To recreate any perspective distortion, it is best to position this graphic using a corner pin node.

1 In the node editor, select the brightness contrast node.

You'll place the corner pin node directly after the brightness contrast node but before the logo is merged with the freeze frame.

2 In the effects library, select tools > warp, and then click on the corner positioner tool to add it to the node editor.

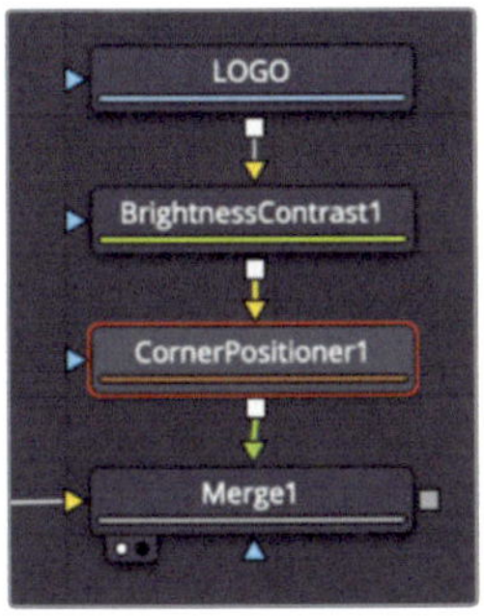

The entire logo has been squeezed onto a tiny area of the van based on the corners of the frame. To fix that you need to position each corner of the logo near the corresponding corner of the frame. You can do this easily by dragging each corner into position or entering values in each of the four corner position controls in the inspector.

What Coordinates Do I Use When Positioning Elements?

The coordinate system Fusion uses starts in the lower-left corner where the X position is equal to 0, and the Y position is equal to 0. The upper-right corner places X = 1 and Y = 1. The center of the screen is therefore always X = 0.5 and Y = 0.5.

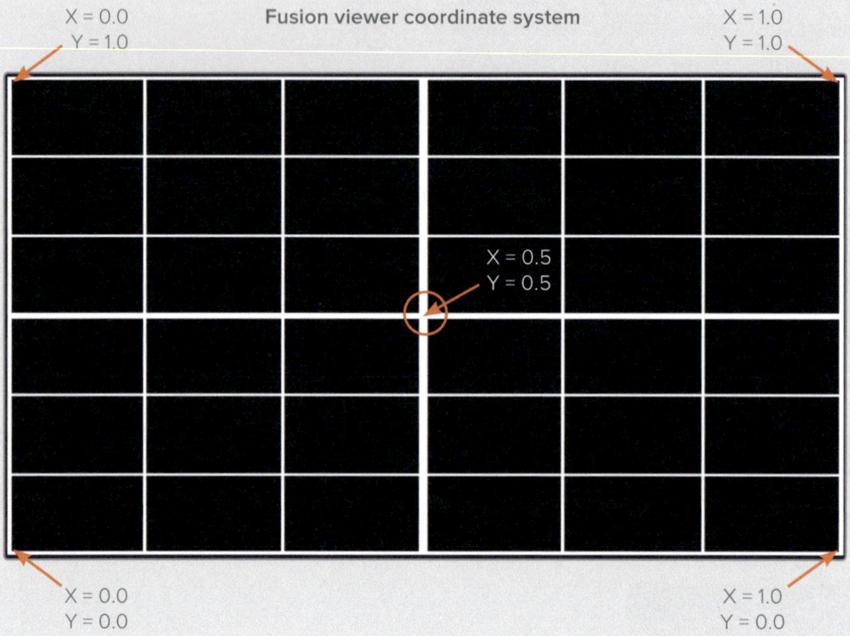

3 In the viewer, drag the upper-right corner of the graphic corner pin outline to the upper right corner of the frame, or enter 1 for X and 1 for Y in the corresponding inspector fields.

4 Drag the lower-right corner of the graphic corner pin outline to the lower-right corner of the frame, or enter 1 for X and 0 for Y in the corresponding inspector fields.

The two corners on the left will frame the graphic best if they are located about 25% in from the left edge.

5 In the inspector, set the top left X value to .25 and the Y value to 1, placing the corner at the top of the frame and 25% in from the left edge.

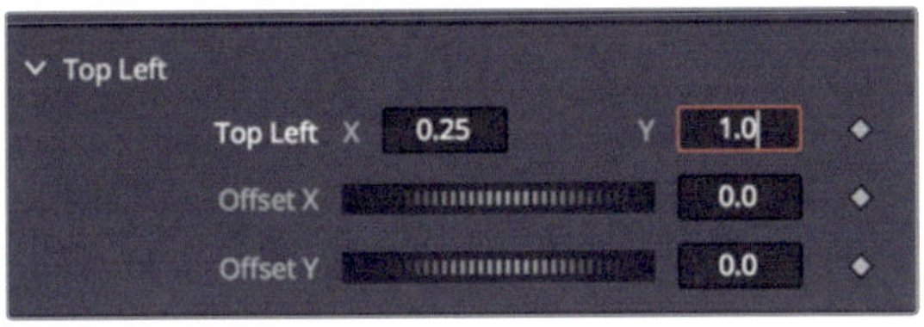

6 In the inspector, set the bottom left X value to .25 and the Y value to 0, placing the bottom at the bottom of the frame and 25% in from the left edge.

With these values entered, the graphic is placed over the van but positioned a bit too close to the roof. Fine tuning the left corners to be lower in the frame will pull the graphic away from the top of the van.

7 In the viewer, drag the top left and bottom left corner pin control points slightly down in the frame until the graphic is more realistically located on the van.

The placement and appearance of the graphic is perfect, but it is still just placed on a still frame without any movement. Hang in there; you'll fix it in the next exercise.

Combining Mattes and Images

So far, you have a freeze frame of the entire shot, but you need to isolate the side of the van with the logo to composite it. This is where the rotoscoping technique you learned earlier comes in handy. You'll create a matte that isolates just the side of the van.

1 Go to frame 65, then from the toolbar, drag the polygon tool into an empty area of the node editor near the merge tool.

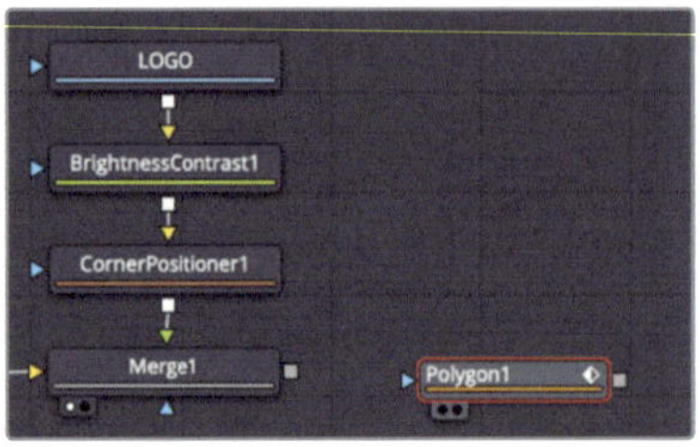

2 With the polygon tool selected, draw a shape around the side of the van using the creases by the door and roof to guide your matte. Be sure to close your matte by connecting the last and first control points or pressing Shift-O to connect them automatically.

You now have a matte, but you have nowhere to connect it into the node tree. All the available blue effect masks inputs are not the type of mask input you need. An effect mask limits the area of an effect. What you are doing here is combining an image with a matte, effectively cropping the image. This type of mask is often called a garbage matte. You need to add a node that can accept a garbage matte input. The most commonly used node for this purpose is called a matte control node.

3 In the node editor, select the merge 1 node, and then, in the effects library select the matte category.

4 Click the matte control tool from the effects library to add it after the merge 1 node.

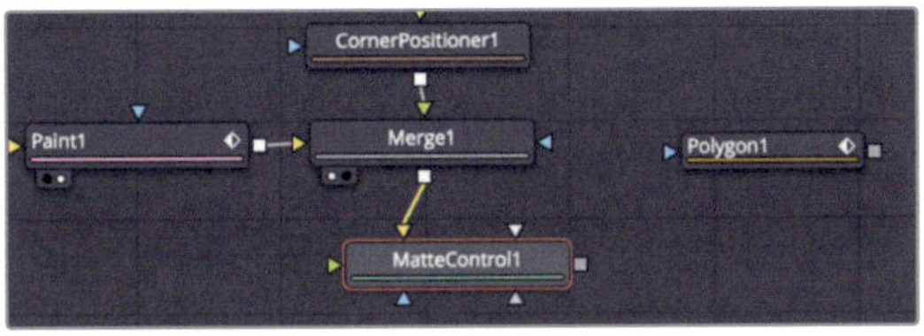

5 Press 1 to see the matte control in the viewer.

The matte control is typically used to copy or combine mattes from a foreground to a background. In this situation, you'll use it to copy a matte to the background freeze frame. However, because the matte control has a number of possible inputs, you need a way to ensure that you select the correct one.

6 Option-drag (macOS) or Alt-drag (Windows) from the polygon output to the matte control node and release the mouse button.

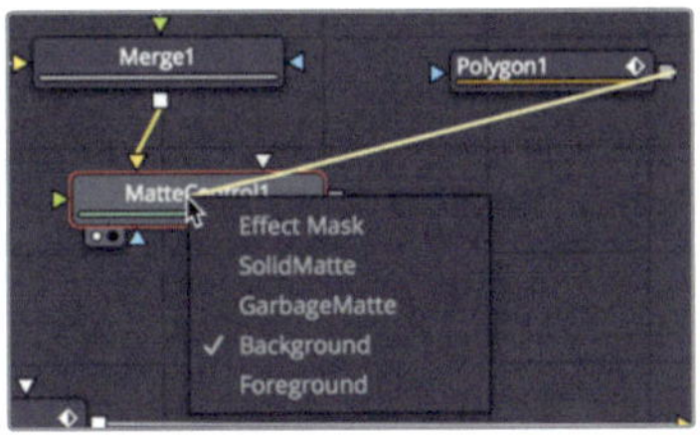

When you release the mouse button, a pop-up menu appears listing all the possible inputs of the matte control. This menu makes it easier to select the correct input. For this task, you want to use a garbage matte.

7 In the pop-up menu, click the garbage matte to connect the polygon.

The viewer now shows a hole where the side of the van used to be. Your matte is used to cut out the side of the van rather than isolating it. You can use the inspector to reverse this operation.

8 With the polygon node selected, in the inspector, click the invert button.

The side of the van with the logo is the only part seen from the freeze frame. You can now composite this still image over the top of the moving image using the planar tracker data to move it in sync with the full-motion shot.

Match Moving with the Planar Transform

For simple corner pinning images, you can connect directly into the planar tracker. However, when you are working with irregular polygon matte shapes or anything other than images of the same aspect and resolution of the composition, the more appropriate method is to use the planar transform.

1 In the node editor, select the planar tracker node.

2 At the bottom of the inspector, click create planar transform.

A planar transform node is created that contains all the transform and perspective distortion data captured by the planar tracker. You can apply this data to any input image or matte, thereby saving lots of time compared to rotoscoping objects.

3 Connect the output of the matte control to the yellow input of the planar transform.

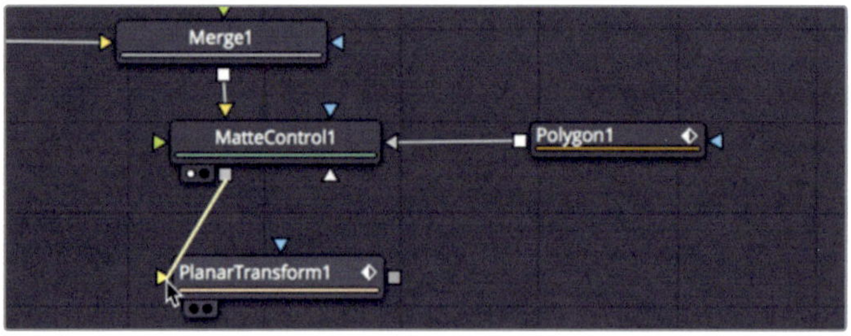

4 Press 1 to view the planar transform in the viewer.

5 Drag through the render range to preview the match-moved side of the van.

The freeze frame of the van's side now follows the driving motion. The next step is a simple merge to composite the van side over the moving van shot.

6 Select the planar tracker and press delete.

Once you have the planar transform, there is no need to keep the planar tracker. All the data captured from the tracking is now stored in the planar transform.

7 From the toolbar, drag the merge node over the connection line between the media in 1 and the media out 1 nodes to insert it.

8 Drag the output from the planar transform node to the green foreground input of the merge 2 node.

9 Select the merge 2 node, and press 1 to see it in the viewer.

10 Press the spacebar to play the entire shot.

So far, you have fairly convincing results. In the next exercise, you'll address a few little cleanup areas to finish this shot.

Finalizing the Composite

If you look carefully at later frames (such as frame 90), you'll see a clear shift in the shadows between the masked region and the moving van.

1 Move the playhead to frame 90.

This is a common problem resulting from the shadows changing in the live shot. Your freeze frame lighting is baked in and doesn't update with the moving shot. A simple blur could disguise the seam. However, because the clean van goes right to the top of the roof, you'll use the variable soft-edge on the polygon tool only where needed, toward the base and back of the van.

2 Move to frame 65.

> **TIP** It's important to always perform changes to a matte on the original frame where the adjustments were made. Doing so will avoid accidently adding keyframes and interpolating between adjustments.

3 Select the polygon node, and at the top of the viewer, click the make double poly button.

4 Right-click over one of the control points in the viewer and choose Controls > Select > Polygon 1: Outer Polygon

5 Drag the middle control point along the right edge of the shape to create a soft transition edge.

> **TIP** If you need to move a second control point, first click off into the gray area of the viewer to deselect the initial control point. Then hold cmd (macOS) or Ctrl (Windows) as you drag the second control point.

6 Select the media out 1 node to hide the on-screen controls for the polygon node and play the clip to preview the results.

The matte now looks clean. The final touch to make our logo appear more natural is to fix the "strobed" motion that the logo appears to have. This is due to the original camera footage having motion blur as the van gathers speed and our graphic having none.

7 With the planar transform node selected, in the inspector, click the settings tab.

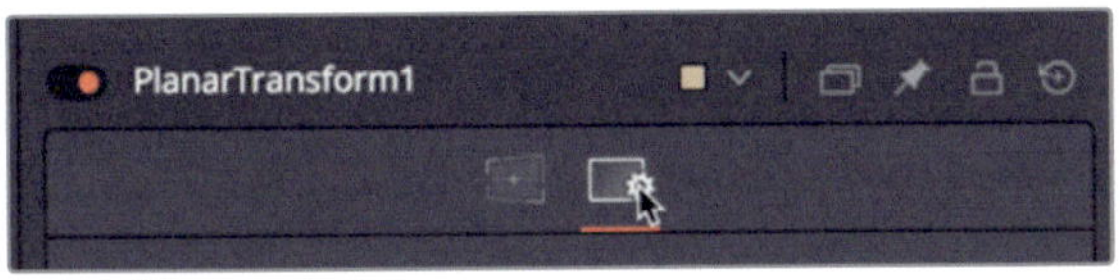

8 Enable motion blur.

The default blur seems a little strong for the shot, so you'll dial back the shutter angle. You'll also increase the quality of the blur to remove any visible stepping.

9 Adjust the shutter angle down to 130.0, and the quality up to 5.

With that, the shot is complete. The technique you learned here can be used to replace many different objects in a shot, from simple street signs to touch screens and even tattoos. Creating the clean surface with the time stretcher, paint tool, and polygon spline gives you a very simple but powerful recipe for common problems that may arise in a shot.

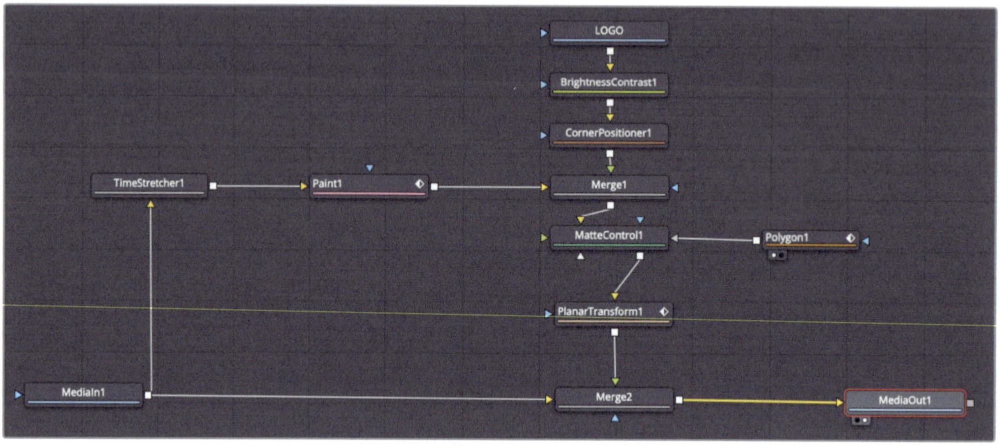

Completed node tree for Lesson 7.

Practice Exercise

This practice exercise is almost identical to the van exercise, although there is no paint required. The goal is to replace the twitching closed eyes on an actor with a freeze frame of his eyes so they remain static throughout the shot.

1 Go to the next to last clip in the timeline.

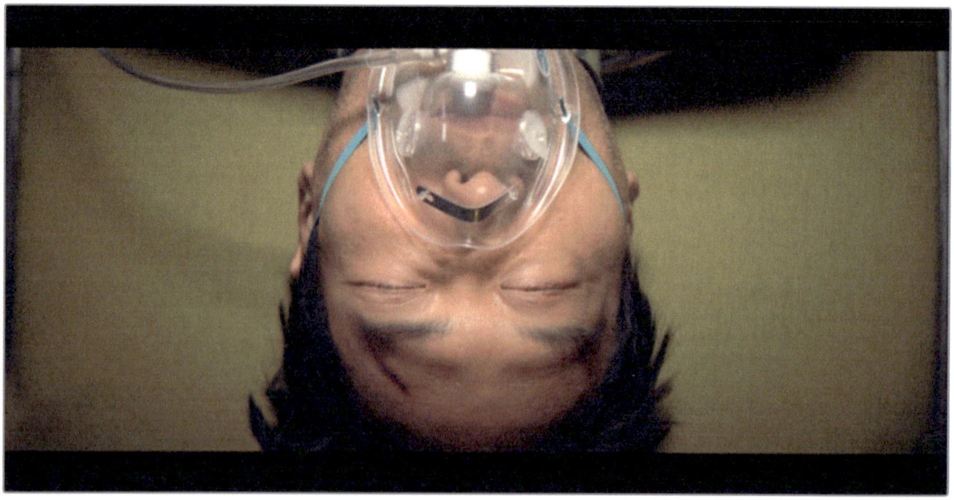

The man on the gurney moves his eyes too much in this shot. This would be an easy fix if the camera didn't rotate as it looked down on him. However, whether it be the eyes, a piece of the set, or a few people, the job of removing patches from a frame or covering up unwanted areas is a common task that can help save a shot. This particular shot is a very quick and easy procedure almost identical to what you have completed with the van shot.

2 Planar track the face of the man.
3 Use the time stretcher node to freeze a frame.
4 Add the matte control and a polygon tool to isolate the eyes.
5 Create a planar transform.
6 Use the planar transform to move the isolated eyes.
7 Composite the eyes over the moving face.
8 Soften the edges of the polygon node.

Lesson Review

1 True or false: You must click the set button before you begin planar tracking.
2 True or false: When drawing a shape around a surface for planar tracking, you should include as much of the background as possible.
3 Which of the following make for good planar tracking surfaces:
   a) Side of a building
   b) Billboard
   c) A bouncing ball
   d) All of the above
   e) None of the above
4 True or false: The planar transform can be used to track planar surfaces and composite the results.
5 True or false: When cloning with the paint tool, you hold the option (macOS) or Alt (Windows) key and click to select the clone offset source, and then keep the option or Alt key pressed while you paint over the destination.

Answers

1. True. Clicking set identifies the reference frame for the planar tracker.
2. False. When drawing a shape around a surface for planar tracking, you should not include any of the background.
3. A and B are ideal for planar tracking. C, the bouncing ball, is not a planar surface and therefore not a good option.
4. False. The planar transform uses the tracking data from a planar tracker and applies to it its input image. The planar transform has no compositing capabilities.
5. False. When cloning with the paint tool, you hold the option (macOS) or Alt (Windows) key and click to select the clone offset source, and then release the option or Alt key when you begin painting over the destination.

Lesson 8

Compositing Green Screen Content

Green or blue screen keying is the classic visual effects work that comes to mind for most people. A foreground subject is shot against a bright blue or green screen which is then keyed to make it transparent, thereby allowing the subject to be placed on a new background.

The keying process is actually a procedural method of generating a matte, rather than the manually drawn mattes you have used in previous lessons. The process of compositing with green screen is really an art unto itself, but a simple workflow can be followed for most keys. The trick is not to try to do everything with a single keying node. Focus the keyer on the fine detailed edges of your foreground subject, and then you can manage other areas with other tools. In the end, combining multiple mattes together will always get you better results quicker.

Time

This lesson takes approximately 60 minutes to complete.

Goals

Completed composite for Lesson 8.

Managing Color for Visual Effects

Before you begin this keying lesson, we must cover some of the more technical aspects of compositing. Color management is a critical part of the visual effects workflow, and the requirements for compositing are slightly different than editing or color grading. You'll start this keying job by setting up a scene-referred color managed workflow.

> **NOTE** The timelines bin includes a backups bin with timelines saved at various stages of the lesson and a completed projects bin with finished compositions. These bins are both available for reference and reverse-engineering the node trees.

1 Open DaVinci Resolve, and in the project manager, open the Fusion 16 Lessons part 2 project you have been using for the past three lessons.

2 From the timelines bin, double-click the **Part 2-START** timeline. Then go to the fourth and final red marker, which is over the green screen of our musician.

3 Click the Fusion page button.

This shot from a Steve Vai music video is typical of a professional green screen shoot. The first thing you might notice is that it appears to have low contrast and low saturation. This is often the case with content acquired on film, or more likely these days, digital film. Digital film cameras apply a very pronounced gamma curve to clips to achieve a higher dynamic range. To a much lesser extent, every HD video clip, Photoshop document, or web graphic, has a *nonlinear gamma* value based on either an sRGB or Rec 709 2.4 color profile.

The problem with nonlinear gamma curves is that compositing operations expect images to have linear gamma, not modified gamma values. Why? Because the math used in almost every compositing operation works more predictably assuming images have linear gamma.

So, if you simply click the Fusion page and start working on your HD or RAW content, you're starting out on the wrong foot.

To composite, you first need to remove the nonlinear gamma curve from this green screen shot and convert it to linear. Then, you'll reverse the process and ensure that the output from the Fusion page will apply the correct nonlinear gamma according to your output format. Sound confusing? Don't worry; DaVinci Resolve handles all of this with very little effort from you. You just need to enable DaVinci YRGB color management.

4 Choose File > Project settings, and in the sidebar, click the color management category.

5 In the color science menu, choose DaVinci YRGB color managed.

Why Use Linear Color Space?

The different gamma curves in a post production workflow, from your camera, the displays, and even the way the human eye perceives light, are not helping when it comes to compositing. Why is that? You can blame it all on math.

The majority of compositing tools calculate their effects expecting linear gamma. When given images with nonlinear gamma, you can get unexpected results. When images are composited using linear gamma, each step in brightness represents a uniform change of intensity rather than of a change in your perceived brightness. Let's look at a simple black to white gradient to help explain this. In the real world (not the one you perceive with your eyes), light is linear. Think of brightness as being a scale that starts with black at 0 and white topping out at 1.0. Based on this linear tonal range, 0.5 is half the intensity, or middle gray.

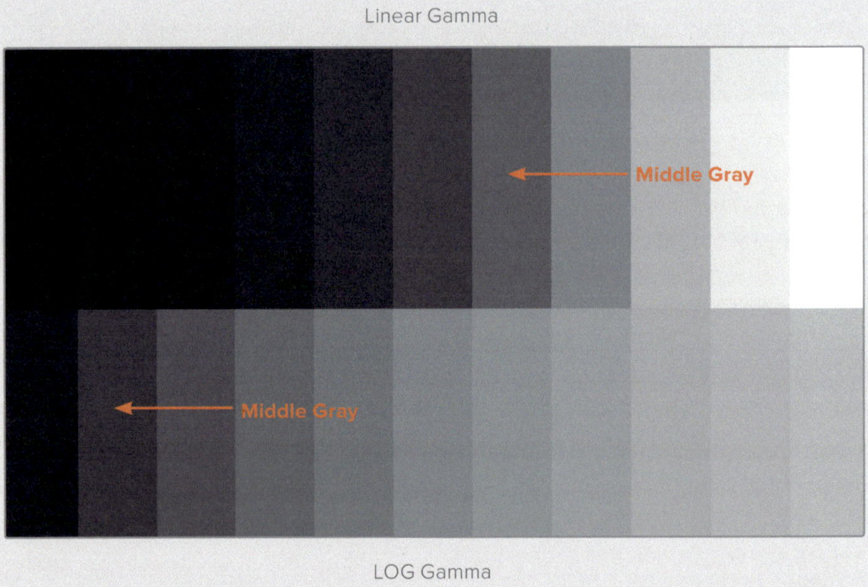

When you apply a LOG gamma curve, similar to images from a digital film camera, middle gray is moved and ends up being somewhere around 0.2!

Your middle gray value isn't actually middle gray any longer and no one has told all the color correction tools of this change! These tools still assume middle gray is based on linear gamma, and they will operate based on that assumption. Of course, you cannot have different color correction tools for each and every gamma curve and image format, so the easiest thing to do is remove all nonlinear gamma curves and work with linear gamma.

In a more practical example, below we have applied the same color correction on this split-screen RAW image. We raised the gamma by 25% and boosted green and yellow in the midtones as well. The lower left side of the split screen shows the adjustment done using the original LOG gamma, and on the upper right the adjustment is done with linear gamma. Both images were then converted back to sRGB with a split wipe for display on the monitor and print.

LINEAR Gamma

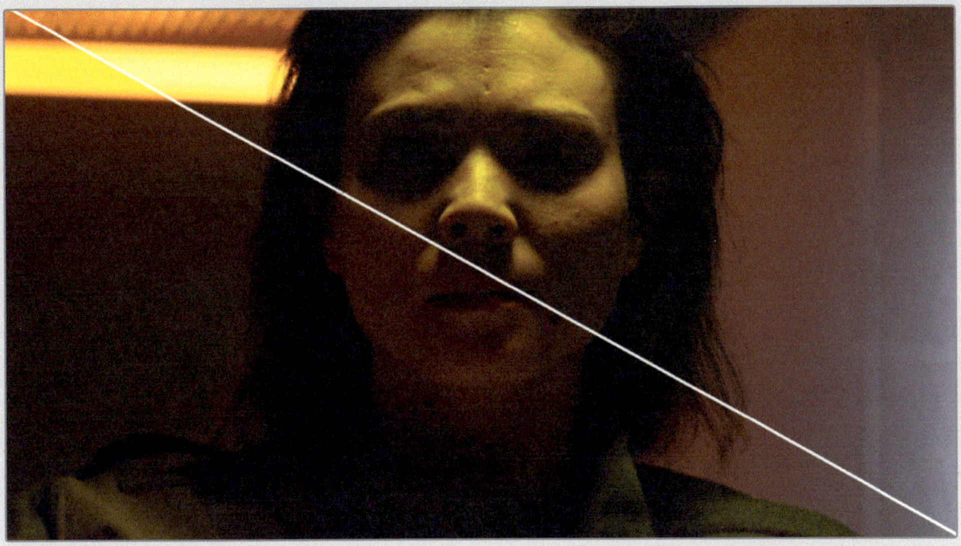

LOG Gamma

Notice the lack of contrast on the lower left face. There is also a noticeable midtone color shift on the log image with a lot more red appearing on the skin and light.

Similar differences happen even with HD content, although the further you move away from linear when compositing, the more likely you will experience issues. All of this is due to the fact that most compositing operations do not handle nonlinear gamma curves. They expect that light is linear, so when you add +2 to brightness, the shadows, midtones, and highlights are increased equally. Color correction is not applied differently to the midtones than to other regions.

The correct way to composite images is to use a scene-referred color management system like DaVinci YRGB Color Management or ACES and convert any nonlinear gamma curve to linear. You can then composite without worry and allow the scene-referred color managed system to handle the conversion back to your final output display. This is how DaVinci Resolve and the Fusion page can work to give you more accurate compositing results.

6 Because this image came from a Blackmagic Design URSA camera, in the input color space menu, choose Blackmagic Film 4.6K GEN 3.

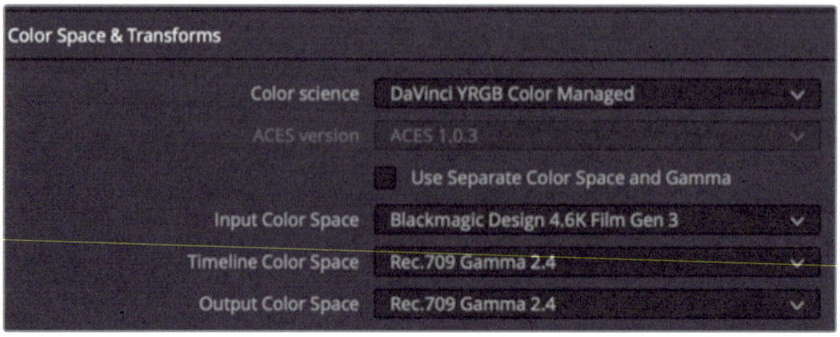

Every clip in the project is now assumed to have come from a Blackmagic 4.6K camera that used the film 4.6K gamma setting. (Of course, this assumption can be changed on a clip-by-clip basis, if necessary.) The output and timeline color space settings are set to the default Rec 709 for HD displays, which will work for this exercise.

TIP If you do not have a calibrated HD display connected to your computer when creating visual effects, it is more common to set the output color space to sRGB to match your computer display.

7 Click save to close the window.

The image in the viewer now appears with more saturation but remains very dark overall. That is because you are looking at the linear color space clip.

This image would be rather unpleasant to work with, so you can correct it in the viewer using a viewer LUT (LookUp Table). The viewer LUT makes for a better viewing experience, while still allowing you to composite correctly using linear gamma images.

8 In the upper-right corner of the viewer, click the down arrow next to the viewer LUT button.

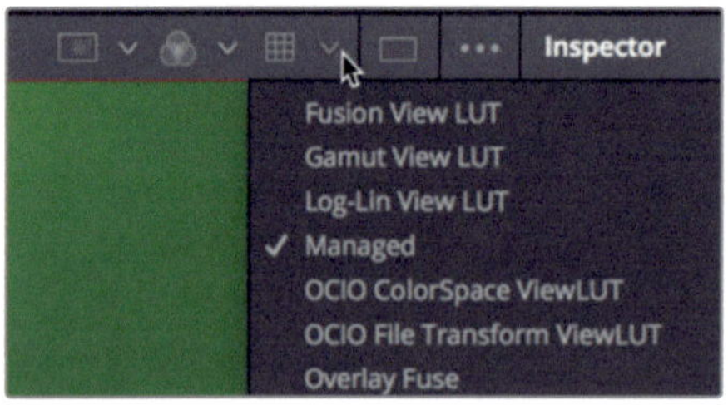

9 From the list of LUTs, choose managed to enable the viewer LUT. This option will use the configuration from the color management project settings.

TIP If you do not use color management, you can add a gamut or Cineon LOG tool after every media in node to convert it to linear gamma. Then, add a gamut or Cineon LOG tool just before the media out node to covert back to the final output gamma setting.

A viewer LUT is a simple color adjustment applied to the viewers in the Fusion page. The image itself is not changed, only its display in the Fusion viewers. Rather than showing the image with linear gamma, Resolve enables the viewer LUT to convert the linear gamma image to the output color space identified in the color management setting.

You are now able to composite using images that look correct, but more importantly, will act correctly during compositing. And when you switch to the edit or color pages, all the gamma curve corrections are managed automatically, based on the output color space setting that you configured in the color management project settings.

Pulling a Green Screen Key

With the color management correctly set up, you can begin the keying process. As you know, combining shots requires a matte: a grayscale image that identifies parts of the foreground as transparent and parts as opaque. Unlike computer-generated images, this live-action green screen shot does not include an alpha channel. So, it is up to you to create the matte through keying. It is often termed *"pulling a key."*

NOTE In this lesson, the shot you will work on uses a green screen, but the keying process works the same for blue screen content.

1 In the node editor, select the media in 1 node, and rename it to GREENSCREEN.

2 In the upper-left corner of the interface, click the effects library button.

All the tools for keying are located in the tools > matte category.

3 Click the tools disclosure arrow, and select the matte category. Click the Delta keyer.

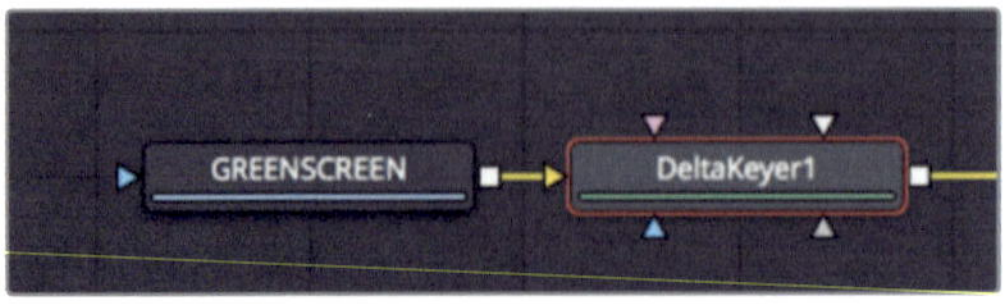

Because the GREENSCREEN node was selected in the node editor, the Delta keyer is connected to the GREENSCREEN output. Whenever you are keying, it is helpful to use two viewers. One where you can see the final output, and the other where you can view the quality of your matte.

4 If you are not already in a dual viewer setup, from the upper-right corner of the viewer, click the single/dual viewer button to display two viewers.

5 Click the effects library button to close the panel and open up more room for the viewers.

The Delta keyer is the primary tool used for green screen and blue screen keying in the Fusion page. As powerful as it is, it is also very simple to use.

TIP The term chroma key is a specific, simplified process of extracting a matte based on a range of hue and saturation in an image. Modern keyers like the Delta keyer use a more sophisticated color difference method to extract the matte.

6 Select the Delta keyer node, and press 1 to display its output in viewer 1.

TIP Make sure both viewers are using the viewer LUT set to managed.

Your first step will be to sample the key color in the foreground image and make it transparent. Doing so tells the keyer whether it is keying a blue, red, or green screen and also dials in a few other optimizations.

7 In the inspector, drag the background color eye dropper and move it over the green screen in viewer 2.

TIP When sampling an area of the frame for a key color, select an area where the audience will be focusing its attention in the final shot. Doing so sets you up for creating an optimal key.

8 Release the mouse button when the eye dropper is near the guitarist's shoulder, screen right.

TIP As you select an area over the screen color, holding cmd (macOS) or Ctrl (Windows) allows you to drag out a rectangle to select an average of an area.

The green background turns mostly black. You've now created a matte for the foreground shot by selecting a green color. Even if this matte looks satisfactory in the viewers, you always need to check it for overall quality.

9 Above viewer 1, click the color controls button, or click the mouse pointer in viewer 1 to make it the active window and press the A key.

The matte for your live-action shot is displayed in viewer 1. A single click of the eye dropper typically produces a matte with lots of gray (semi-transparency).

10 Drag the gain slider slightly up to subtract more value from the green background. Adjusting gain too far will cause some of the white foreground to become transparent, so you only need to increase it a small amount. Staying around 1.3 should be safe.

11 Drag the balance slider to the right, until it is between .7 and .8. The balance slider takes more or less information from the colors other than the screen color—in this case, red or blue. Dragging it to the right causes blue to be more transparent and red to be more opaque. Since there is a lot of red in this foreground subject, dragging to the right removes some of the transparency in the foreground.

Although this is not nearly good enough for the final result, this is a good start since it retains all the fine hair detail.

Even with the cleanest of green screen keys, such as you have here, you must refine the matte a bit to ensure that areas intended to be opaque appear as solid white, and the areas intended to be transparent appear as solid black.

Refining a Matte

Refining your matte is done within the Delta keyer. Taken in stages, the first stage is to ensure you have selected all the various shades of green you can, without causing any damage to the fine details in the matte. This pre-matte selection refinement is done in the pre-matte tab of the inspector.

1 In the inspector click the pre-matte tab.

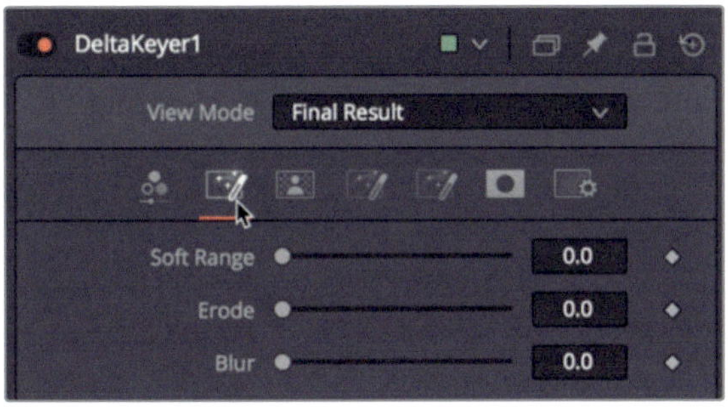

The goal in the pre-matte tab is to select areas of your green screen that were not made transparent with the single-click eye dropper. Selecting other areas will even out the color selection of the green screen.

TIP You can use the clean plate node in place of the pre-matte tab for more control over the pre-matte process.

2 In viewer 1, drag a selection rectangle around the dark gray area to the left of his hat.

When making these pre-matte selections, be careful not to get too close to the fine hair details you are trying to preserve.

It may seem as if you have a pretty clean matte, but your computer screen can play tricks with the gamma in your image. To offset the computer screen gamma setting, you can make a temporary gamma adjustment to the viewer that will allow you to check the matte for unwanted semi-transparent areas.

3 From the options menu in the upper-right corner of viewer 1, choose gain/gamma from the list.

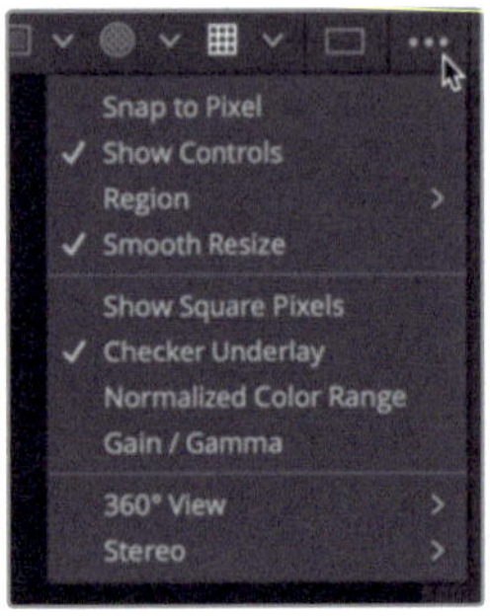

The gain/gamma overlay allows you to temporarily adjust the viewers without affecting your final output. This will allow you to find holes in the matte that may be hidden by the color profile of your computer screen.

4 Drag the gamma slider up about halfway to view any gray areas in the transparent background of the matte.

The gamma adjustment usually exposes some small areas that can be removed.

5 Drag a selection rectangle across the white specks that appear near the guitarist's left leg (screen right).

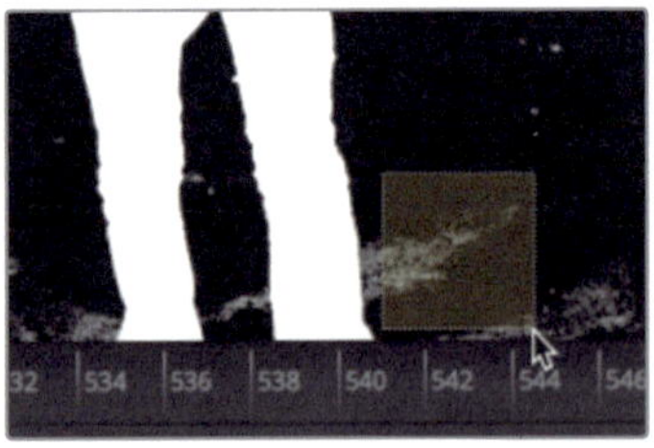

During the pre-matte selection process, and during other matte refinements you'll make later, do not concern yourself with parts of the screen that are not near our foreground subject. Those areas are best handled by other mattes. But, conversely, do not get too close to the fine hair details to prevent them from eroding.

The pre-matte selection can be used to expand the green selection and then be eroded to fill in any holes created in the foreground.

6 In the inspector, drag the soft range slider up to around 0.02 to expand the pre-matte selection.

7 Drag the erode control up to around 0.002 to eat away at any holes created by the soft range.

Both the soft range and the erode are small adjustments that you may decide to come back to later in the process if the edges around the subject become too pronounced.

To refine the matte density further, you can switch to the matte tab of the Delta keyer.

8 In the inspector, click the matte tab.

The matte tab contains parameters for modifying the density and edges of the matte. It is arguably the most important tab in the Delta keyer because the quality of your matte determines the quality of your key. By adjusting the threshold sliders at the top of the matte tab, you clamp the black and white cut-off points. Values that fall below the low threshold setting are considered as pure black (transparent), and values that fall above the high threshold setting are considered as pure white (opaque).

TIP As you make adjustments to the matte, zoom in to, and/or expand the viewer size to get a clearer look at the matte, especially around the fine hair details.

9 To clamp the black, transparent areas of your matte, drag the low threshold slider to the right until there are very few gray spots in the background. Again, pay extra attention to the hair details. The more you increase the low threshold, the more you eat away at the thin hair strands. In this shot, if you drag above 0.2 you have probably lost some hair detail.

10 To clamp the white areas of the subject, drag the high threshold slider to the left until the white areas filling in the foreground are mostly pure white. Decreasing the high threshold too much below 0.75 can cause the fine details to merge together.

TIP When refining the key, you are primarily concerned with getting good quality edges for your subject. The black transparent areas and the core of your foreground subject can be handled using various types of mattes.

The next steps are very subtle adjustments using the clean foreground and clean background controls. You generally apply these with an extremely light touch because they tend to cause harsh matte edges, but they can also "fill in" the small holes that appear in the black and white parts of the matte. As you adjust them here, you'll use a combination of the keyboard and the slider to make very small scale adjustments.

11 To fill in any black holes in the white areas of the matte, hold the cmd key (macOS) or the Ctrl key (Windows) and drag the clean foreground slider to around 0.0005, and then do the same for clean background.

12 From the options menu, choose gain/gamma from the list to close the overlay and reset the gamma in the viewer.

It will be impossible to refine the matte to the point where it fixes every pixel of the frame without also sacrificing some of the hair detail. You'll find that for every shot, compositing with green screen requires multiple tools that work together. However, the basic adjustments in the Delta keyer that you have completed here are a useful starting point for any key.

Rotoscoping Auxiliary Mattes

Your matte still has unwanted areas such as the gray or semi-transparent areas around the lower portion of the guitar and a few white specks in the black transparent areas. Sometimes, you won't be able to key out everything, so you need to pull out the digital duct tape, more formally known as auxiliary mattes.

Auxiliary mattes are mattes other than the main core matte created by the key. These auxiliary mattes help to patch matte holes that are impossible to patch otherwise. Two auxiliary mattes are used regularly on every keying job. One is the garbage matte that removes areas of the set not covered by the green screen. The second is a hold out matte, The hold out matte covers up unwanted semi-transparent areas in the foreground that the keyer didn't catch. Let's start with creating a garbage matte.

1 From the toolbar, drag a B-spline tool into the node editor under the Delta keyer.

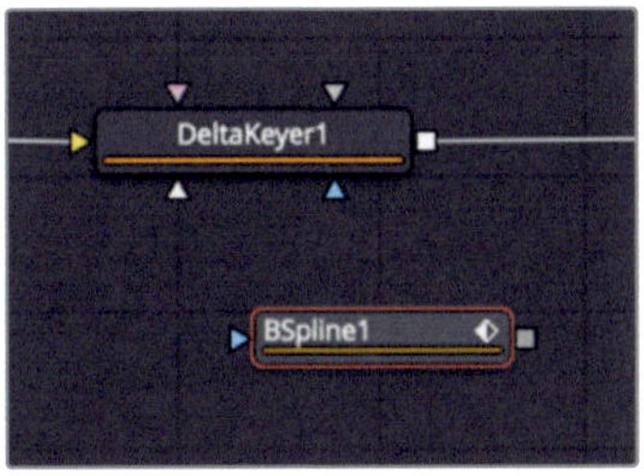

2 Select the B-spline node, and rename it to G_MATTE.

When drawing a matte, it is sometimes easier to use a B-spline tool rather than Bezier splines as you have been doing. B-splines produce smoothed curved edges without the need to manage handles. For consistently smooth curved shapes (when you are not concerned with making extremely detailed shapes with lots of corners) B-splines can be easier to work with.

Let's use the B-spline tool to draw around the area you want to keep, and then invert the matte.

> **TIP** When you draw a shape with a B-spline curve tool, the control points you set influence only the shape of the curve. The control points do not define the location of the actual spline curve. That being the case, it is best to start by drawing a very loose shape, and then adjust the control points to create the curve you need.

3 In viewer 1, draw a wide, rough shape around the guitarist.

TIP Similar to drawing a polygon spline, remember to close the matte by clicking on the first control point you added or by pressing Shift-O.

4 Once you finish drawing the matte, Option-drag (macOS) or Alt-drag (Windows) from the G_MATTE output to the Delta keyer node.

5 Release the mouse button, and in the input pop-up list, choose garbage matte.

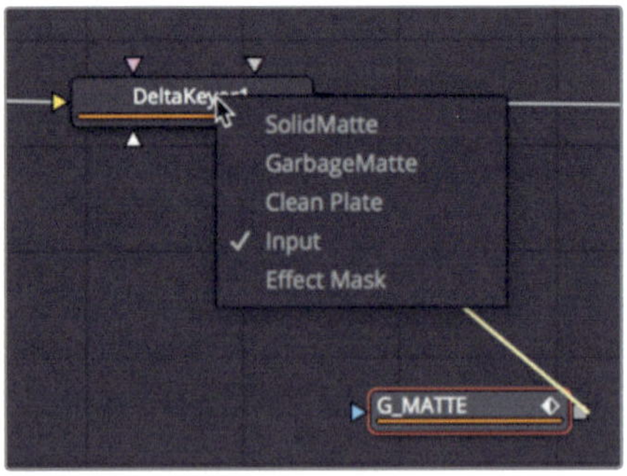

Because you drew this garbage matte around the subject, the guitarist's silhouette is removed from the shot. You need to invert it.

6 Select the G_MATTE node, and in the inspector, click the invert checkbox.

This matte is done only for a single frame, and this guitarist likes to move as he plays. The drawing of polygon or B-spline mattes and animating them over a series of frames is called rotoscoping. A basic rotoscoping technique is called divide and conquer. Divide and conquer bifurcates the clip with keyframes and continually divides each section with keyframes until the motion of the object is covered. This ensures you only add keyframes when absolutely necessary.

7 Move to the end of the render range.

By default, after you close a polygon shape, any change you make to a control point adds a keyframe. Changes to the shape on different frames are interpolated. This behavior makes it incredibly efficient to animate small changes to your matte as the subject in your clip moves.

8 Refine the shape to better fit the around the guitarist by adjusting any of the control points along the B-spline.

> **TIP** You can adjust the smoothness of a B-spline's curve by holding down the W key, selecting the control point for the curve, and dragging left or right to increase or decrease the smoothness.

When using the divide and conquer technique, the idea is to set keyframes at the start, end, and middle of your animation. You then continue to divide those segments by adjusting the shape to fit at halfway points between keyframes until the shape's movement in each segment accurately matches the object you're trying to rotoscope.

9 Move to the middle of the render range, around frame 570.

10 Adjust the control points to fit around the guitarist.

11 Repeat the process by dividing the keyframed sections from frame 530 through frame 570 so that the animated shape fits the guitarist throughout the moving shot.

12 When you come to a point where you must move the entire shape, click the select all points button in the viewer toolbar and drag a control point to move all the control points and set a keyframe.

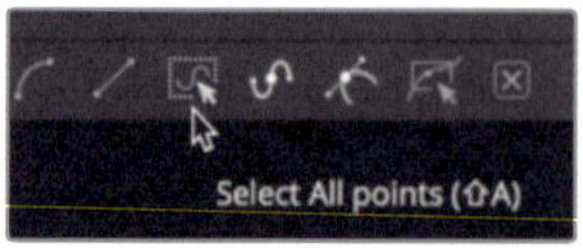

TIP Pressing Option left and right arrow (macOS) or Alt-left and right arrow (Windows) will move the playhead to the next and previous keyframe, respectively, to help you more quickly refine the polygon matte.

13 Once you finish with the first half of the shot, divide the second half starting at frame 570 through frame 603. Adjust the shape wherever you feel a keyframe is needed, continually dividing sections as you go.

Rotoscoping a Hold Out Matte

Another auxiliary matte you will typically create does the opposite of the garbage matte. A hold out matte fills in any holes that appear in the white matte of the foreground subject. Some bright areas of the white guitar have caught green bounce light from the screen. To correct this area, you'll draw another polygon shape to cover the dark gray area where the body of the guitar is located.

1 From the toolbar, add a polygon node into the node editor, next to the G_MATTE node, and rename it to **HOLDOUT**.

2 In viewer 1, zoom into the matte and draw a shape around the midsection of the matte to cover up any gray areas.

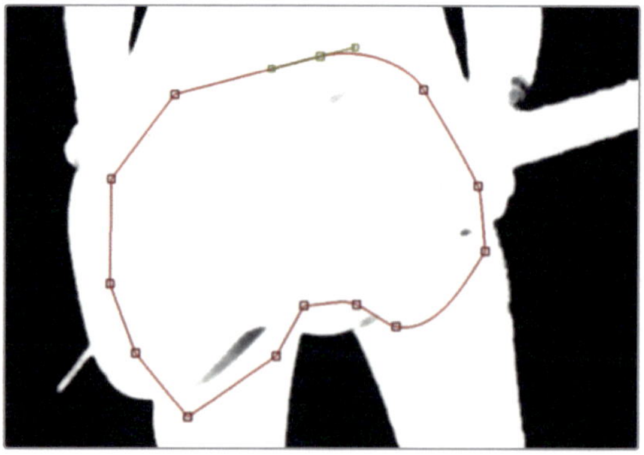

The hold out matte must be connected to a different input of the matte control than your garbage matte. A hold out matte is connected to the solid input on the Delta keyer.

3 Option-drag (macOS) or Alt-drag (Windows) from the HOLDOUT output to the Delta keyer node, and in the input pop-up menu, choose solid.

4 Again using the divide and conquer technique, make adjustments to the polygon Bezier matte for the remaining frames in this shot. Use the select all points button above the viewer or draw a selection rectangle around a group of control points to move them, then refine any single points as needed.

A note about garbage and hold out mattes: Often people unaccustomed to creating green screen composites attempt to do everything in the keyer. Let me dispel that myth right now. Using auxiliary mattes is not an admission of a failed key. The use of auxiliary mattes means you are being smart about your time and aware of the entire process. Use a keyer for what it is good at: creating soft edges and extracting the fine hair detail. Use the auxiliary mattes to avoid wasting time fiddling with keyer controls for items easily done with a spline shape.

Lining Up the Background

You can only go so far adjusting the matte against a black background. Eventually, you must view the foreground over the actual background clip to get a complete picture of the matte quality.

1 In the upper-left corner of the interface, click the media pool button, and from the keying bin, drag the BKGD HD clip into an empty area of the node editor.

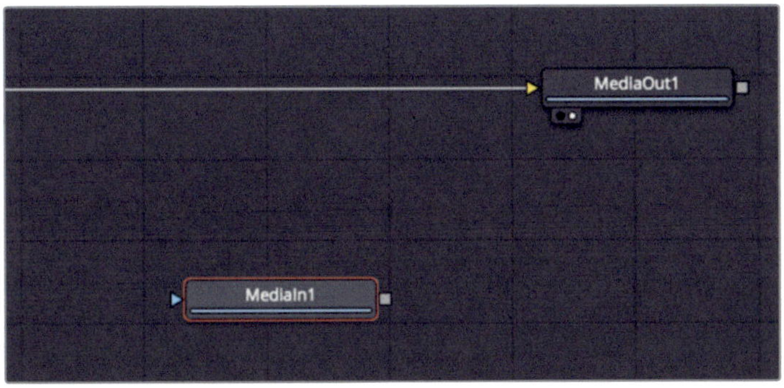

2 Press 2 to see the clip in the viewer.

The viewer is empty, as if there is nothing to see. Remember that the comp has a start and end time based on the source clip from the edit page timeline. If you view the render range, it starts at frame 530. Unless the background clip you brought in is over 530 frames long, it will not be visible in this composition. It is helpful to view the keyframe editor when bringing in new clips to see where they are located along the time range.

3 In the upper-right corner of the window, click the keyframes button.

By default, the keyframe editor shows only tracks for animated controls. You'll need to switch the filter menu to show all the tracks.

4 From the filter menu in the upper-right corner of the keyframe editor, choose animated to disable the filter.

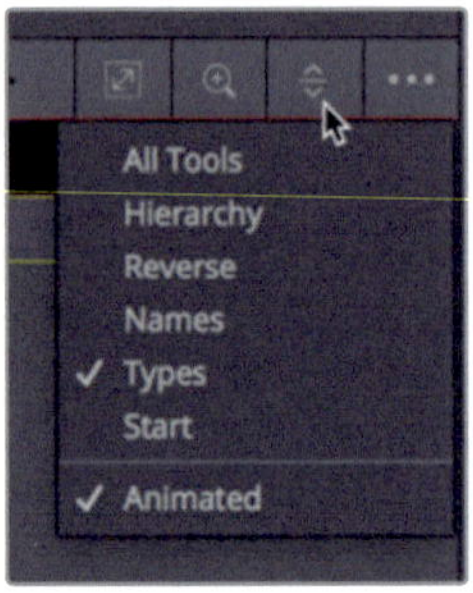

The keyframe editor now shows all the tracks, including the background node. It is clear now why the viewer is empty. The background node clip begins at frame 0 and ends at frame 73. This is a full 437 frames before the comp starts. You can use the global in/out controls in the inspector to slide the background clip into place.

5 Select media in 1 node in the node editor.

6 At the top of the inspector, position the mouse pointer between the two global in/out handles.

7 Drag the global in/out bar to the right until you get near the 530 start frame.

8 When the global in is near 530, it can become easier to zoom into the keyframe editor and slide the clip into place there.

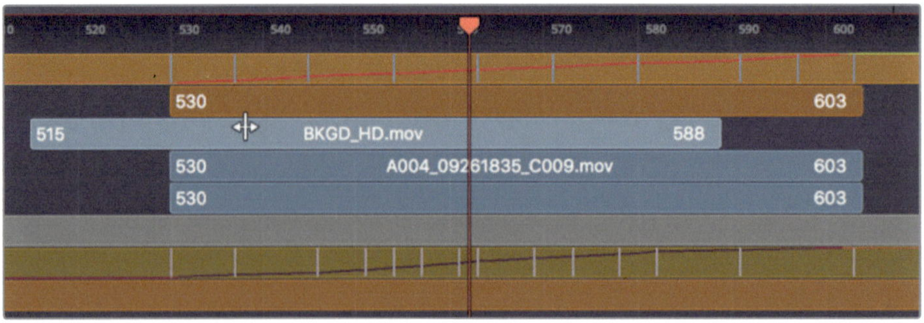

Now you are able to view the clip in viewer 2.

9 With the media in 1 node selected, press F2 and rename the clip to BACKGROUND.

To view the key over the background, you'll use a merge node.

10 In the upper-right corner of the window, click the keyframes button to close the panel.

11 Click the connection line near the yellow input of the media out node to disconnect it.

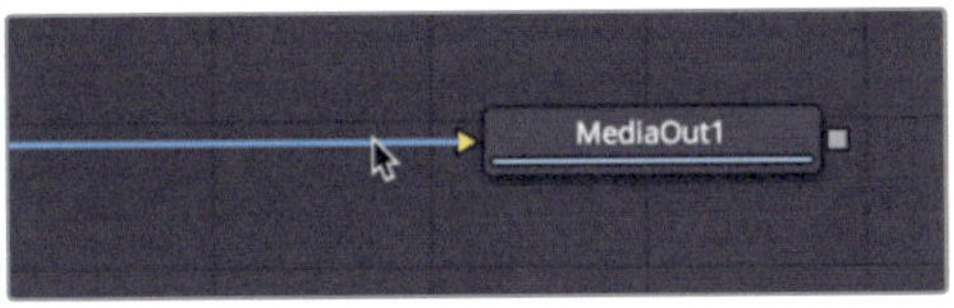

12 Drag the output of the Delta keyer to the output of the BACKGROUND node to create a merge node.

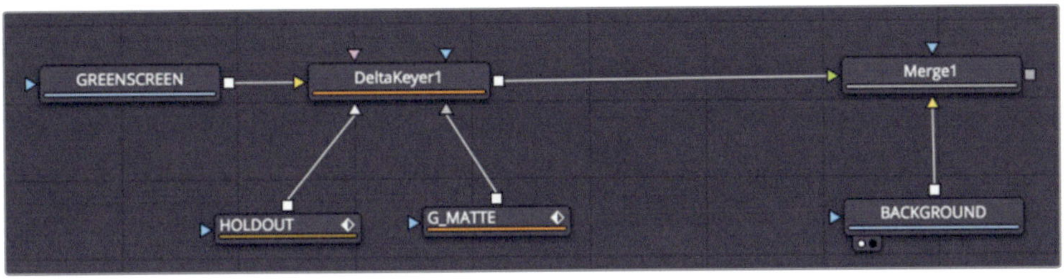

13 Drag the output of the merge node to the media out node.

14 Select the merge node, and press 2 to display it in viewer 2 to see the initial composite.

Although this looks good for the amount of work we have done so far, the foreground and background still look like very different images. The next step is to color correct both elements so they appear as if they are in the same location.

Color Correcting Elements

Color correcting the background and foreground is done for two reasons. The first is to remove any green color that remains on the foreground subject, and the second is to match the foreground and background so they realistically appear to exist in the same setting. Let's tackle the green tint issue first. The replace color at the bottom of the matte tab in the inspector already includes a fair amount of spill suppression.

1 Select the Delta keyer, and in the inspector, select the matte tab.

  Spill is created from bounced light reflecting off the green screen that spills onto the foreground subject. Since removing that green spill is inextricably linked to the process of extracting a matte, you will sometimes create holes in your foreground matte. This is due to the replace color attempting to remove the green screen color that has spilled onto your foreground. You inevitably must balance the quality of the spill suppression with the generation of the matte. A way around this dilemma is to set the replace color to source.

2 At the bottom of the inspector, set the replace color to source.

  Setting the replace color to source reintroduces an amount of the original green screen pixels instead of trying to remove them. However, you are now left with green spill on the musician.

  You can remove spill fairly easily using a color corrector node directly after the Delta keyer.

3 From the toolbar, insert a color corrector node between the Delta keyer and the merge 1 node.

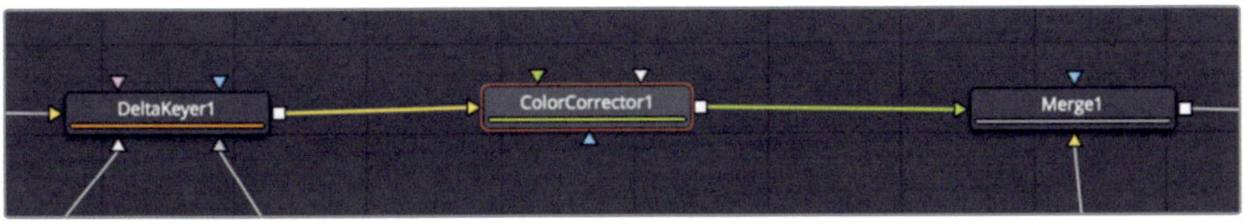

The color corrector node includes a number of modes for correcting highlights, midtones, and shadows, as well as spill suppression.

4 At the top of the color corrector inspector, choose suppress from the menu.

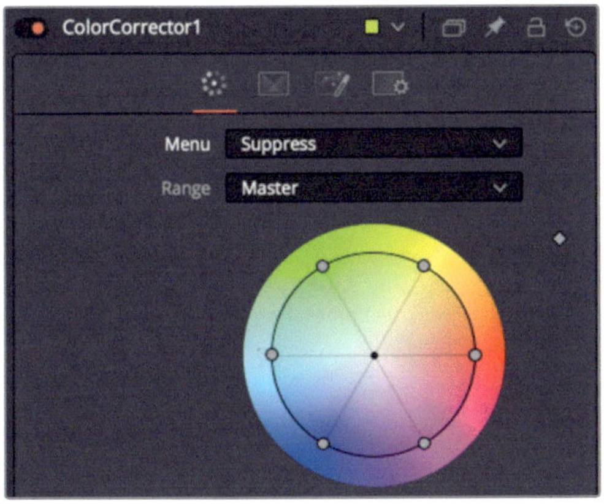

TIP Zoom into the viewer as far as needed to view the changes as you reduce the spill colors.

The suppress wheel allows you to drag the control point for the desired spill color toward the center, thereby reducing its saturation.

5 To reduce the green and some of the bright yellow tint around the edges of the foreground, drag the green and yellow control points in toward the center of the color wheel.

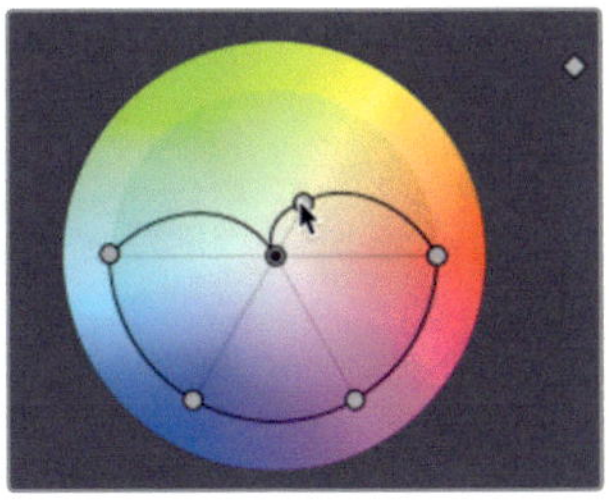

TIP In many cases, spill suppression can produce a noticeable reduction in image brightness. You can counteract this effect by slightly raising luminance.

6 With the color corrector node selected, press cmd-P (macOS) or Ctrl-P (Windows) to disable the spill correction. Then, press that keyboard shortcut again to re-enable the spill suppression.

Now you can focus on actually matching the foreground and background. You can choose to perform the foreground color correction in the existing color correction node, or if you prefer to isolate your adjustments, you can add another color correction node.

Matching Foreground and Background

To match the foreground and background, you'll use both viewers to see both images side by side. In each viewer, you'll display a sub-view to show a histogram. This will help guide you as you match the two elements.

1 Select the color corrector 1 node in the node editor. At the top of the inspector, set the menu from suppress to colors. This will be used to color correct the foreground.

2 Press 2 to see the foreground color corrector in viewer 2.

3 Select the background node in the node editor, and then from the toolbar, click the color corrector tool again. This will add the color corrector to be used to correct the background.

4 Set viewer 1 back to viewing colors instead of the alpha channel, and then press 1 to see the background color corrector in viewer 1.

5 Click the sub-view button in the upper-left corner of viewer 1 to display the histogram sub-view.

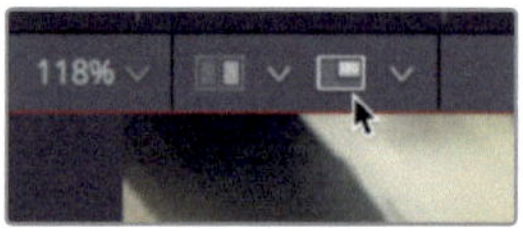

A histogram display helps you see whether your image is exposed properly. Learning to read one will improve your compositing skills when it comes to matching elements.

TIP Other sub-views are available by clicking the pop-up menu to the right of the sub-view button.

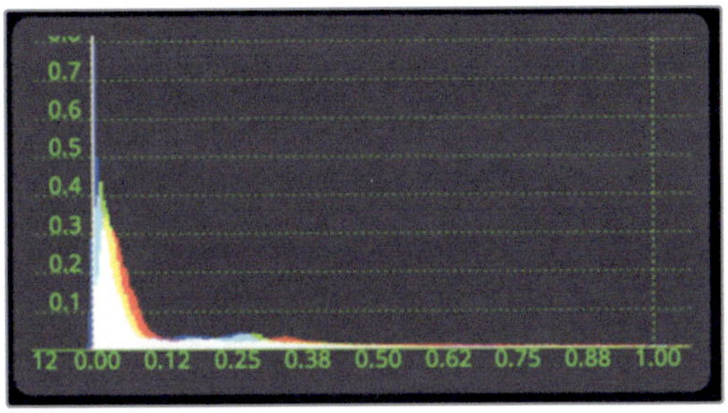

A histogram displays the brightness distribution of pixels in an image. Black, at a value of 0.0, is on the left, while white, at a value of 1.0, is on the right. The number of pixels falling into any given tonal range is represented by the height of the histogram.

6 Position the mouse pointer in the lower corner of the histography sub-view, and then drag on the corner to expand the graph.

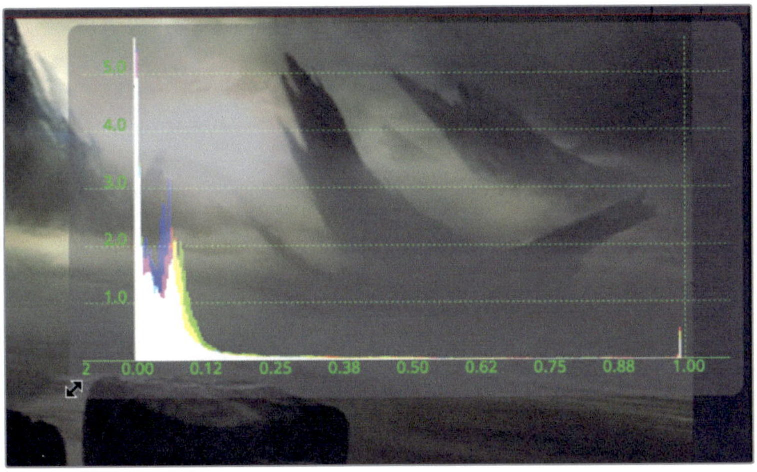

7 Open the histogram in viewer 2 and roughly expand it to the same size as the histogram in viewer 1.

Since most of the background shot is dark, you'll start by making adjustments to the shadows.

8 Select the background's color corrector 2, and from the range menu, choose shadows.

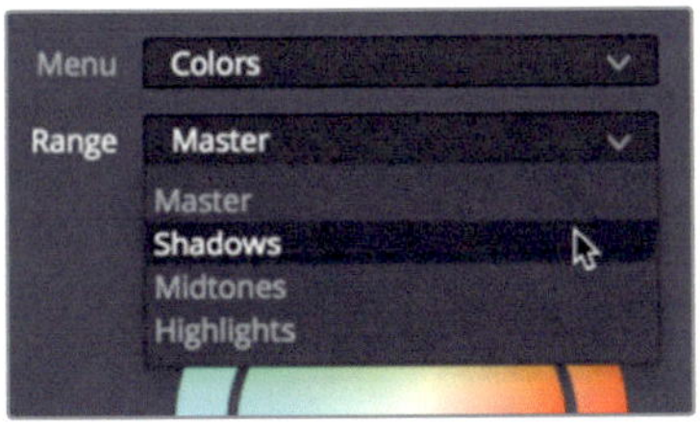

Any adjustments you make now will affect only the darker pixels in this shot.

9 Drag the gamma slider to the right and set it between 1.2 and 1.3. This will bring out a lot of the detail in the rocks.

10 Set the range menu back to master and adjust the gamma slider to the right to set it between 1.1 and 1.2. This will increase the overall midtones of the shot without blowing out the sky or causing the rocks to become muddy and gray.

Now, you'll work on the foreground. The foreground is already combined with its matte in the Delta keyer. Any adjustments to the gamma or lift in the color corrector will need to take into account that you are working with a premultiplied alpha channel.

11 Select color corrector 1 and click the options tab in the inspector, and click the pre-divide/post-multiply checkbox.

12 Click back on the correction tab.

Two observations that can be gleamed from looking at the image and the histogram: The highlights are very hot and the blacks are not nearly as dark as the background.

13 Drag the gain slider to the left to bring down the overexposed highlights on the guitar and the top of his hat. Somewhere between 0.6 and 0.7 should work well.

14 To darken the shadows, drag the gamma to the left, but only a bit; do not go below 0.8.

Let's see how the two images now appear together in the composite.

15 Select the merge 1 node and press 2 to see it in viewer 2.

Although the brightness values seem very close, the colors are very different.

16 Click the sub-view button to close the histogram in viewers 1 and 2.

17 Select color corrector 1 and drag the saturation down between 0.6 and 0.7.

18 From the range menu, choose shadows and drag the color indicator in the color wheel toward teal to match the ambient light in the background.

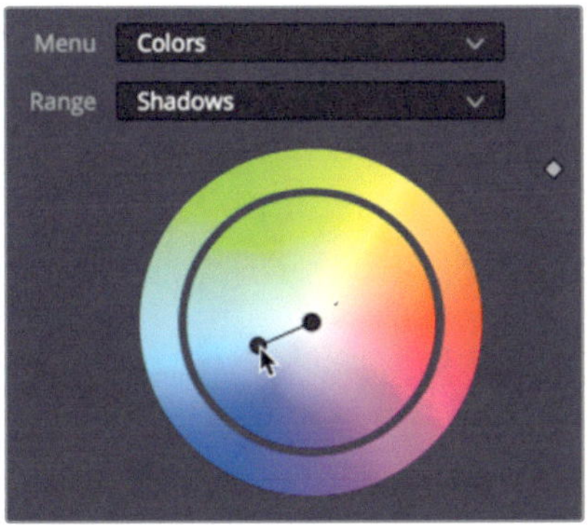

19 From the range menu, choose midtones and drag the color indicator in the color wheel toward yellow/green to match the sunlight coming through the smoky clouds.

TIP It is possible to color correct the foreground and background using the tools in the color page. Adding a second media out node (effects library > tools > I/O > media out) connected to the Delta keyer will allow you to add a new source to the color page. That source will carry the key signal from the Delta keyer into the color page.

The main point of this green screen lesson is not only to teach you the steps in setting up a green screen key, but perhaps more importantly to show you that keying a foreground subject is never completed using only one tool. It is far more common to employ multiple techniques to address separate areas. In the end, if you learn to break down the foreground into its own small problematic areas, you can address each area with different tools to get the best possible results.

If you stopped here, you could be rather pleased with the results we've achieved in such a small amount of time. However, you can also go the extra step and make it really special by continuing on with Appendix A, which will introduce you to an additional color correction step called a light wrap.

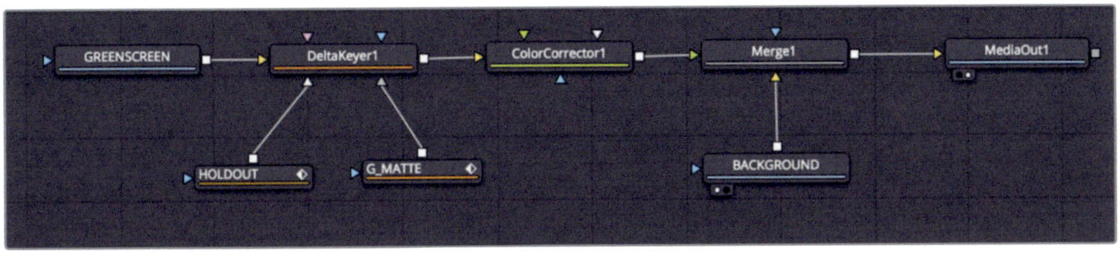

Completed node tree for Lesson 8.

Lesson Review

1. In the Fusion page, what is the primary tool for pulling a green screen key?
2. True or false: a solid matte connects to the Delta keyer in order to fill any holes that appear in the white matte of the foreground subject.
3. True or false: Removing spill or bounce light coming from the screen onto the foreground subject requires that you add a color corrector node.
4. True or false: Adjusting threshold in the Delta keyer suppresses spill on the foreground.
5. What project setting must be enabled for the Fusion page to automatically switch all media in nodes to linear gamma?

Answers

1 The Delta keyer is the primary tool for green screen keying.

2 True. A solid matte connects to the Delta keyer to fill holes in the foreground subject.

3 False. Removing spill or bounce light coming from the screen onto the foreground subject can be done in the Delta keyer or additionally in a color corrector node.

4 False. Adjusting threshold can clip only the black or white levels in a matte. It cannot modify the RGB levels for spill suppression.

5 The color science setting must be set to DaVinci YRGB Color Managed to convert all the media in nodes to linear gamma in the Fusion page.

Part III
3D Compositing

Lesson 9

Setting Up a 3D Scene

Many visual effects are about reproducing the characteristics of the real world such as perspective, atmosphere, depth cues, and occlusion of objects in a scene. And while you could create these effects using 2D compositing, they become much easier to do in a 3D compositing environment.

Fusion's 3D compositing includes the ability to position multiple elements in 3D space. You can add cameras, lights, 3D shapes, and apply a host of tools specifically designed for generating 3D visual effects and motion graphics.

This lesson and the following one combine to introduce you to the basic construction of a 3D scene by creating a broadcast news graphic. You'll learn how to navigate in 3D space and use some of the everyday tools for 3D compositing and motion graphics.

Time

This lesson takes approximately 50 minutes to complete.

Goals

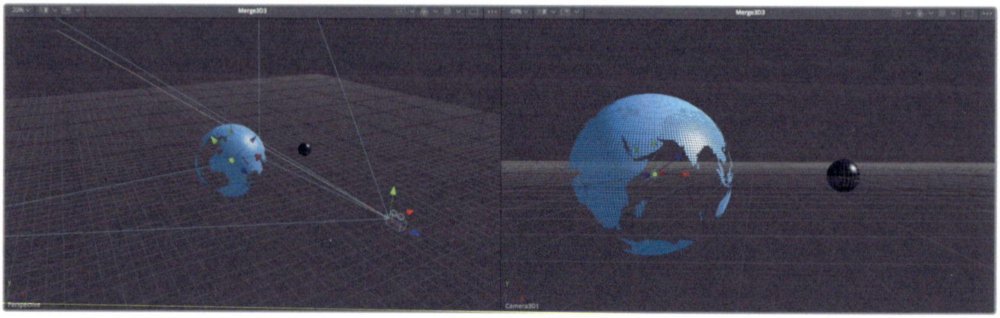

Completed composite for Lesson 9.

Placing Elements on 3D Shapes

Instead of creating a Fusion composition from the timeline, you can create a Fusion composition in a bin. This is often appropriate when creating motion graphics where you may not be ready to edit it into a timeline, but you need to begin creating the graphics for a project.

NOTE The timelines bin includes a backups bin with Fusion comps saved at various stages of the lesson, available for reference and reverse-engineering the node trees.

1 Open DaVinci Resolve. In the project manager, right-click and choose restore project archive.

2 Open R16 Fusion Lessons folder > **Fusion 16 Lessons part 3 dra**.

This project includes all the content required for the remaining lessons in this book.

3 Once the archive opens in the edit page, right-click in the Media > 3D Globe bin and choose new Fusion composition.

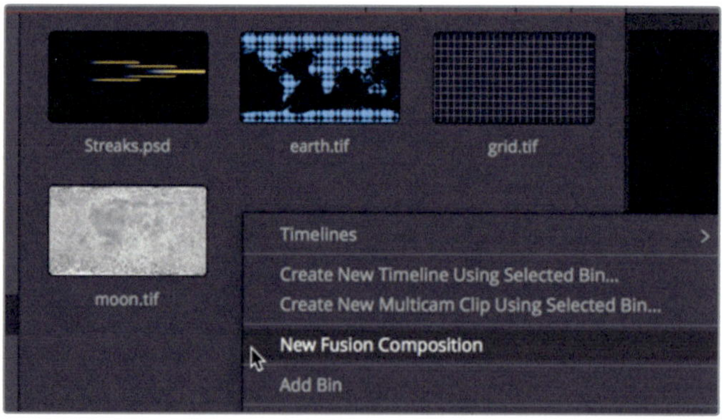

4 In the dialog, name the clip BREAKING NEWS, and click create.

Similar to the Fusion composition generator you used in earlier lessons, this Fusion composition can be opened in the Fusion page without any media associated with it and without creating a timeline.

5 Right-click over the BREAKING NEWS clip in the bin, and choose open in Fusion page.

For this project, you will use elements from the media pool to create a 3D scene.

NOTE It is much easier to work with 3D objects if you have two viewers, so if your user interface is configured with a single viewer, correct that now.

6 In the upper-left corner of the Fusion page, click the media pool button, and select the 3D globe bin.

This bin contains a few still images that you will use to create a stylized globe and a moon for our broadcast news graphic.

7 From the 3D globe bin, drag **earth.tif** into an empty area of the node editor, and press 1 to display it in viewer 1.

This is a stylized map of earth that has an alpha channel, as you can tell from its checkerboard background. Again, let's get organized right from the start by renaming each new element.

8 In the node editor, select the media in 1 node, press F2, and rename this image **EARTH**.

Any video clip, still image, or 2D generator cannot be part of a 3D scene without first connecting to an image plane 3D node or a shape 3D node.

9 With the earth node selected, open the effects library, and from the tools > 3D category, click the shape 3D tool. Then press 1 to view the earth image in 3D space.

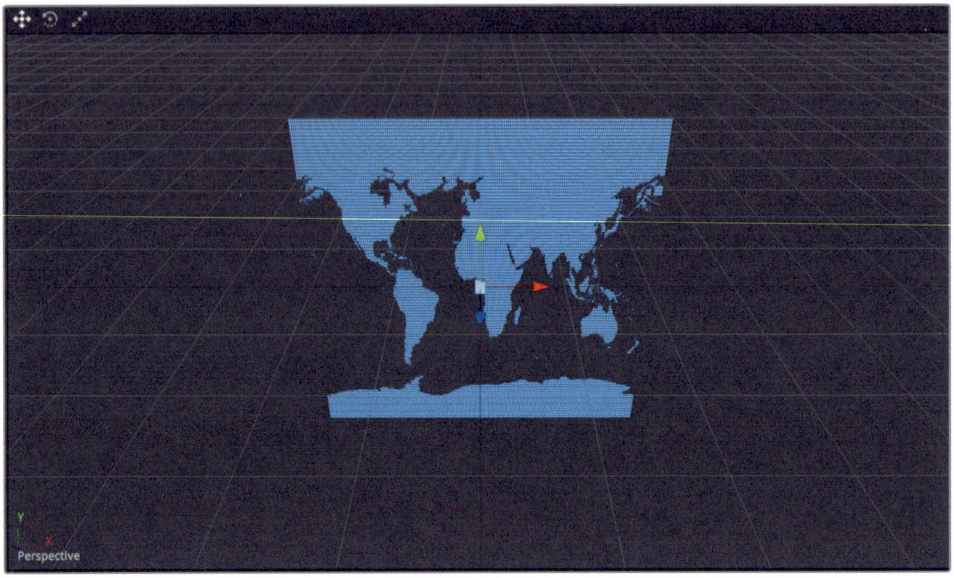

TIP You can use an image plane 3D instead of a shape 3D node. The image plane 3D retains the aspect ratio of the video clip connected to it.

The earth image now appears in a 3D viewer. It is placed on a 3D image plane, and can be rotated and viewed from any angle.

10 While holding down the middle mouse button, Option-drag (macOS) or Alt-drag (Windows) left and right in viewer 1.

By using the modifier key and middle mouse combination when dragging, you can rotate around the image to see it from the left or right side. To be clear, you are not rotating the image itself; you are changing your view, as if you were walking around the image. You can also change the view to see it from the top or bottom.

11 Continue holding down the Option (macOS) or Alt (Windows) key, and while holding down the middle mouse button, drag up and down in viewer 1.

When a 3D node is displayed in any viewer, that viewer becomes a 3D viewer and is set to a perspective view. This perspective view gives you the ability to view the image from any angle in 3D space.

12 Open the media pool, and from the 3D globe bin, drag grid.tif into an empty area of the node editor. Press 2 to display it in viewer 2.

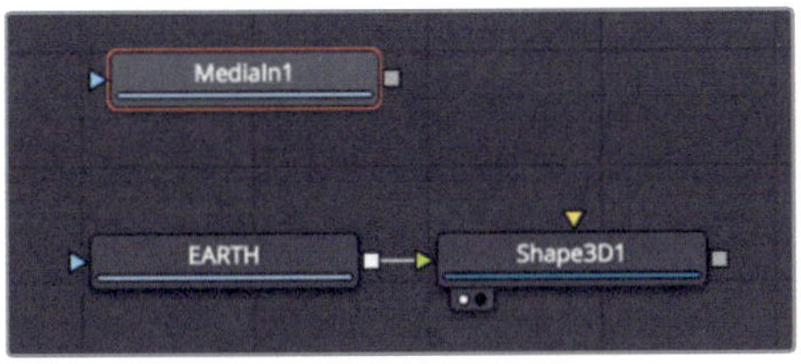

This image is a grid. Before using it, you should rename it appropriately.

13 Select the media in 1 node, press F2, and rename this image **GRID**.

14 With the grid node selected, open the effects library, and from the tools > 3D category, click the shape 3D tool. Press 2 to view the grid in 3D space.

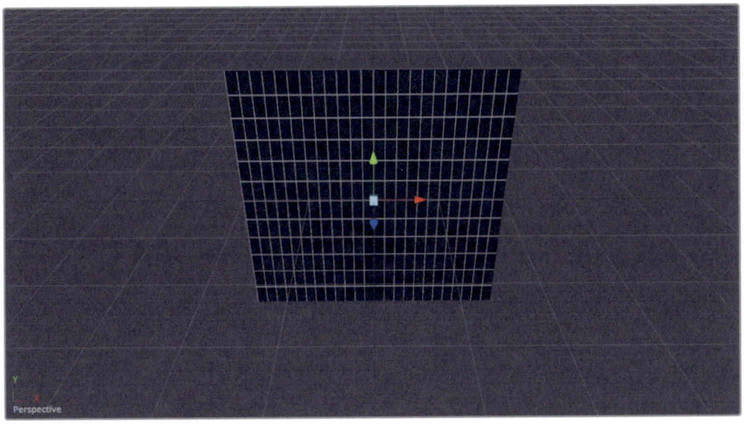

The grid is also displayed in a 3D viewer, but the two images are not connected and cannot interact. Each exists in its own 3D world. However, you can combine 3D images to exist in the same 3D world using a merge 3D node located in the last group of tools in the toolbar.

While the merge node is the fundamental 2D compositing tool, the merge 3D node is the fundamental 3D compositing tool.

15 With the grid's shape 3D 2 node selected in the node editor, click the merge 3D node in the toolbar. Press 2 to view the merge node in viewer 2.

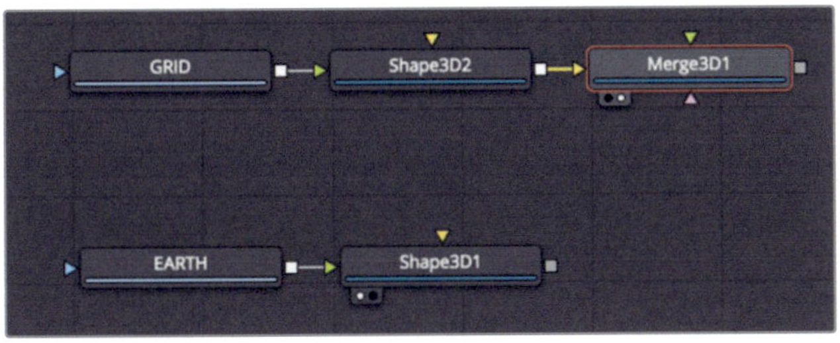

Although you may see no visual difference compared to the grid view, the merge 3D node allows you to connect other 3D images, so they will exist in the same 3D space.

16 Drag the output of the earth's shape 3D 1 node to the green input of the merge 3D 1 node.

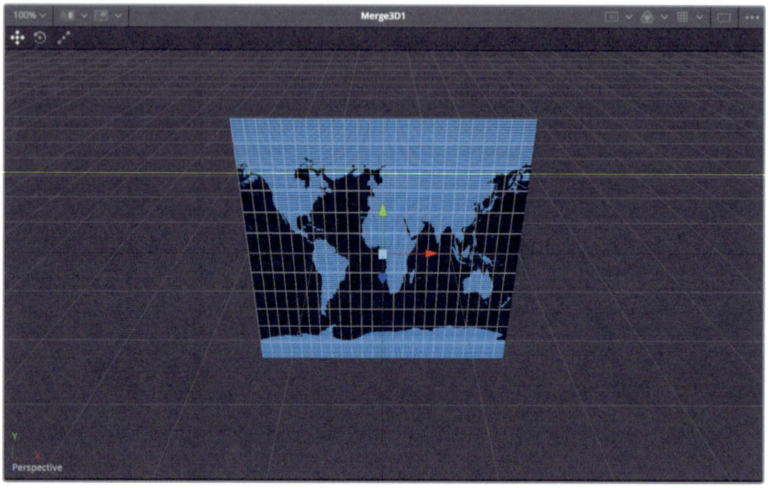

The viewer shows the images on top of each other. To give the appearance of a more realistic, 3D view, you can enable default lights.

17 Right-click in viewer 2, and choose 3D options > lighting.

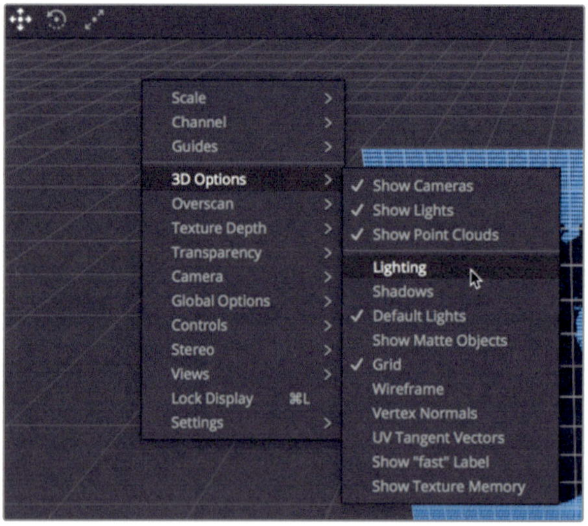

18 Preview the light by holding the Option key (macOS) or the Alt key (Windows) and using the middle mouse button to drag around in viewer 2.

The default lighting provides a soft directional light that comes from the upper right of the 3D world. The light object is not visible in the viewer or the inspector; it merely provides some default shading to the objects so they appear more three-dimensional while you work.

Navigating in 3D

Unlike the standard 2D merge node, the 3D merge node can connect multiple clips and images. The merge 3D node has no foreground and background because the order of elements as seen in the viewer is determined by their Z-position values. This behavior makes it extremely important that you understand how to view and move elements in 3D space.

1. In the node editor, select the shape 3D 2 node. This is the grid's shape 3D node.
2. In the inspector, click the transform tab.

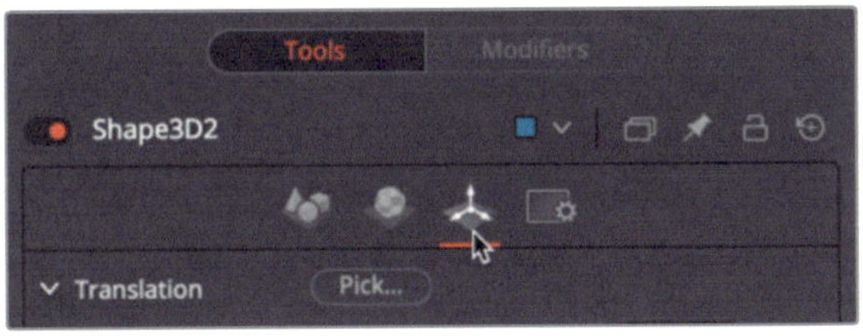

In Fusion's 3D world, the origin point of the coordinate system is in the center of the world (X = 0, Y = 0, and Z =0). Each element you add to the 3D world will start at this 0,0,0 location.

To move an image in 3D space, you use the shape 3D node (or the image plane 3D node), which adds 3D positioning, rotation, and pivot controls to an image.

3. In the inspector, drag the X translation to slightly move the grid to the right.

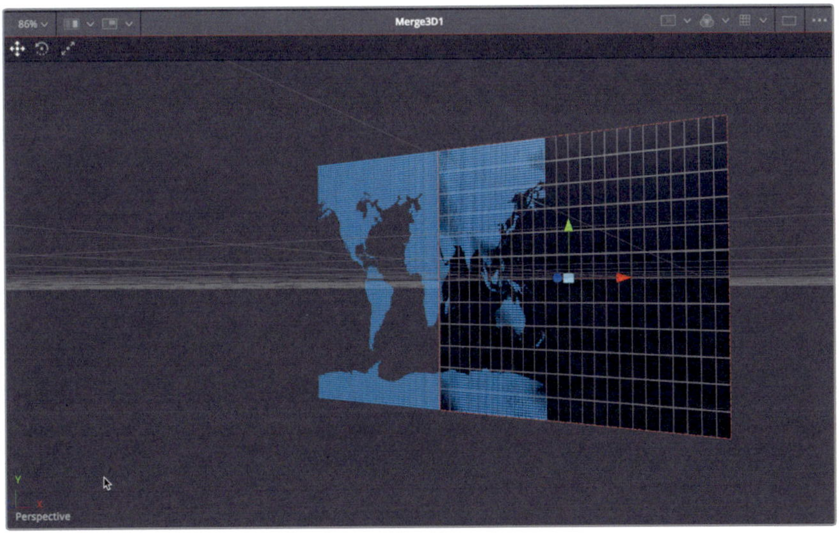

The X translation moves the grid to the left or right. You can perform the same transform operation using the red arrow in the viewer. Other viewer arrow overlays move the object up and down (green, Y translate arrow) or toward and away from you (blue, Z translate arrow).

4 In the inspector, drag the Z translation slightly to the right until the grid is in front of the earth.

Unlike 2D compositing with a merge node, you do not change the inputs to the merge 3D node to get the grid in front of or behind the earth image. That ordering is done by altering the Z translation value and the angle at which you view the composite.

5 In viewer 2, Option (macOS) or Alt (Windows) drag right while holding down the middle mouse button until the earth rotates in front of the grid.

You have changed your viewing perspective; so, even though the Z translation sets the grid in front of the earth, your perspective view of the scene determines the layer order.

The following key and mouse button combinations are extremely useful when working in a 3D viewer.

- Pan: Drag while holding down the middle mouse button.
- Rotate: While holding down the middle mouse button, Option-drag (macOS) or Alt-drag (Windows).
- Zoom in and out: Hold down cmd (macOS) or Ctrl (Windows) and scroll the middle mouse wheel.

Now, let's get back to a default starting point and begin to build the planets.

6 In the inspector, click the gray dot under the X and Z translation controls to reset them to 0.0.

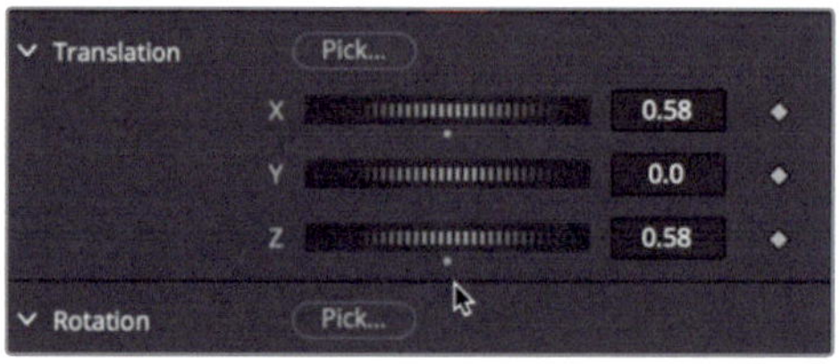

7 In viewer 2, Option (macOS) or Alt (Windows) drag left using the middle mouse button until the earth and grid are facing front.

Dragging in the viewer while set to perspective view only changes the angle from which you are viewing the objects, not the objects themselves. It is similar to moving a camera around your subjects.

TIP You can use the perspective view axis control in the lower-left corner to orient yourself and the viewer. When you move the perspective view so that the green arrow points up, the red arrow points right and the blue arrow points directly toward you, the perspective view is at a default front view in the 3D world.

Intuitively moving the perspective viewer to different angles, and moving objects in 3D space are essential activities in 3D compositing. In upcoming lessons, you'll get more practice with these techniques.

Using Shapes in 3D

The shapes 3D node is a flexible node that enables you to wrap video or images around basic 3D shapes. For this 3D planet scene, you'll apply the grid and the earth images to spherical shapes and practice moving them in 3D space.

1 Select the grid's shape 3D 2 node, and in the inspector, click the controls tab.

The controls tab allows you to select different primitive 3D shapes to wrap the image around.

2 From the shape menu, choose sphere.

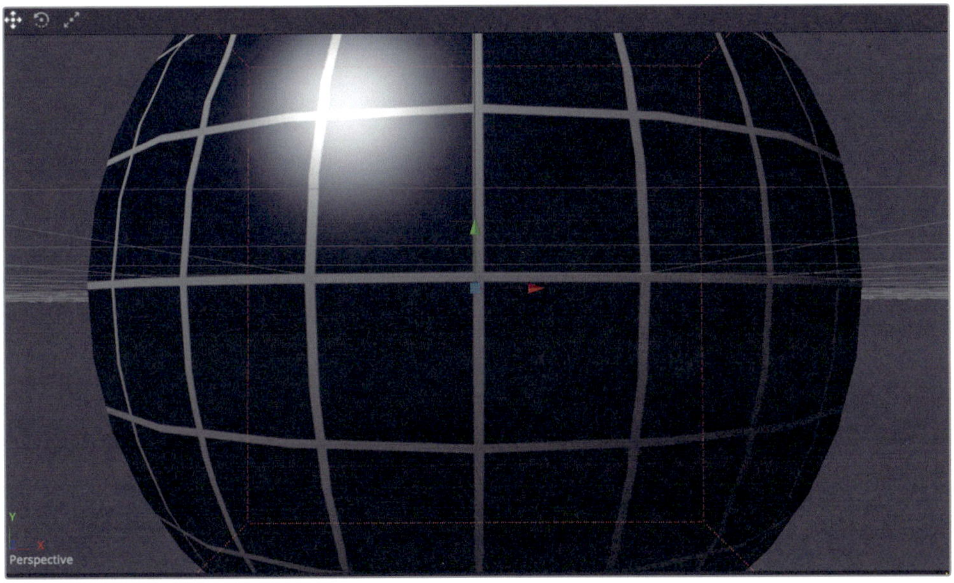

The grid changes from a flat plane to a sphere. This sphere is very large, so it encompasses the entire image of the earth and obstructs it from view.

3 Drag the radius slider to around 0.2.

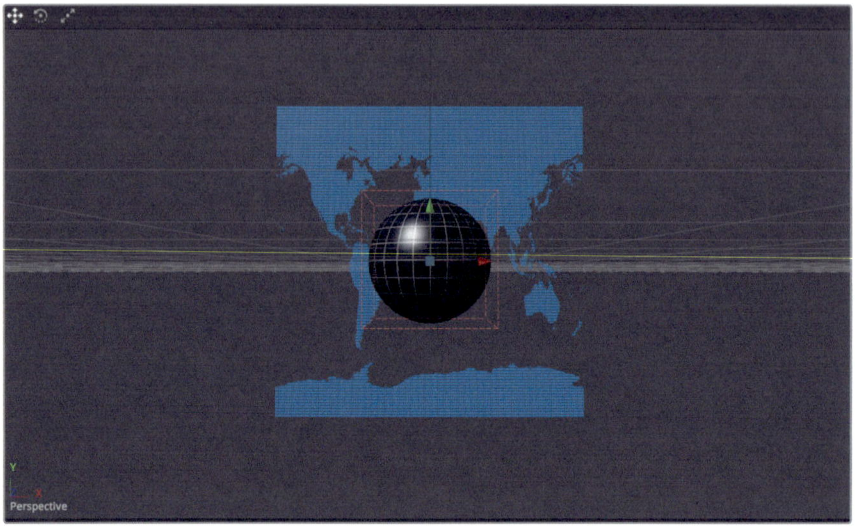

The grid is now small enough to reveal the earth image.

4 In the node editor, select the earth's shape 3D 1 node, and in the inspector's controls tab, in the shape menu, choose sphere.

The earth is mapped onto a sphere shape that surrounds the smaller grid. The alpha channel contained in the earth image creates some transparent areas where the grid can be seen.

5 In viewer 2, hold down the cmd (macOS) or Ctrl (Windows) key, and scroll the middle mouse button to zoom the viewer out a bit until you can clearly see the top and bottom of the earth sphere.

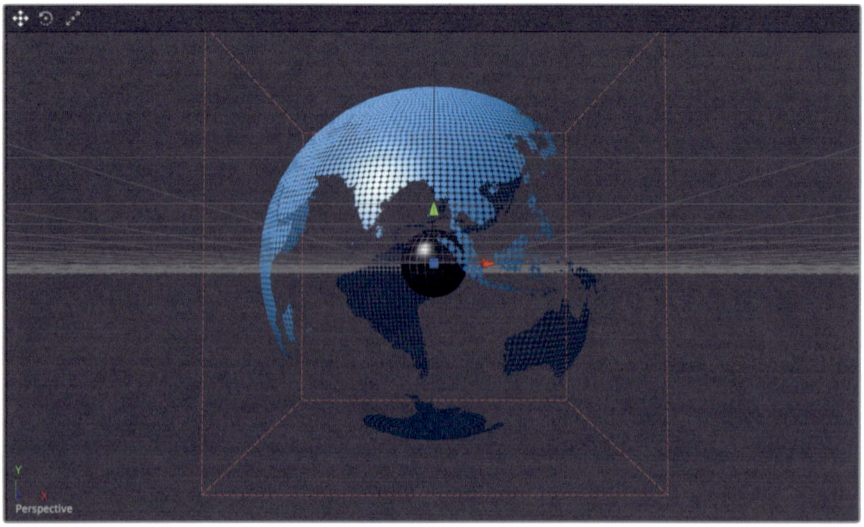

6 While holding down the middle mouse button, Option-drag (macOS) or Alt-drag (Windows) right and left in viewer 2 to rotate your view around the earth and grid. Then, return to the original perspective view position.

Fusion provides a true 3D environment with 3D shapes, rotation, and object depth. At the moment, the grid is located inside the earth sphere because both objects occupy the center location of the 3D world at the X, Y, and Z translation coordinates of 0,0,0. Let's move the grid into a more practical moon-like position.

7 In the node editor, select the grid's shape 3D 2 node.

8 Using the arrow overlays in the viewer, drag the red X translate arrow to the right until the grid reaches the edge of the viewer.

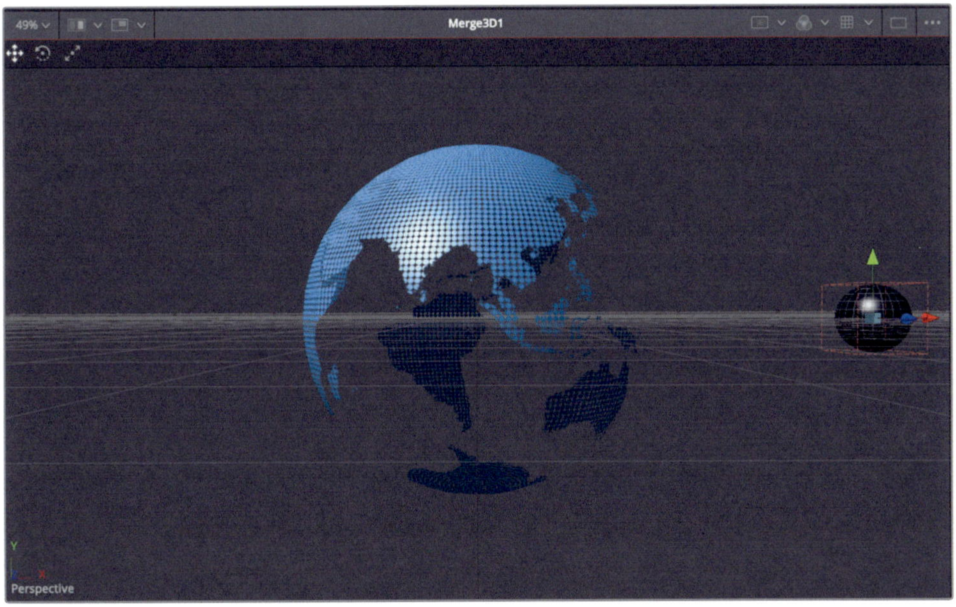

You now have a modern, high tech earth with its grid moon in perfect position to begin animating them.

Entering Simple Expressions

To suggest reality, the planets need to rotate. The earth needs to rotate around its own axis, while the grid needs to rotate around its own axis while also orbiting around the earth. Let's start with the earth rotation but instead of using keyframes to create the rotation, we'll use a simple expression.

1 Select the earth's shape 3D 1 node, and in the inspector, click the transform tab.

Expressions can be simple conditional statements, mathematical equations, or links to other parameters that generate animation. The pick whip you used in Lesson 4 to animate the dotted line is a shortcut to writing a simple expression. When you used the pick whip, you exposed an expression field that was populated by the parameter name you chose. For this exercise, you'll enter a simple expression directly in the expression field to create a rotation animation. The expression will simply create a link based on the current frame in the project to the degree of Y rotation applied to the earth.

2 In the Y rotation number field, type = (equals sign), and then press enter to open the expression field.

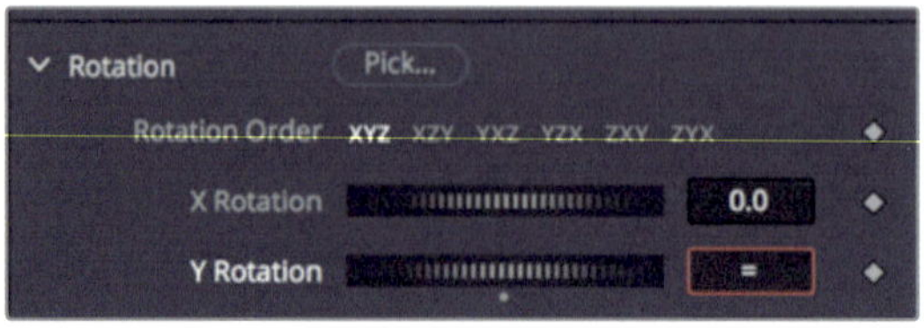

In the expression field you can enter a simple expression to control the Y rotation.

3 Click in the expression field, type **time*10**, and press enter.

TIP The time expression is always written in lowercase type.

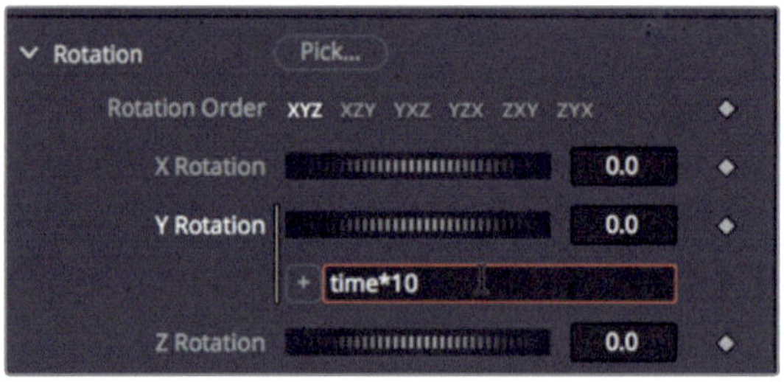

Although this procedure may seem cryptic at first, it is easy to understand. Because this field is for Y rotation, all of its values refer to degrees of rotation. Entering time tells the Y rotation parameter to change its value as the playback advances frame-by-frame. At frame 1 the rotation value will be 1 x 10 degrees, and at frame 10 the rotation value will be 10 x 10 = 100 degrees of rotation around the Y axis.

TIP A list of additional expression examples can be found in the DaVinci Resolve Reference manual.

4 Play the animation and see that the earth is rotating on its Y axis.

This movement is a bit fast, so you can easily change the expression and try new speeds just by changing one number.

5 Click in the expression field, and change the number from 10 to 1. You can do so even as the animation is playing, making it very easy to try several different values.

Now, you can animate the grid. The easiest way to do so is to link the rotation of the grid to the rotation of the earth, as you did back in Lesson 4 with the plane and the dotted line.

6 In the upper-right corner of the inspector, click the pin button to prevent the earth's shape inspector from closing when you select another node.

By pinning the earth's shape inspector, you can select another node and see the two nodes' controls at the same time.

7 In the node editor, select the grid's shape 3D 2 node.

The grid's shape 3D 2 controls appear at the top of the inspector, while the earth's shape 3D 1 controls remain open in the inspector below.

8 In the grid's shape 3D 2 controls, click the transform tab. In the number field to the right of the Y rotation slider, enter an = (equals sign), and press enter to open an expression field under the rotation control.

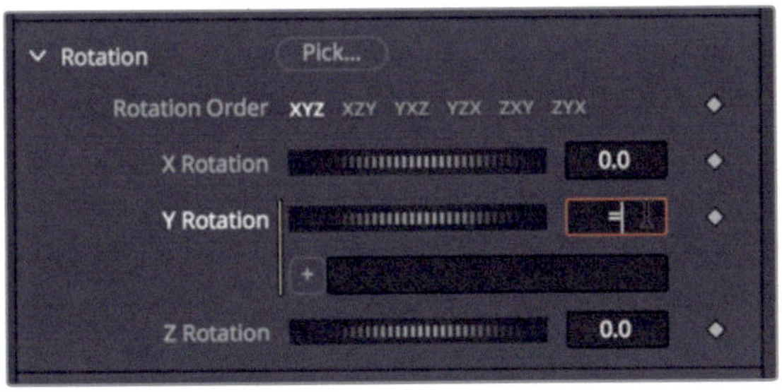

Instead of entering an expression, you'll use a pick whip to link the two parameters together.

9 At the top of the inspector, in the grid's shape 3D 2 controls, drag from the plus icon on the left of the expression field to the Y rotation label in the earth's shape 3D 1 controls.

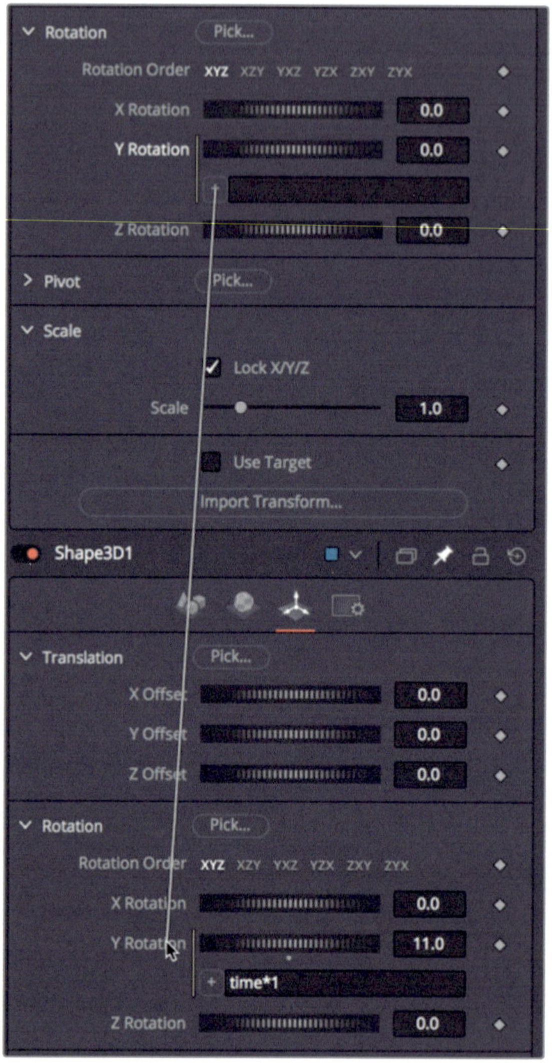

10 Play the composition to see the linked animated shapes.

Although that worked, the grid is rotating slowly. We would like it to rotate a bit faster than the earth's rotation to make it more visible. It is easy to do so by modifying the expression that you added with the pick whip.

11 In the expression field, insert ***2** to the end of the expression. The * sign represents multiplication. So by multiplying the expression by 2, you are increasing the grid's rotation to be two times faster than the earth's rotation.

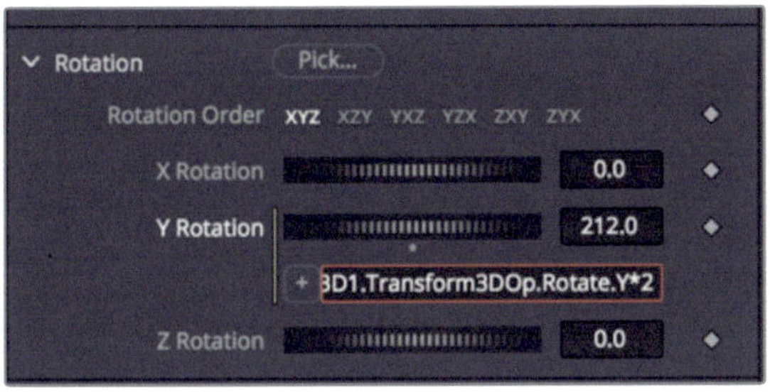

12 Play the comp to see the shapes rotating.

These rotating elements look good so far. Since you won't be linking any more parameters right now, you can unpin the earth's shape 3d from the inspector.

13 In the node editor, select the earth's shape 3D 1 node.

14 In the upper-right corner of the inspector, click the pin button to release the earth's shape inspector. The earth's controls will close when you select another node.

Animation via expressions allows you to animate without keyframes. So, no matter how you change your composition, the rotation never stops; it just recalculates and continues along its course, counting upward and rotating.

Creating Multiple Axes of Rotation

The grid also requires an orbital center as well as the axis of rotation. The orbital center will be used to rotate the grid around the earth.

To create a second point of rotation to orbit the grid around the earth, you'll need to add a new pivot point for the grid. The 3D equivalent of the transform tool that you used when 2D compositing is the transform 3D. It includes a complete set of 3D transforms, including rotation and pivot.

1 Select the grid's shape 3D 2 node, and then, from the effects library, go to the tools > 3D category and click the transform 3D tool.

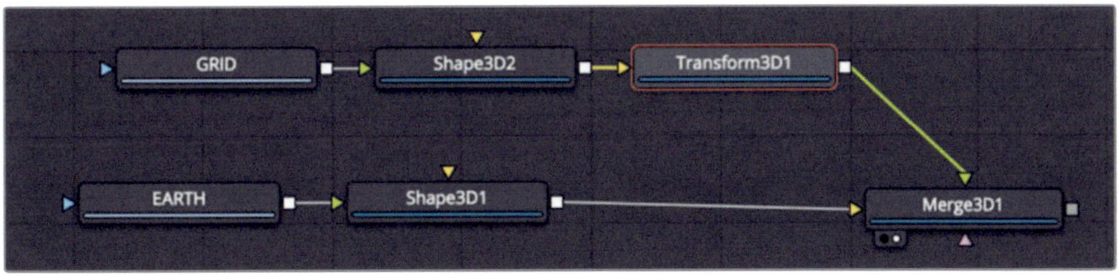

Inserting a transform 3D node adds a second *pivot point* to the grid. The grid has its own pivot point based on the shape 3D node that acts as the grid's axis of rotation. The transform 3D node adds an additional pivot in the center of the 3D world that will act as the grid's orbital center.

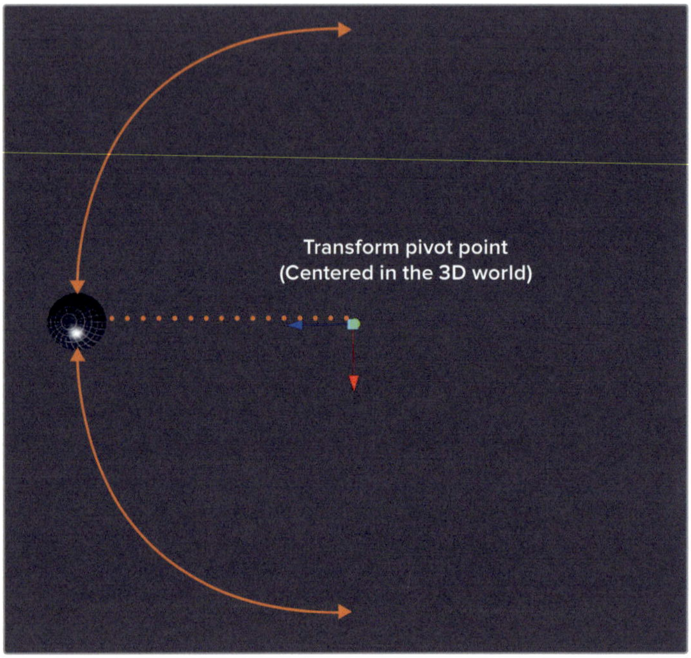

Transform pivot point centered in the 3D world.

2 With the transform 3D selected, drag the Y rotation slider back and forth to see the results of the added pivot point. Then reset the Y rotation to 0.

Connecting the two transforms to an object creates a hierarchy, so that one transform can control or influence the other. A transform that controls others is called a parent, while a transform that is controlled is called a *child*. In this comp, the parent is the transform 3D node, while the child is the shape 3D node. Moving or rotating the transform 3D node will move or rotate the grid, but moving or rotating the shape 3D node will not impact the transform 3D node.

TIP When you are rotating only one object on a single axis, you can use the pivot controls to offset the object's axis of rotation.

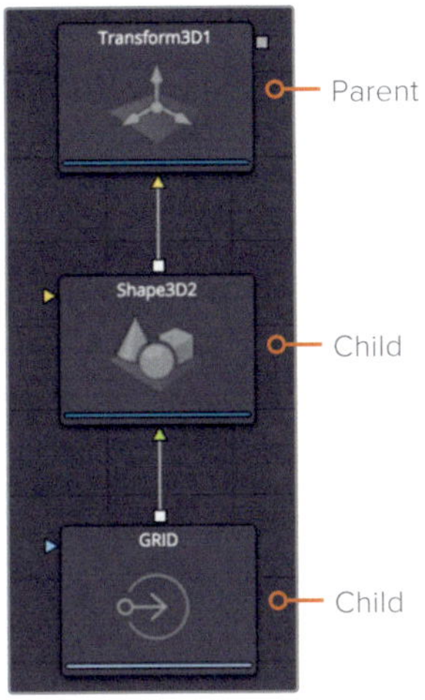

3 Select the grid's transform 3D node, and then in the transform tab, double-click the Y rotation field and press = (equals sign). Press enter.

4 In the expression field, enter **time*1.5**, and then press play to see the orbital rotation.

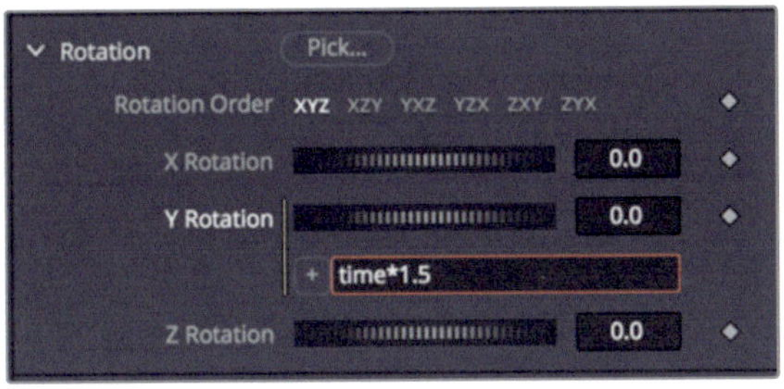

Based on the location of the transform 3D's pivot, the grid swings around behind the earth but also still rotates around its own pivot.

You can change the starting point of the orbit just by adding or subtracting the degrees to modify the position. For instance, if we subtract 90 degrees from the starting point of the orbit, it should start with the grid in front of the earth.

5 At the end of the expression field, enter **time -90**, and then press enter.

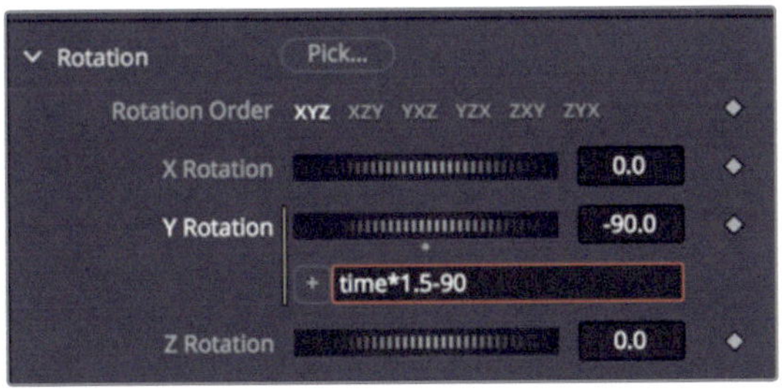

6 Play the comp to see the spheres rotating.

As you review this animation, look over the node tree and think about what each node is actually controlling. Where are the textures coming from, and what is driving the animation? Think about having to extend this beyond two planets and adding a third. Examining a node tree is one of the benefits of the node tree interface: You can see exactly how each object is processed.

Adding Lights and Cameras

3D scenes in the Fusion page can include cameras and lights. You use the cameras and lights much like you would in the real world to frame and illuminate objects how you see fit.

The perspective view and the default lighting you have been using in the viewers are automatically present whenever you create a 3D composite. However, they are primarily there so you can see objects and view them from different angles and are not intended to be used in your final rendering. Although you do not need to add more lights or a camera to a 3D scene, doing so adds significant creative control over the final appearance.

1 From the toolbar, drag a new merge 3D node to an empty area of the node editor.

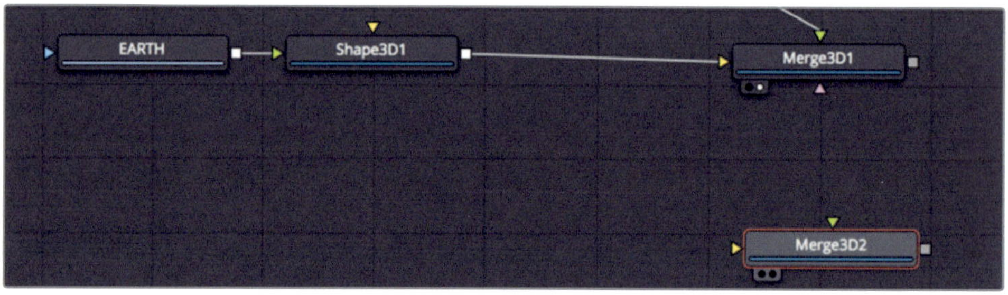

Although you can connect everything into a single merge 3D, things can get a bit tangled and unwieldy if you choose that path. Even on simple composites, as we have here, it is wise to keep things organized by using multiple merge 3D nodes to group items together. We'll use a new merge 3D to connect our lights. Similar to standard 2D merge tools, you can connect merge 3D nodes together, creating a more organized and flexible, yet still singular 3D world.

2 Select the merge 3D 2 node, and in the effects library, from the tools > 3D category, select the light subcategory.

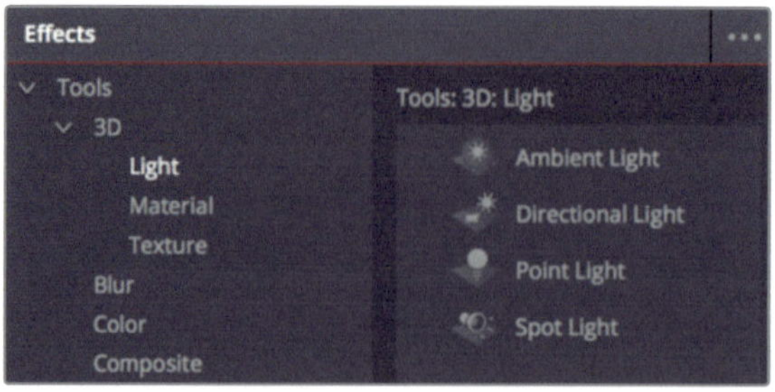

3 In the light subcategory, click the directional light tool to add it to the node editor, connected to the merge 3D 2 node.

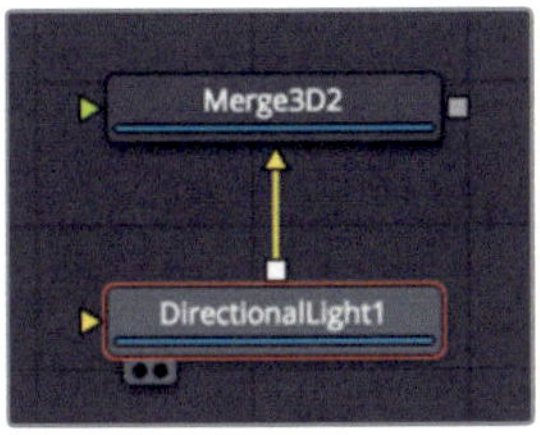

The directional light is added to the scene and connected to the merge 3D 2 node. Now, you'll connect the merge 3D 2 node and the merge 3D 1 node.

4 Drag the output of the merge 3D 2 node to the merge 3D 1 node to connect it.

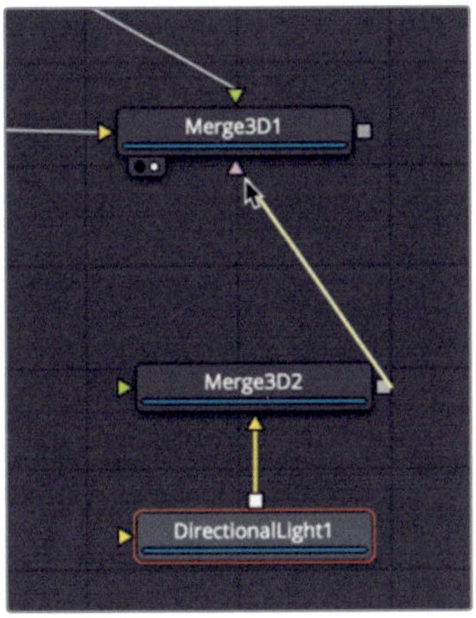

> **TIP** Holding the Option key (macOS) or the Alt key (Windows) and clicking on a connection line will add a router that allows connection lines to bend instead of having diagonal lines overlap other nodes.

When a light is added to a scene (directly to a merge 3D or indirectly through a connecting merge 3D), the default lights in the scene are disabled and only the lights connected to the node tree illuminate the objects in the scene. However, we do not see any lights illuminating our scene, because, by default, lights are not passed from one merge 3D to another. This default setting allows you to combine 3D objects into one scene but light them independently.

5 Select the merge 3D 2 node, and in the inspector, click the pass through lights checkbox.

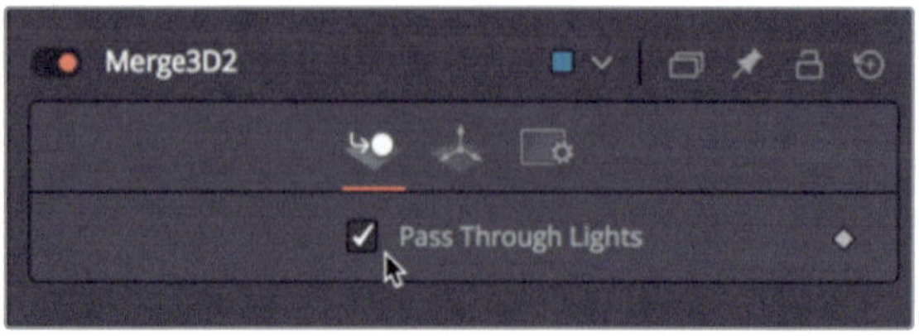

Now any lights you connect to merge 3D 2 will illuminate all the objects in merge 3D 1.

6 Select the directional light node, and in the transform tab of the inspector, use the Y and Z rotation controls to set the light as if coming from the upper-right corner of the frame.

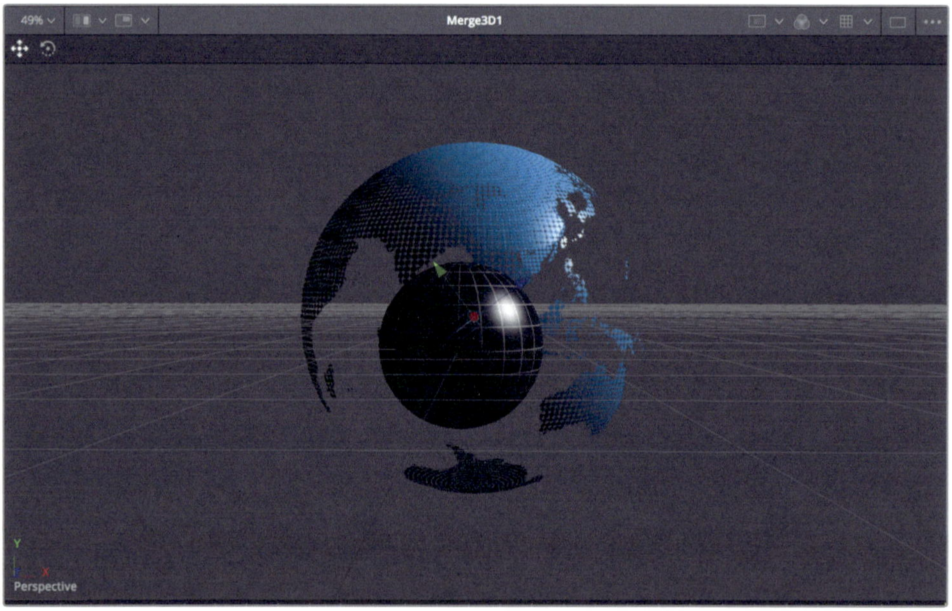

The directional light made the scene too dark in areas located opposite to the light's direction. You can add an ambient light to fill in those dark areas.

7 Select the merge 3D 2 node again, and in the effects library, click the ambient light tool.

8 Select the ambient light node, and lower the light's intensity to around .300.

Your entire 3D scene is now lit using only two lights. With that in place, you can start to work on framing your composite by introducing a camera.

Why Are There Different Light Types?

The Fusion page includes four lights that you can add to a scene, each with its own characteristics.

Ambient light illuminates an entire scene equally—similar to adding a gain brightness—because it has no position or rotation. It is primarily used to fill in areas that other lights may leave too dark.

Directional light has a clear direction but lacks a specific source location. You do not control its position, but you can use rotation controls to indicate from where in the scene the light appears to be coming. This light is akin to sunlight, because no matter how far away an object may be, there is no light fall-off.

Point light has a clear position in space that emits light in all directions therefore only its position affects the light, not its rotation. A light bulb is a good example of a point light. Unlike both ambient and directional lights, a point light may fall off with distance.

Spot light comes from a specific point and has a clearly defined cone with fall-off to the edges of that cone. This is the only light capable of casting shadows.

Setting Up and Animating a Camera

Even without a camera, you can set up the perspective view to see your scene from different angles. This can often stimulate ideas on how ultimately you may want the scene framed. To show as much of the arc animation from the moon as possible, let's position the earth slightly to the left using just the standard perspective view.

1 In viewer 2, hold down the middle mouse button and drag to position the frame so the earth is against the left edge.

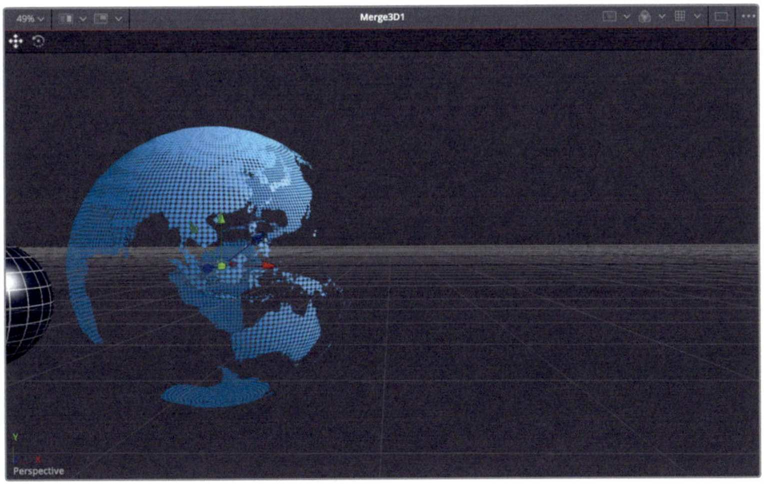

You can now add a camera and connect it to the merge 3D 1 with the moon and earth objects.

2 Select the merge 3D 1 in the node editor.

3 With the merge 3D 1 node selected, click the camera tool in the toolbar to add it to the merge 3D 1 node.

You can set the viewer to display the camera output, using the perspective axis label in the lower left corner of the viewer.

4 In the lower-left corner of viewer 2, right-click the perspective axis label, and choose camera 3D 1.

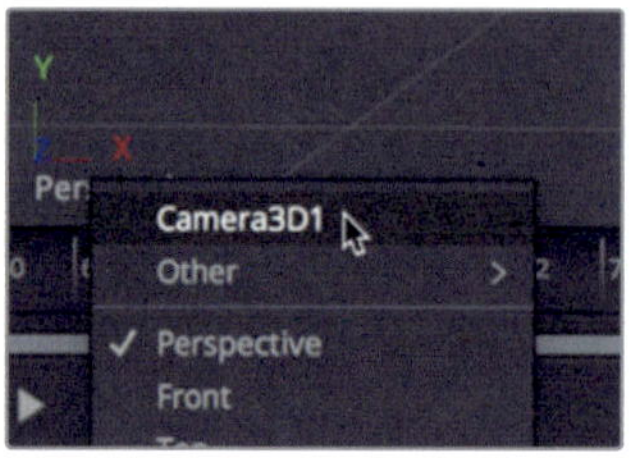

Now you are able to see the output of the camera in viewer 2. It is similar to looking through the camera's viewfinder.

Since we never framed the camera, it is located at the center of the 3D scene inside the earth sphere, which is not where we want our camera.

Rather than tweaking a lot of parameters to essentially match what we have in the perspective viewer, you can copy the perspective view to the camera.

5 In the lower-left corner of viewer 2, right-click the perspective axis label and choose perspective from the menu.

6 Right-click the perspective axis label again, and choose copy pov to > camera3D1, and then switch back to the camera 3D 1 view.

Copying the viewer's perspective point of view (POV) to the camera repositions the camera so it matches the perspective framing.

This is where having two viewers comes in handy. It will be easier to animate the camera if viewer 1 shows us the perspective view where we can manipulate the camera's position, while viewer 2 shows the camera output.

7 In the node editor, select the merge 3D 1 node and press 1 to display it in the viewer.

8 In viewer 1, right-click the perspective axis label again and choose perspective.

For our animation, we'll create a simple pull back on the camera to expose more space on the right. We'll use this space in the next lesson when we add 3D text.

9 Move to the start of the render range.

10 Select the camera 3D 1 node, and in the inspector, click the transform tab.

11 In the transform tab, click the keyframe button to the right of the Z translation parameter.

This sets our starting keyframe. Now, we'll set an ending keyframe and move the camera back a bit.

12 Move to the end of the render range.

You can use the inspector to move the camera or change the camera's position directly in the viewers.

13 Position the mouse cursor over viewer 1, hold the cmd key (macOS) or the Ctrl key (Windows) and scroll the middle mouse wheel until you can clearly see the earth and the camera in the frame.

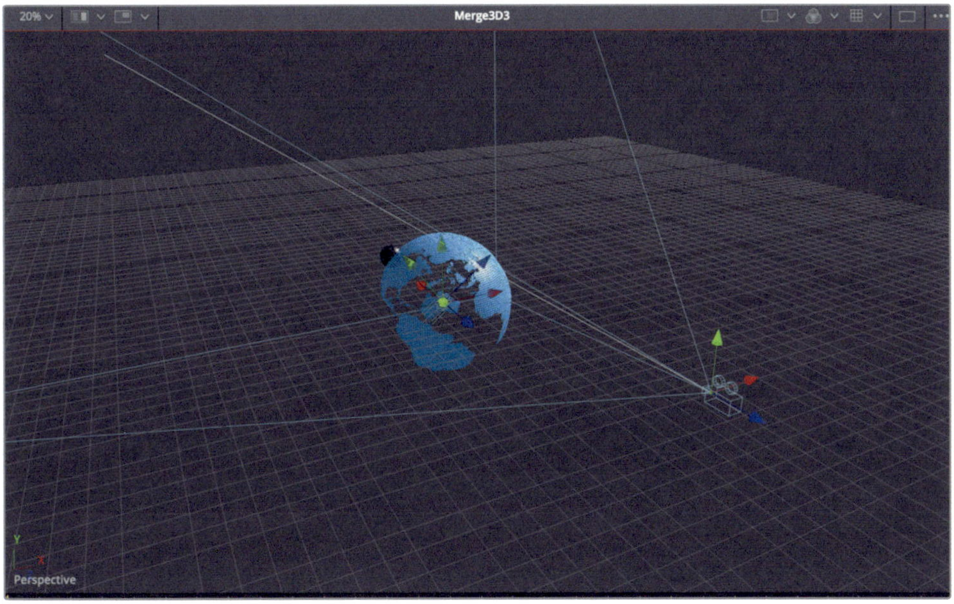

The camera, like all objects in a 3D scene, includes three arrows as translation controls in the viewer. The green arrow moves the camera up and down on the Y axis, the red arrow moves it side to side on the X axis, and the blue arrow moves the camera in and out on the Z axis. Since we want to move the camera back a bit, we'll drag the blue arrow in viewer 1.

14 In viewer 1, drag the camera's Z translate blue arrow back, away from the spheres until the inspector's Z translation is set around 4.5.

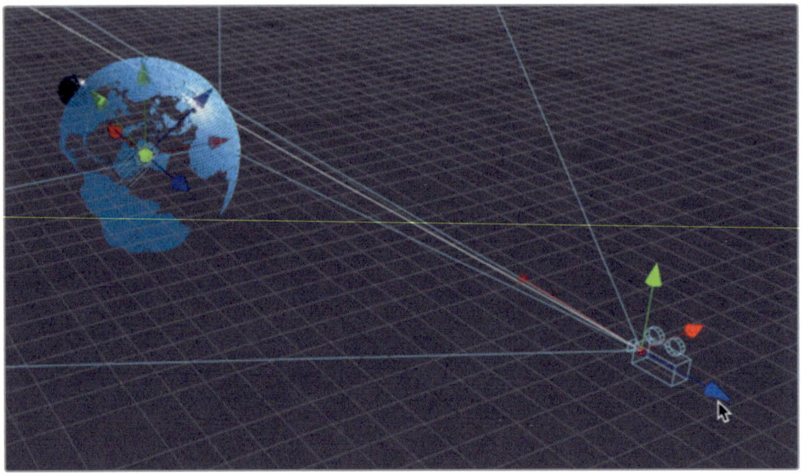

Moving the Z translate adds an ending keyframe for the animation.

15 Play the comp to see the camera animation.

For this lesson, everything is now in place. You have planets in orbit, a well-lit scene, and an animated camera move. In the next lesson, you'll finish this broadcast news design by adding 3D text. You'll also learn how to render your 3D scene and add 2D image-processing effects to add the final polish.

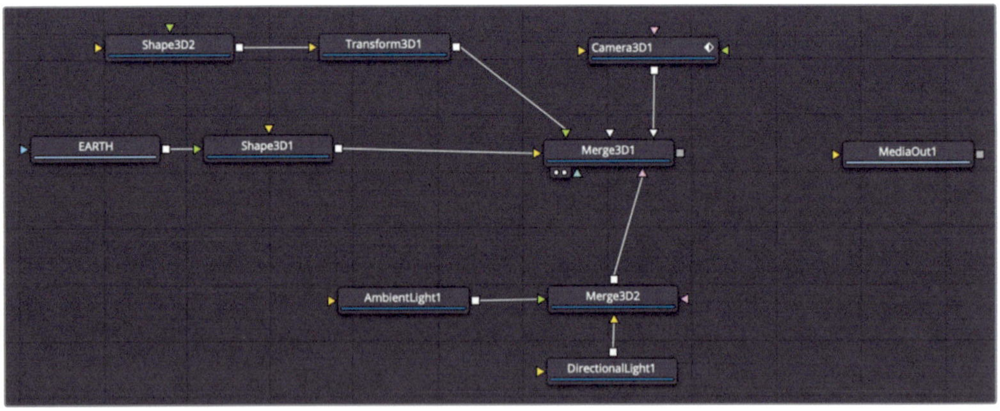

Completed node tree for Lesson 9.

Practice Exercise

Let's try expanding this universe with other elements. Located in the 3D globe bin are two additional images, a **moon.tif** and a **Streaks.PSD** file. We'll add both these textures to our projects and animate them based the instructions below.

1 Replace the grid texture with the moon.tif texture. Keep the same animation and size.
2 Use the streaks image on another sphere that is slightly larger than the earth's radius, and make the streaks appear to travel around the earth's outer atmosphere.

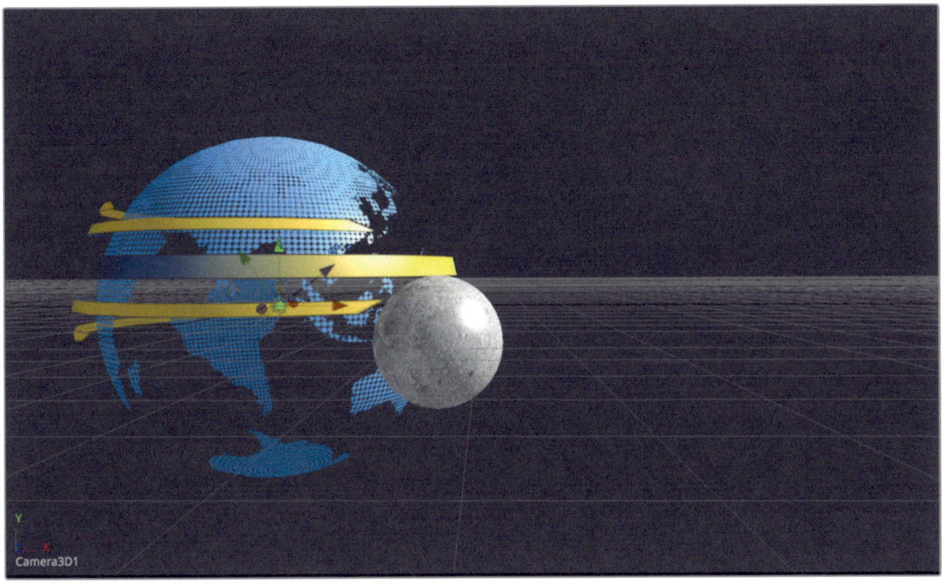

TIP The streaks may not look angular once added to the sphere shape. To correct this, increase the amount of detail, or tessellation, of the streaks by adjusting the base subdivisions and height subdivisions in the inspector.

Lesson Review

1 True or false: A media in node must go through a shape 3D node or an image plane 3D node before it can connect to the merge 3D node.
2 True or false: The center of Fusion's 3D scene uses the following coordinates: x = 0.5, y = 0.5, z = 0.5.
3 True or false: To link two different parameters together, you enter a + (plus) sign in the value field to expose the expression field.
4 True or false: Lights must be added to a scene using their own merge 3D node. They cannot be connected into a merge 3D that already has objects connected.
5 True or false: In the viewer, a green arrow on a camera, light, or 3D object moves the object along the Y axis.

Answers

1 True. A media in node must go through a shape 3D node or an image plane 3D node before it can connect to the merge 3D node.

2 False. The center of Fusion's 3D scene uses the following coordinates: x = 0, y = 0, z = 0. A 2D scene uses x = 0.5 and y = 0.5 as the center of the viewer.

3 False. To link two different parameters together you enter an = (equals sign) in the number field to expose the expression field.

4 False. Lights, 3D objects, and cameras can all be connected to the same merge 3D node, but it makes for better organization and flexibility if you separate them.

5 True. In the viewer, a green arrow on a camera, light, or 3D object moves the object along the Y axis, up and down.

Lesson 10

Designing 3D Broadcast Graphics

From movie titles to commercials and broadcast news, 3D text and graphics are used in so many ways to convey a project's feelings and energy. In this lesson, you'll continue with our 3D scene from the previous lesson. We'll add a compelling 3D broadcast news title using a material *shader* that will produce a shiny metallic surface for our text. In addition to the title, you will explore how to combine 2D and 3D scenes, and how to take the broadcast graphic we create and use it as a template in the edit page.

Time

This lesson takes approximately 70 minutes to complete.

Goals

Completed composite for Lesson 10.

Adding 3D Text to a Scene

With motion graphics, the animation is central, so it is often a good idea to first build your animation and set the timing of the graphic using simple versions of your design objects. Later, you can apply the final materials and look. That's the method you'll follow here: In the first half of the lesson, you'll animate very simple 3D text and shapes, and then you'll refine their look in the second half.

> **NOTE** The timelines bin includes a backups bin with Fusion comps saved at various stages of the lesson, available for reference and reverse-engineering the node trees.

1. In the project manager, open the Fusion 16 Lessons part 3 project.
2. In the edit page timelines bin, right-click over the **Lesson 10 START** and choose open in Fusion page.

   This composition opens into the Fusion page with our globes, lights, and camera animation from the previous lesson. We've started this composition as if you completed the practice exercise from Lesson 9, with the added yellow streaks and the new moon look.

The standard 2D text+ node cannot be connected into a merge 3D composite; it only connects to 2D composites. When you begin a 3D composite, as you did in the previous lesson, you must use a text 3D node to add text.

3 Select the merge 3 node, and press 1 to see it in viewer 1.

4 From the toolbar, drag the text 3D tool into an empty place in the node editor. Press 2 to display that node in viewer 2.

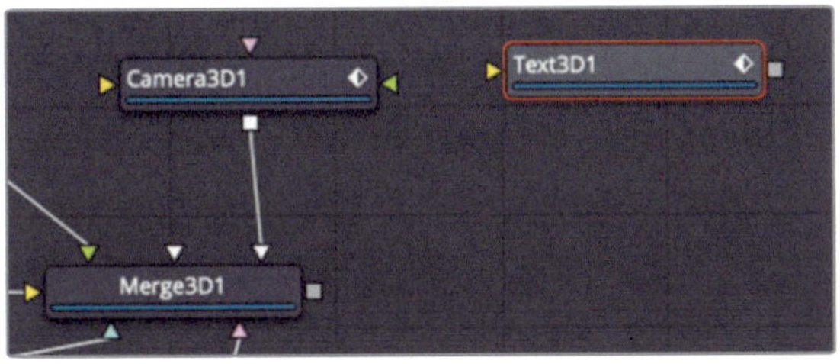

You now have a 3D viewer with access to 3D text controls, but we'll proceed with steps that are very similar to using a standard 2D text node.

5 With the text 3D node selected, go to the inspector, and in the styled text field, type **BREAKING NEWS**, using two lines of text.

As with 2D text, you can assign a font, size, and other text properties.

6 Set the typeface to Open Sans Bold Italic and the size to 0.5.

You also have your choice of 3D controls, such as extrusions and bevels.

7 Scroll to the bottom of the inspector, and click the disclosure arrow to open the extrusion parameters.

These parameters can add depth to text and give titles a greater sense of weight and substance.

8 Set the extrusion depth to 0.1, and the bevel depth and bevel width to around 0.02.

9 Right-click in viewer 2 and choose 3D options > lighting to enable the default shading in the viewer.

> **TIP** You may need to zoom in to the viewer to see the changes in the bevel depth and width.

10 While holding down the middle mouse button, Option-drag (macOS) or Alt-drag (Windows) in the viewer to view different angles of the text in the perspective view.

The text is lit using default lights that more clearly show the results of the extrusion and beveled edges. That's all the effort you will put into the look right now.

Animating 3D Text

You can animate 3D text using keyframes, expressions, and modifiers just as you can any object. We'll create a simple keyframe for our 3D text and then connect it into the 3D scene.

1. Move the playhead to frame 40, and in the transform tab of the text 3D inspector, click the keyframe button next to the X rotation parameter.
2. Using the X rotation slider, drag to the right until the text is almost face down, with an X rotation value of around 60 degrees.

3. Move the playhead to frame 60 and set the X rotation to 0 so the text is rotated up right.
4. Play the animation to review it.

   This will create a quick tilting animation, but the hinge of the rotation is at the bottom of the text. It would look nicer if the text rotated around a pivot between the two text lines. You can change the rotation pivot point using the pivot controls.
5. Move the playhead somewhere after frame 60. This will make it easier to view the pivot point position.
6. In the inspector, open the pivot controls.

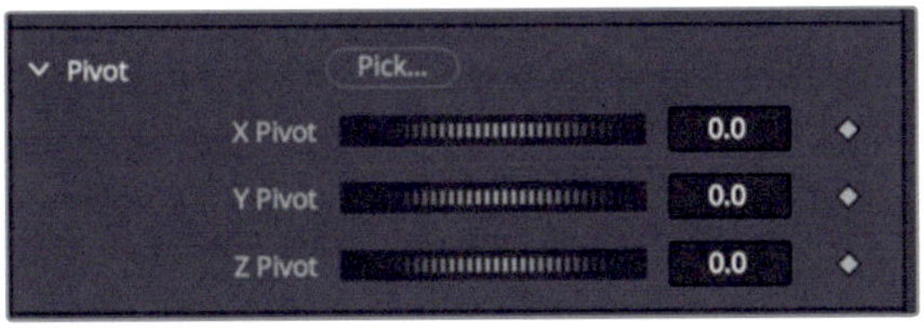

7 Using the Y pivot slider, drag to the right until the pivot point in the viewer sits between the two lines of text.

8 Play the animation to review the new rotation pivot point.

The text now rotates using a pivot point located between the two lines of text for a more symmetrical rotation.

Animating Text in a Scene

The rest of the animation can best be set by viewing it in context of the camera and the 3D scene. So, we'll need to connect the text into the main node tree.

1 Select the merge 3D 1 node in the node editor.

Instead of going to the effects library to get a tool, if you know the name of the tool you want, the quickest way to add it to the node editor is to use the select tool dialog.

2 Press Shift-spacebar to bring up the select tool dialog.

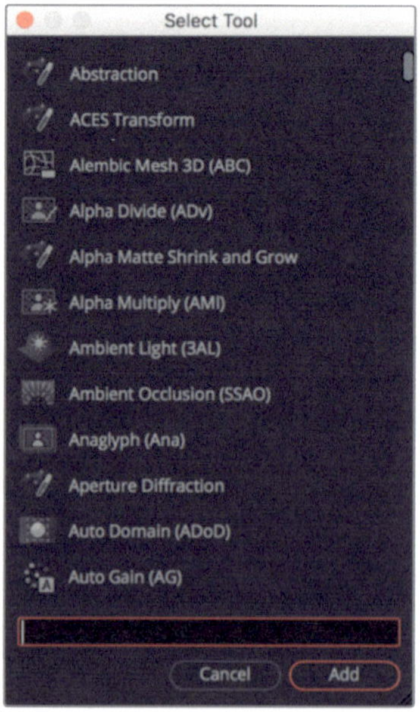

The select tool dialog allows you to search for any tool in the Fusion page by typing a few characters or the tool's shortcut.

3 Type in **merge** to search for the merge 3D tool.

4 From the list of tools, select the merge 3D tool, and click add to add the tool directly after the selected merge 3D 1 node.

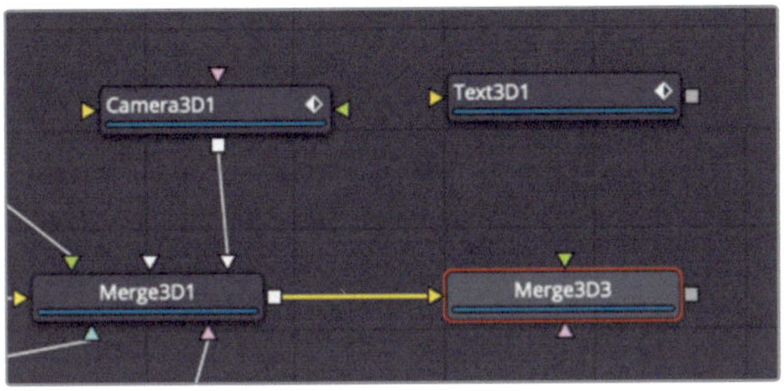

A new merge 3D 3 node can now be used to connect the text into the composition.

5 Drag the output from the text 3D node to the input of the merge 3D 3 node.

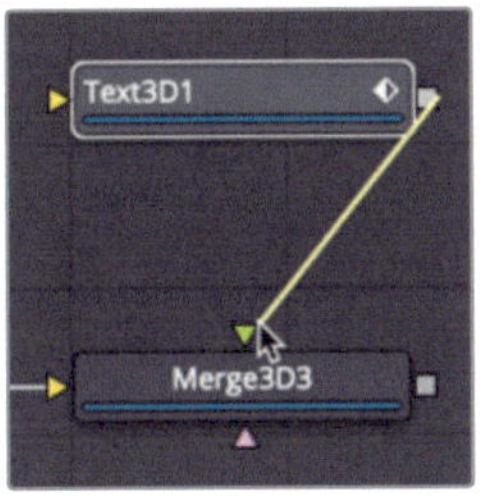

6 Select merge 3D 3, and press 2 to see it in the viewer.

7 Right-click over the perspective axis control, and choose camera 3D1 to view the camera output in the viewer.

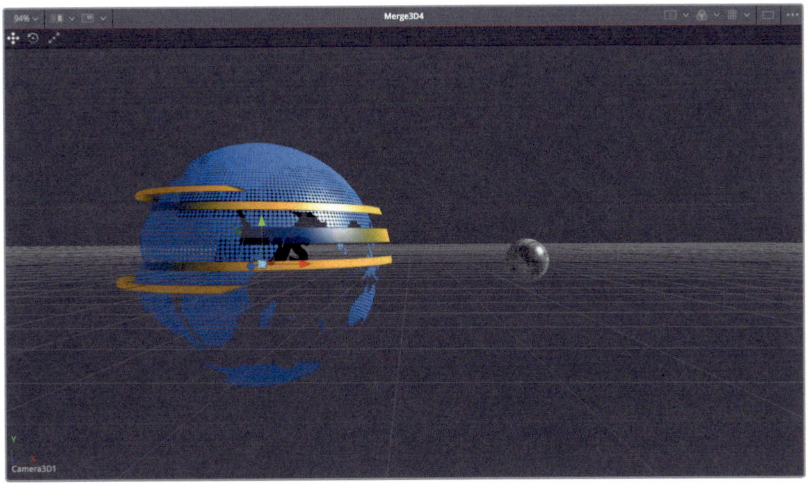

The text is obscured by the earth and needs to be moved off to the right, as well as brought forward toward the camera, so it doesn't run into the rotating moon.

8 Select the text 3D node and drag the X translation slider to the right to move the text out from inside the earth sphere, around 1.5 to 2.0.

9 To bring the text forward, drag the Z translation slider to bring the text forward, around 2.5.

The text appears black in the frame because the lights are not being passed through the merge nodes. Before you continue animating the text, let's enable the lights to pass through to merge 3D 3.

10 Select the merge 3D 1 node, and in the inspector, click the pass-through lights checkbox.

Now that the text is illuminated, you can continue animating. The current position of the text is where the text should end up at the end of our animation, so you can set a keyframe on frame 60 for the Z value. The X value will remain at the same value for the duration of the shot, so there is no keyframing needed on that parameter.

11 Move the playhead to frame 60, select the text3D node, and click the keyframe button for the Z translation parameter.

Now, you'll move back to frame 40 and position the text forward until it is out of frame behind the camera. To make it easier to see exactly what the text is doing, you'll view the perspective view in viewer 1, while you see the camera's output in viewer 2.

12 Select the merge 3D 3 node in the node editor, and press 1 to see it in viewer 1. Then set the perspective axis pop-up to view the perspective view, if needed.

13 In viewer 1, position the perspective view so that you can see the entire scene with the camera.

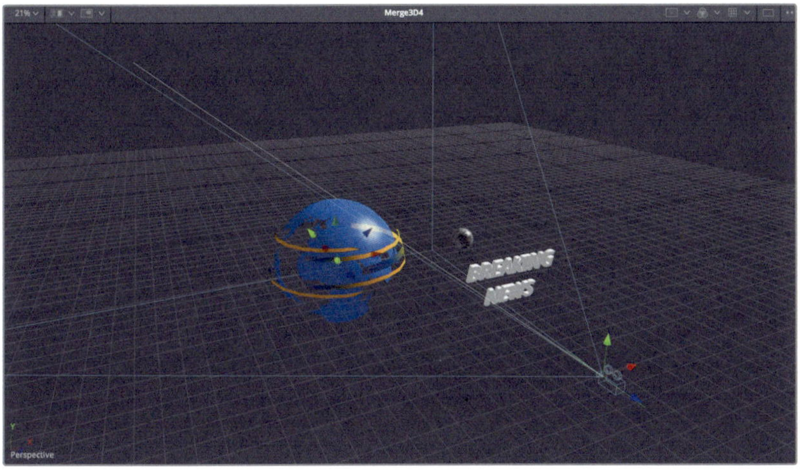

14 Move the playhead to frame 40, and drag the text 3D node's Z translation slider until the text is behind the camera, with the Z translation value around 5.0.

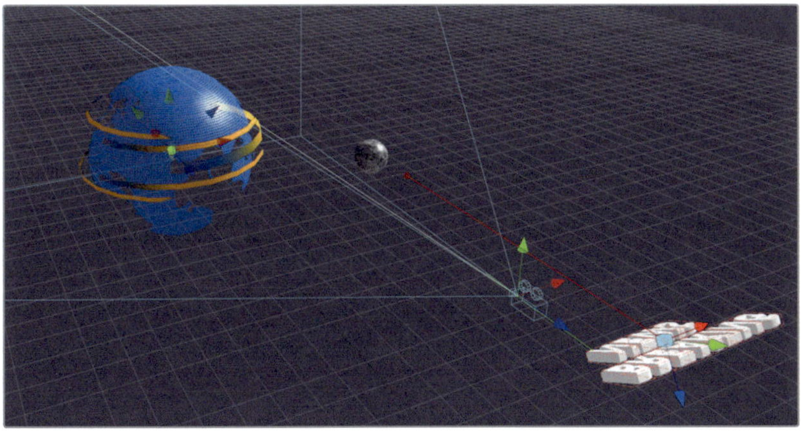

15 Play the animation to review it.

Now you have a hard-hitting breaking news animation that flies in from behind the camera. You can give the graphic a bit more visual interest by adding some other animated 3D elements.

Playing Around with 3D Shapes

The next addition to your broadcast design will be to add some 3D rectangular shapes that will fly by the camera. Adding design elements can contribute a sense of movement to a composition as well as kinetic visual interest. You will start with only one shape, and then use the duplicate node to place multiple offset, cloned shapes in your scene. Like the text, you'll first create these elements independent of the main composition and then merge them in later.

1 Click in an empty area of the node editor to the right of the text node.

  We'll use the select tool dialog to add a shape 3D node. By clicking in an empty area of the node editor, you are identifying where the node will be located when you add it.

2 Press Shift-spacebar to bring up the select tool dialog, and type in **shape**.

3 From the list of tools, select the shape 3D tool, and press add to add the tool to the node editor.

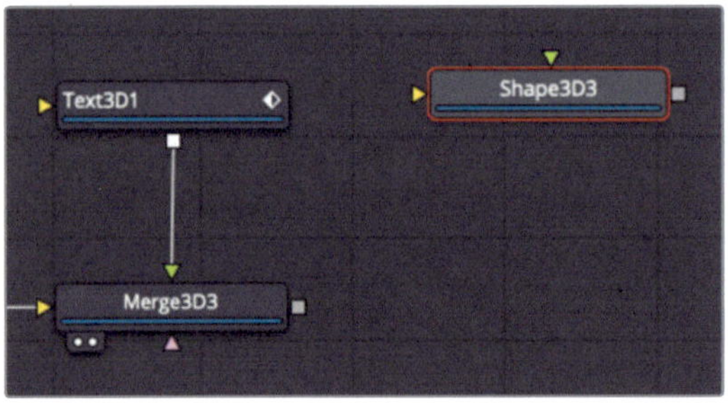

4 Press 1 to view the shape 3D in the viewer, and then click in the viewer and press the F key to fit the selected shape in the viewer.

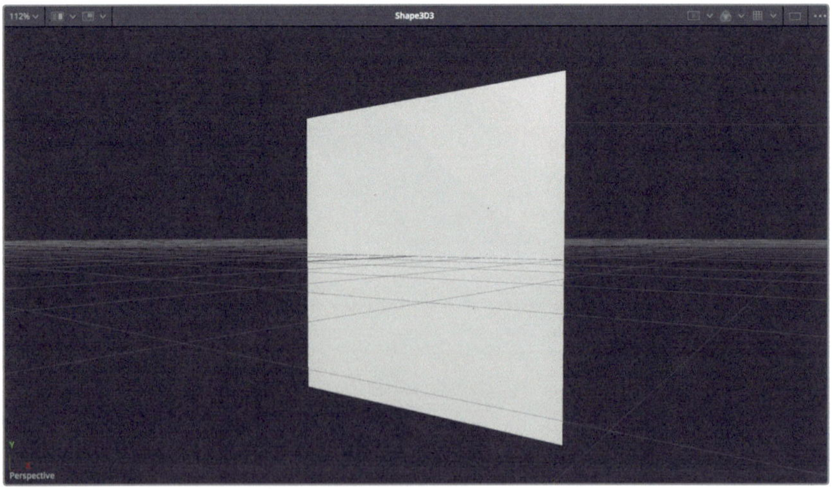

Using this shape 3D, you'll deform a cube to make it into a flat rectangle.

5 In the inspector, use the shape menu to change the shape from plane to cube.

Next, you'll flatten the cube but leave a small amount of depth to make it a slab.

6 In the inspector, select the lock width/height/depth checkbox, and then lower the depth parameter to 0.2 to create a thin rectangular slab shape.

7 Right-click in viewer 1 and choose 3D options > lighting to enable the default shading.

Bending, Twisting, and Shearing 3D Shapes

Motion graphics is a design process and is experimental in many ways. You often return to various parameters at a later time and tweak them as the project grows. Let's experiment with giving this static rectangle a bit more punch by deforming it a bit.

1 In the effects library, choose tools > 3D, and click the bender 3D tool to add it to the shape 3D 3 tool.

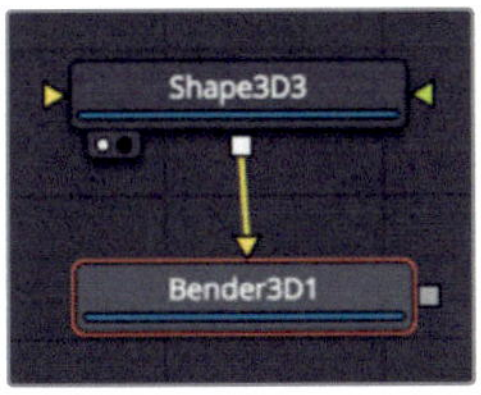

2 Press 1 to see the bender 3D node in the viewer.

The bender 3D tool is used to bend, taper, twist, or shear the geometry it is connected to—in our case, the flat rectangular shape.

3 In the inspector, drag the amount slider back and forth.

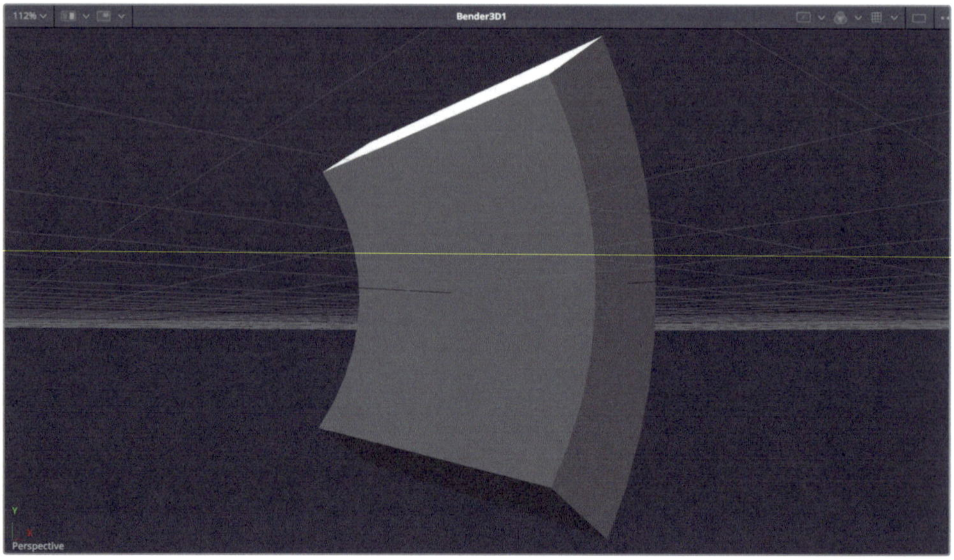

The default configuration is set to bend the geometry along the Y axis, but there are other settings that may be more appropriate for our broadcast design.

4 Reset the amount slider, and then change the bender type menu to shear and drag the slider to the left until it is set around -0.5.

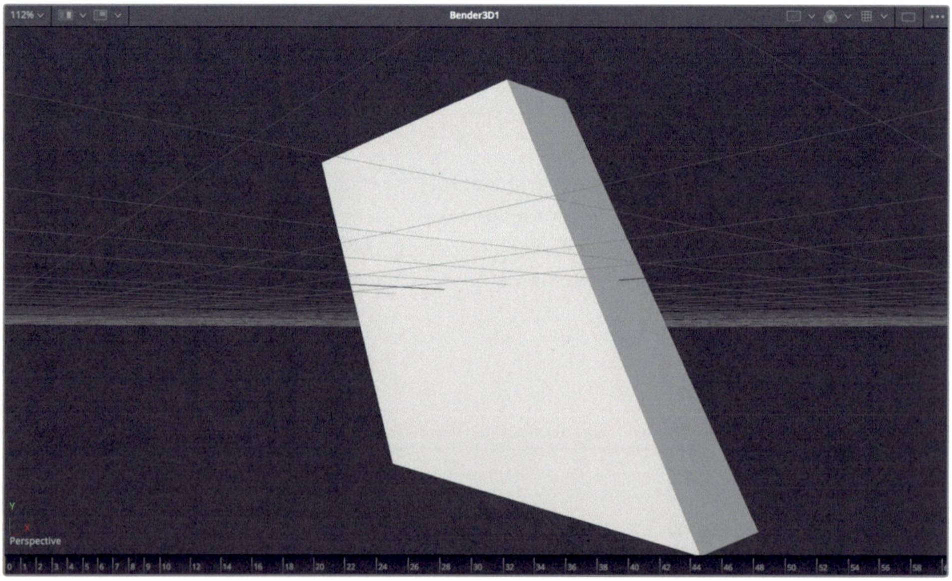

This is a shape that has some attitude to it. We'll use this as the base shape for our design element.

Duplicating 3D Shapes

From this one 3D element, you will create duplicates and offset them in regards to position. Doing so will fill out your scene and add more visual interest than just the spheres and the text. The easiest way to duplicate an object multiple times is to use the duplicate 3D node.

1 Select the bender 3D 1 node, and from the effects library, click the duplicate 3D tool to add it to the node editor. Then press 1 to see it in the viewer.

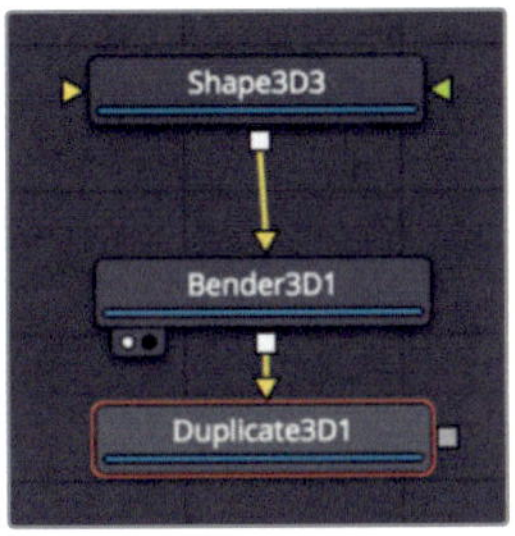

You can use the duplicate 3D node to quickly duplicate the geometry connected to its input. Once you set the number of copies, you can create repeating patterns with an array of objects.

2 At the top of the inspector, in the last field, type **10** to create that number of copies.

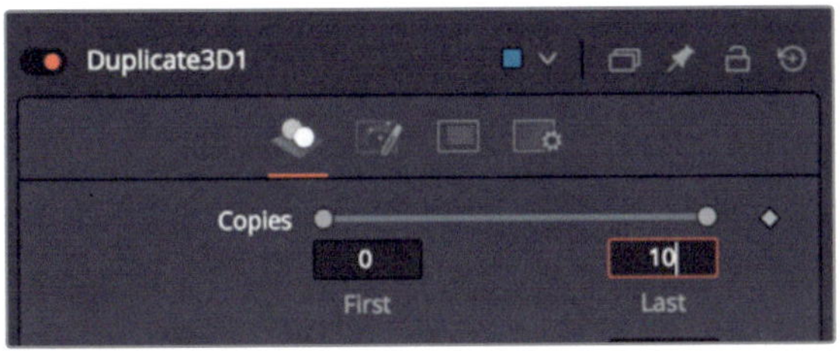

TIP Setting the first field value higher than 0 will exclude the original shape and show only the copies.

3 Drag the X offset to the right to spread out the 10 copies until there is space between each one, around an X offset value of 1.3.

The X offset creates a line of 3D rectangles.

4 With the duplicate 3D node selected, click in an empty area of viewer 1 and press F to fit all the replicated rectangles in the viewer.

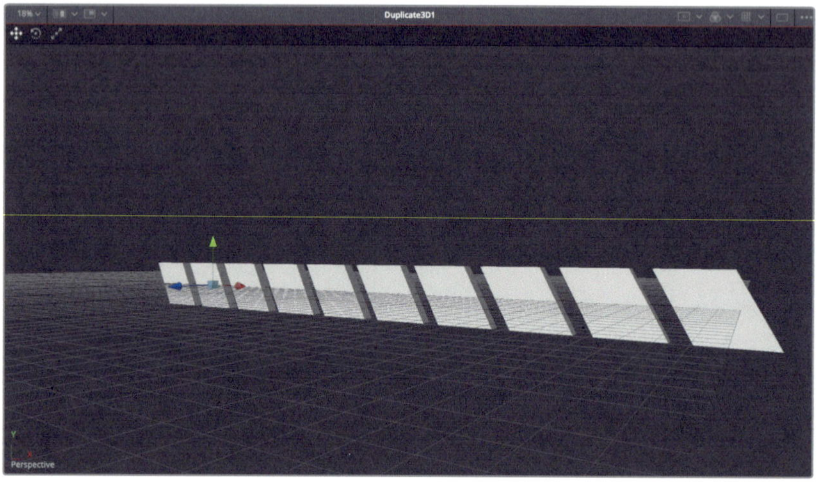

You have ten copies of a rectangle spread out. Each copy is a copy of the previous copy, so the parent is copied, then the copy is copied, and the copy of the copy is copied, and so on. By default, all the copies line up in a row; however, the jitter tab includes controls for offsetting each copy.

5 At the top of the inspector, click the jitter tab.

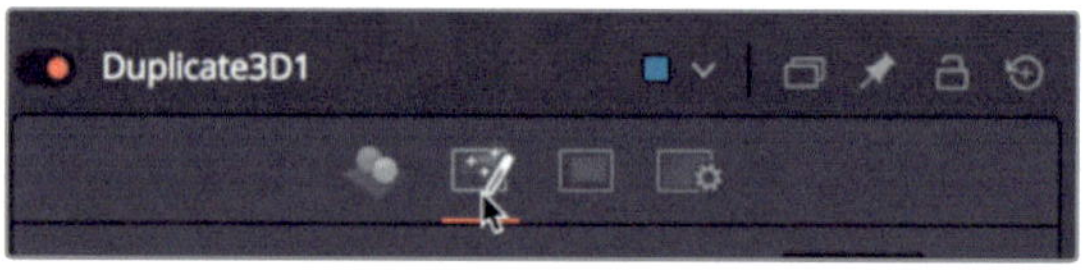

6 In the inspector, drag the Y and Z translation sliders to about 0.5.

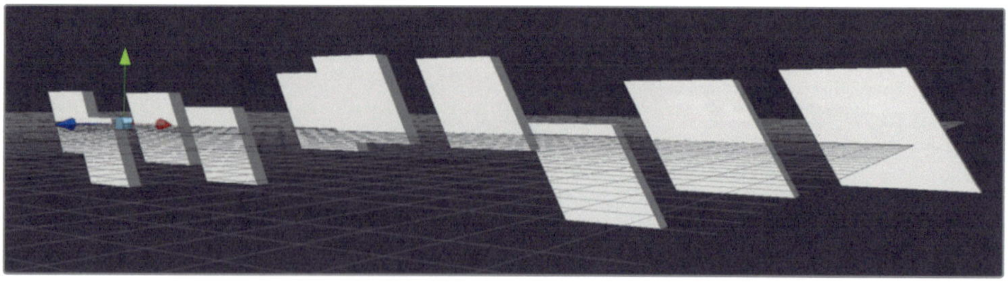

You now have ten 3D rectangles that have more of an energetic look. To do any more with these rectangles, such as animating them, it will help to see them in context of the rest of the composition.

7 Press Shift-spacebar to bring up the select tool dialog and add a merge 3D tool.

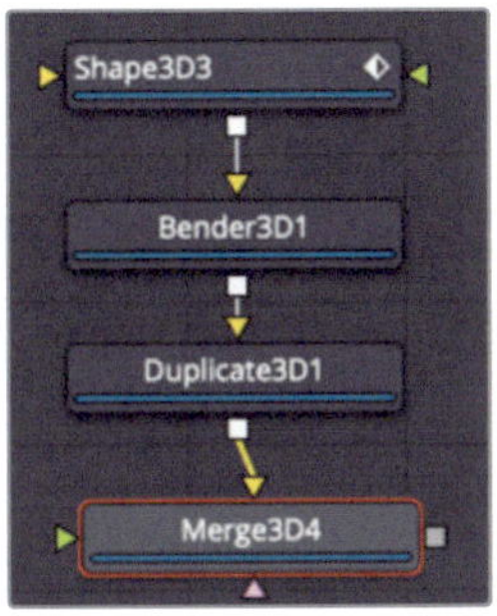

8 Press 2 to view the new merge 3D 4 in the viewer.

9 Drag the output of the merge 3D 3 to the new merge 3D 4 node.

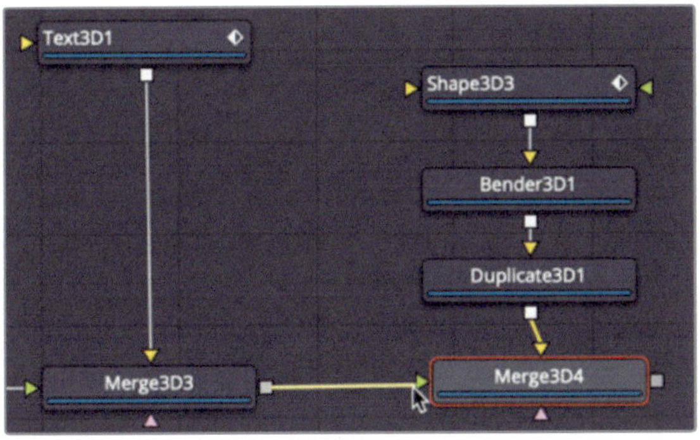

The rectangles appear dark in the viewer because the merge 3D 3 node is not passing the lights through to the new merge node.

10 Select the merge 3D 3 node, and in the inspector, click the pass-through lights checkbox.

Now that we can see the rectangles in context of the other objects and the lighting, we can correctly position and animate them.

The duplicate 3D node can only offset the copies. To move the entire row of duplicated rectangles, you move the original object.

11 Select the shape 3D 3 node and use the on-screen controls to position the shape in the lower-left corner of the frame.

12 To ensure they do not intersect with the earth sphere, in the inspector, click the transform tab and drag the Z translation to 1.5.

The move forward has made the rectangles a bit too large for this scene, so you'll need to scale them down.

13 Drag the scale slider down between 0.6 and 0.7, and then use the Y offset to position them slightly higher in the frame.

Now, using some simple keyframing, we can slide the 3D shapes through the frame.

14 Select the shape 3D 3 node, and move the playhead to frame 60.

As soon as the text is in place, on frame 60, we'll have the rows slide in from the right.

15 In the inspector, drag the X translation slider until the rows are completely off-screen, around 5.0.

16 Click the keyframe button to the right of the X translation slider to set the first keyframe.

17 Move the playhead to frame 110, and drag the X translation slider to the left until the rows are off-screen, around -15.0.

18 Play the composition to see the animation.

Each rectangle now slides across the screen and eventually out of frame.

The duplicate 3D node is a powerful tool and is fun to experiment with. You can even connect multiple duplicate 3D nodes together to create larger arrays of objects. The 3D motion graphics possibilities are endless.

Applying and Customizing Materials

In the first part of this lesson, you focused on your 3D scene animation, but now you will focus on improving its look. 3D objects, including text, use *illumination materials* called shaders, to give surfaces the appearance of real-world materials. Fusion comes with a few shaders for use in your projects, but you can also modify these templates or build your own. For this lesson, you'll use one of the templates in the effects library to give the text and rectangles a chrome look. Let's get started.

1 In the effects library, select the templates category, and then select shaders.

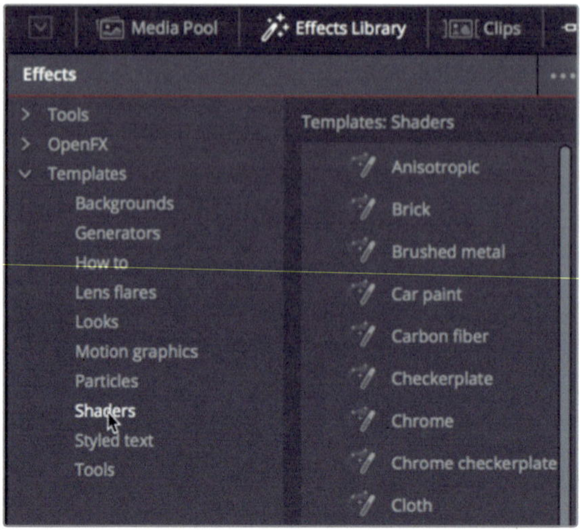

TIP You may want to pan the node editor by dragging while holding down the middle mouse button to give yourself more room on the right side as you build out the material.

2 Drag the chrome shader to an empty area of the node editor above the shape 3D 3 node.

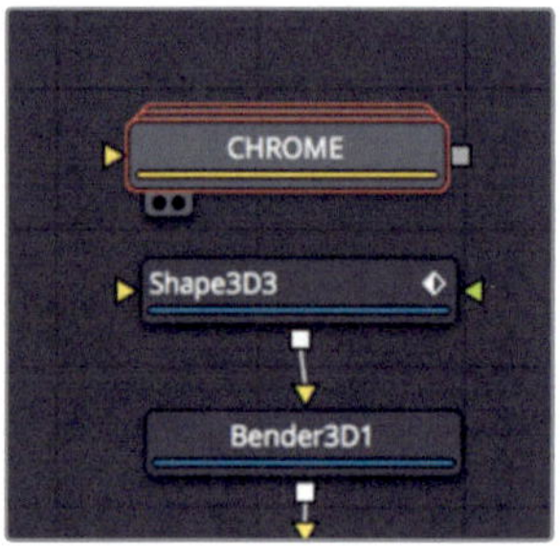

3 Press 1 to see the shader in the viewer.

A shader is represented as a sphere in the viewer, but that shape has nothing to do with the actual project. It is just intended to help you visualize how the material will look in a 3D environment. Fusion also adds a light to the viewer, so you can see how the material's surface responds to light.

TIP If you want to see the material displayed as a different object, right-click in the viewer, and choose shape to select from a sublist of 3D primitives geometry.

4 Drag the output of the shader to the green material input on the shaped 3D 3 node.

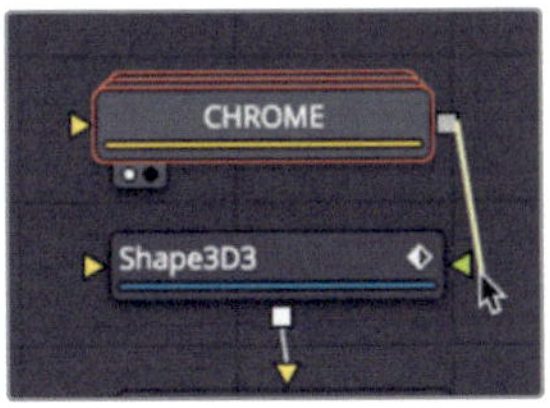

The rectangles now get a reflective chrome material applied.

Adding material to 3D text requires a slightly different technique. The 3D text node requires you to put a replace material 3D node after it to apply the material to the text.

5 Select the text 3D 1 node in the node editor, and from the effects library, select tools > 3D category, and click a replace material 3D tool to add it to the node editor.

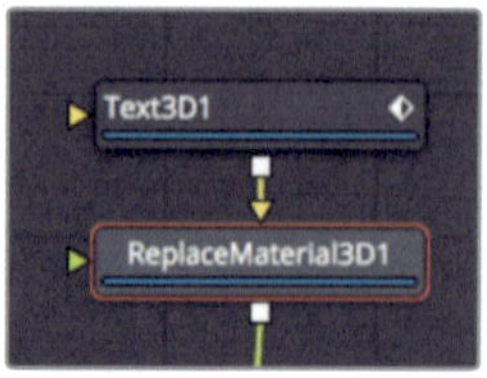

The replace material node has only one connection other than the yellow input where the text is connected. The green replace material input is used as the input for the material you want to use.

6 From the templates category in the effects library, drag a second chrome shader to the node editor, just to the left of the text 3D 1 node.

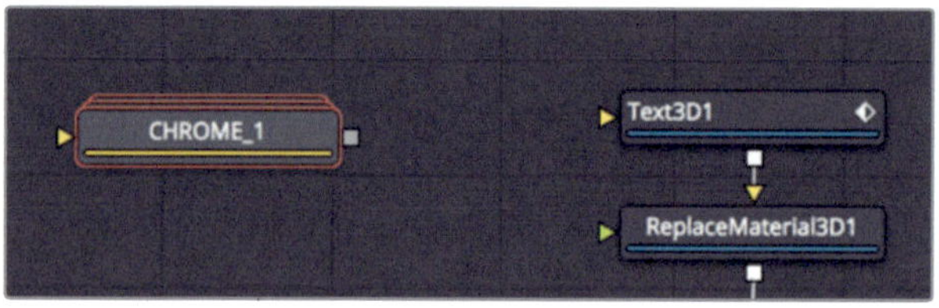

7 Drag the output from the chrome 1 node to the green material input of the replace material 3D node.

The text now has the same chrome appearance as the shapes.

8 Play the animation to see it with the new reflective chrome material.

Modifying the Material

When creating materials or using the templates, you can modify them to get the precise look you want. We will look at two different ways we can change the appearance of the shaders in this composition.

1 Select the chrome node connected to the shape 3D 3 node.

Every shader template includes a subset of controls in the inspector. The inspector for the chrome template includes color swatches for the diffuse color and the specular color. The diffuse color is the main color of the object when even light is hitting it, while the specular color is the color of the brightest highlights.

2 Drag the color indicator in the diffuse color swatch to a rich red color.

The red color is applied to the chrome look.

Although the controls in the inspector may seem a bit limiting, each template is really a grouped node tree, collapsed to take up less room. You can expand the group to make adjustments to any node used to make up the chrome shader.

3 Select the chrome 1 node connected to the replace material node.

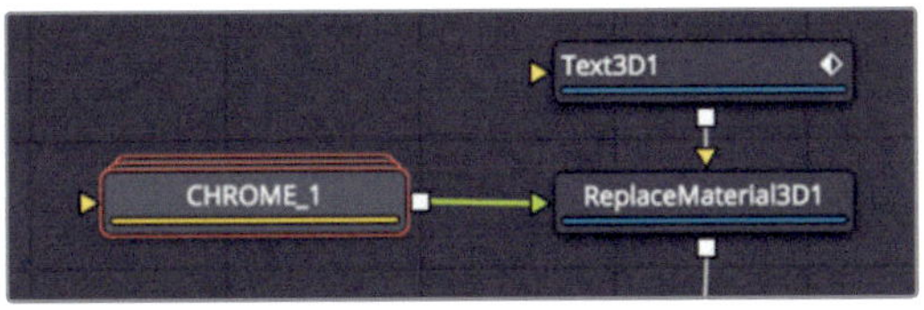

Notice the appearance of the chrome node in the node editor. It has a unique stacked icon to indicate it is a group.

4 Double-click the chrome 1 group node to expand it.

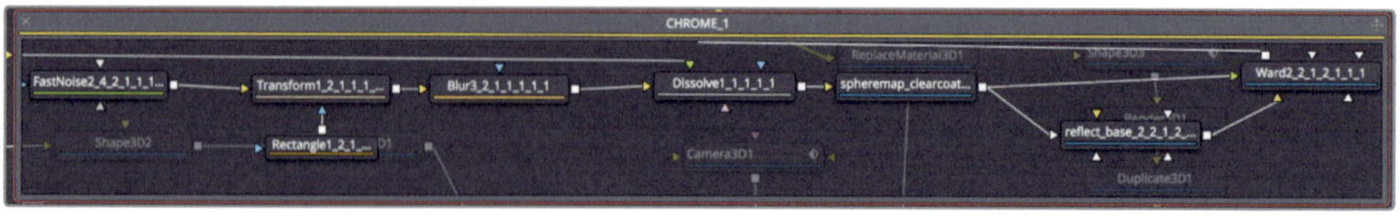

The expanded group is displayed in a floating node editor window. These are all the nodes used to make up the chrome shader. The exposed nodes allow you to change any aspect of this material.

5 Select the fast noise node on the far left of the floating window, and press 1 to see it in the viewer.

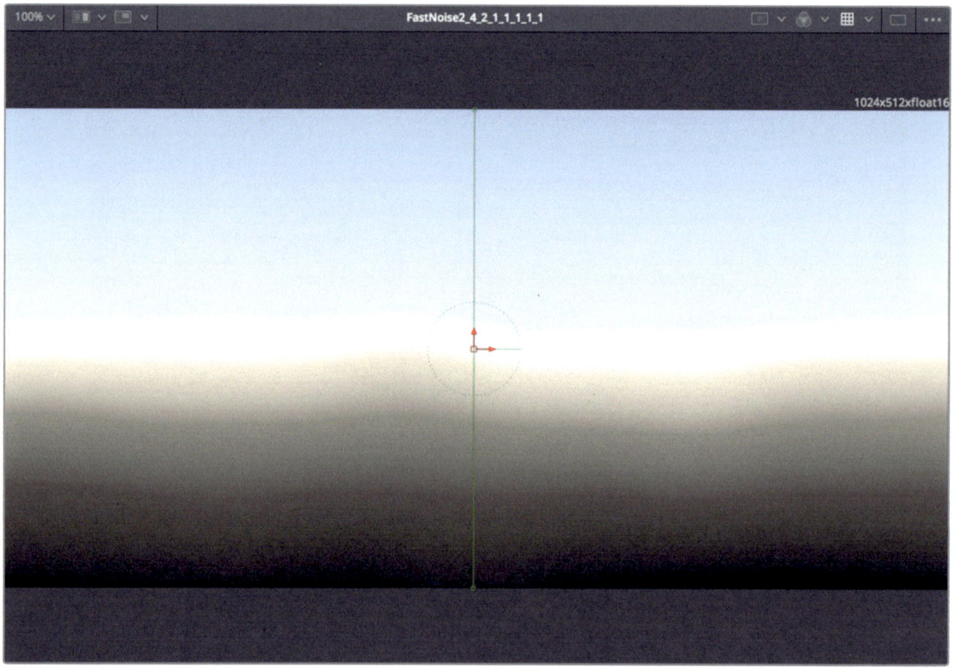

This node is creating the reflective texture of the material. You can replace it with any image from the media pool.

6 Press delete to remove the fast noise node.

7 Open the media pool, and from the 3D globe bin, drag the Italian night HDR image in the same location where the fast noise node was located.

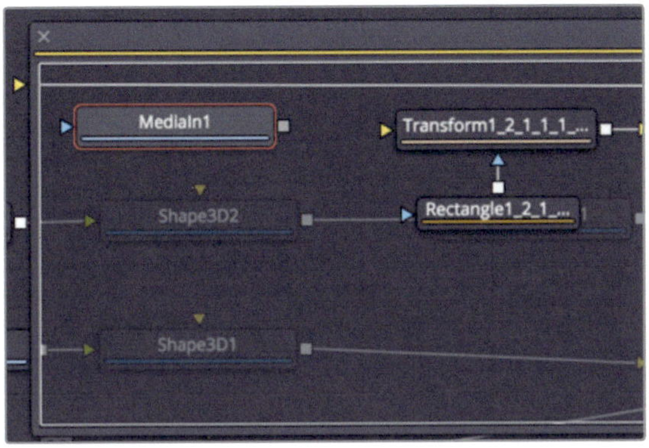

Adding the new node inside the floating node editor window will add it to the window and not the main node tree.

8 Press 1 to see the image in the viewer.

This is an HDR (high dynamic range) equirectangular image that can be used as a reflection map for surface materials created in Fusion. You can replace the fast noise with this image to create a new look for this chrome texture.

9 Drag the output of the media in 1 node to the yellow input on the transform node.

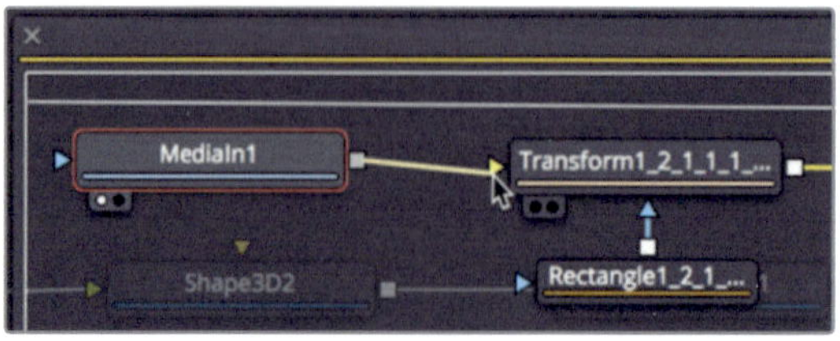

10 In the upper-left corner of the floating window, click the close button to collapse the group.

11 Play the animation to see it with the new reflective chrome material.

This shows just how flexible the templates are for making custom materials.

Take the time to explore other template shaders that come with DaVinci Resolve, because they will help you understand how shaders are made and allow you to eventually begin creating your own shaders for your motion graphics.

Converting 3D into a 2D Image

Every 3D scene ends with a render 3D node that converts the 3D environment into a 2D image. Once you add the render 3D node, additional 2D image processing can be inserted, and you can render the output directly into the edit page timeline via the media out node. The render 3D node is not just a conversion node from 3D to 2D; it also includes several render processes that can enhance the look and quality of your comp. The most significant of those is adding depth of field.

1 In the node editor, select the merge 3D 4 node, and from the effects library, select tools > 3D category, and click the renderer 3D tool to add it to the node editor.

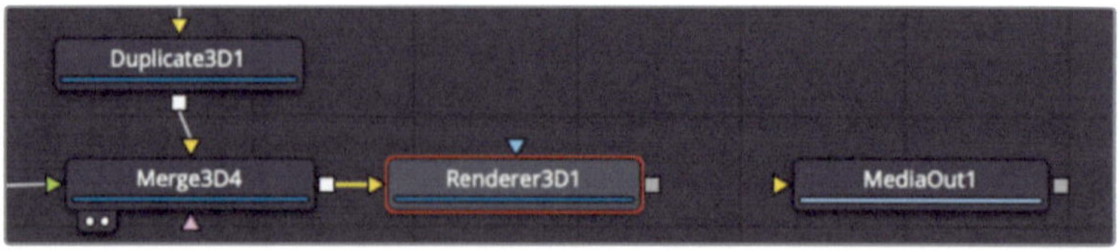

2 Press 1 to see the renderer 3D output in viewer 1.

The renderer 3D is set to render the default perspective view camera, so the first item you will want to take care of is to choose the camera you want to render.

3 In the inspector, choose camera 3D1 from the camera menu.

The renderer 3D node includes two render engine options:

The software renderer engine uses only the system's CPU to produce the rendered images. It is usually much slower than using the OpenGL renderer engine but produces consistent results on all computers.

The OpenGL renderer engine employs the GPU processor on the graphics card to accelerate rendering. Using this method, the output may vary slightly from system to system, depending on the graphics card in use. The increased speed of the OpenGL renderer engine makes it possible to customize supersampling and 3D depth-of-field options. For these reasons, the OpenGL renderer engine is most commonly used.

4 In the renderer type menu, choose OpenGL.

You should notice little if any difference in the viewers because this 3D composite hasn't taken advantage of any special compositing operations that would differentiate the two render engines. However, you should see a difference between the perspective view in viewer 2 and the renderer 3D output in viewer 1. Lighting is not initially enabled for the render 3D node, as it is in the viewer. You'll need to enable lighting to get the shading you see in the perspective view.

5 In the inspector, select the enable lighting checkbox to see the effects of your directional and ambient lights.

One last check is to make sure the render 3D is set to output the correct resolution for your shot. In this case, the project resolution is set to 1920 x 1080, so that is what the renderer 3D should be set to as well.

6 In the inspector, click the image tab, and make sure the width and height values match the project resolution of 1920 x 1080.

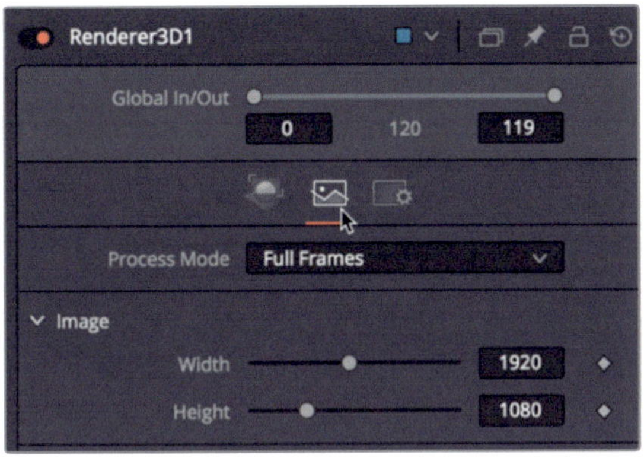

Right now, your project looks nice, but still a bit flat. You can remedy that by adding depth of field.

Configuring Depth-of-Field Effects

To give your project more photo-realism, you can simulate a camera's shallow depth of field setting. Depth of field is the range before and after the focal plane that appears acceptably sharp. Areas outside that range are increasingly out of focus.

The first step in setting up depth of field is to enable it in the renderer 3D node. You'll then set up the camera's focal plane.

1 In the renderer 3D inspector, click on the controls tab and select the enable accumulation effects checkbox, and then set the quality to 20.

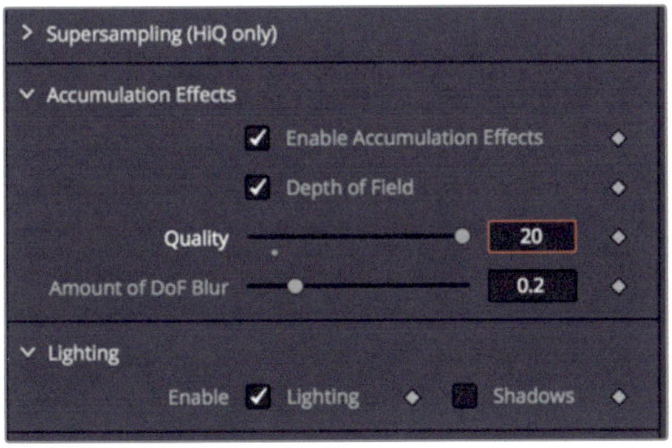

The higher the quality, the better the depth of field (DoF) will look, but the longer and harder the computer will have to work to process the effect. The amount of DoF blur changes the size of the in-focus area. The larger the number, the shallower the depth of field.

2 Set the amount of DoF blur to 0.1, and play through the animation to preview it.

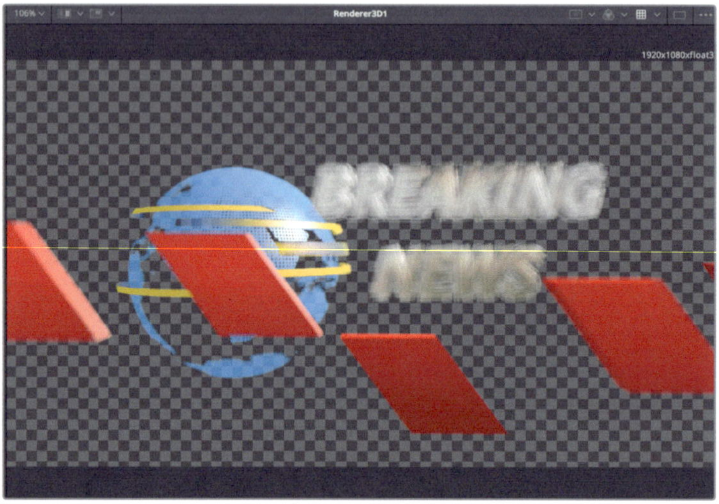

The scene is clearly blurred, but the text never comes into focus. Just as with a real-world camera, you need to set the focal point so that the camera is focused on the area where the text lands.

3 When you are finished previewing the animation, stop playback, and right-click the camera axis control in viewer 2 and select the top view.

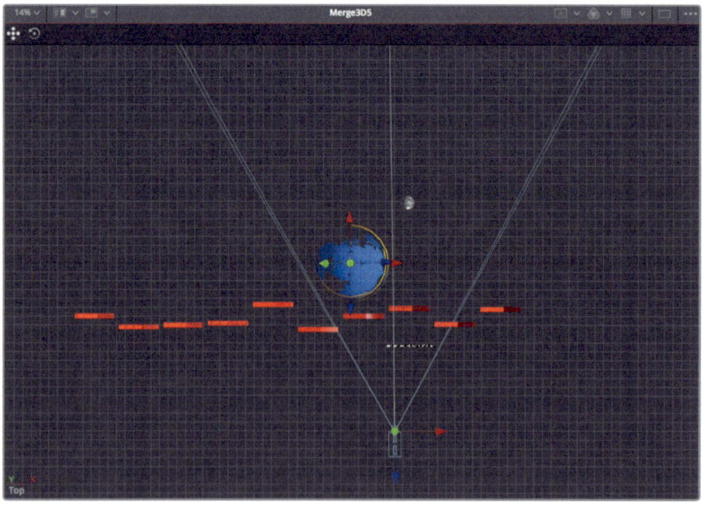

TIP If you need to, in viewer 2, hold down the cmd (macOS) or Ctrl (Windows) key, and scroll the middle mouse wheel to frame the viewer so you can clearly see the text, spheres, and the camera.

The top view will make it easier to precisely focus the camera on the text.

4 In the node editor, select the camera 3D 1 node, and in the inspector, scroll down to open the control visibility section.

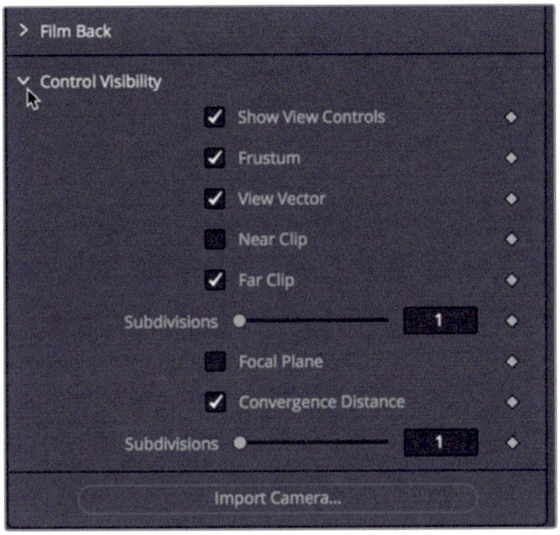

You can determine where the camera is focused by using a focal plane in the top view. The focal plane is like the focal point in a camera. It tells the camera where to focus in the 3D scene.

The focal plane is not visible by default, but you can enable it and set its position in the camera node's inspector.

5 Select the focal plane checkbox.

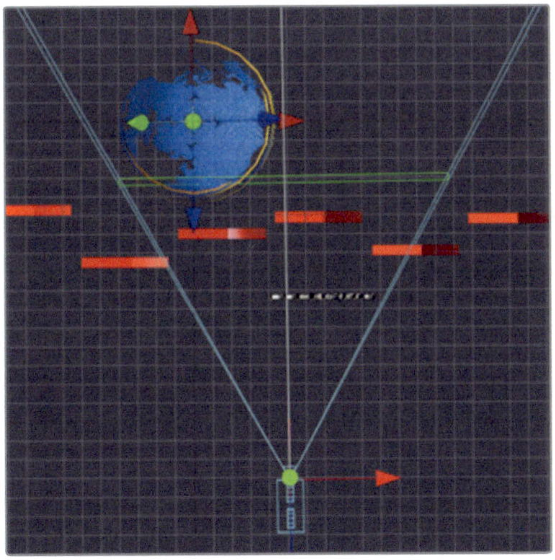

In viewer 2, a green rectangle is displayed in the camera's angle of view outline to represent the focal plane.

6 Move the playhead to frame 60 of the composite. This is where the text is in its final landing position.

To focus on the text at this frame, you can position the focal plane directly over your text.

7 In the inspector, adjust the focal plane slider until the green focal plane intersects with your text in viewer 2.

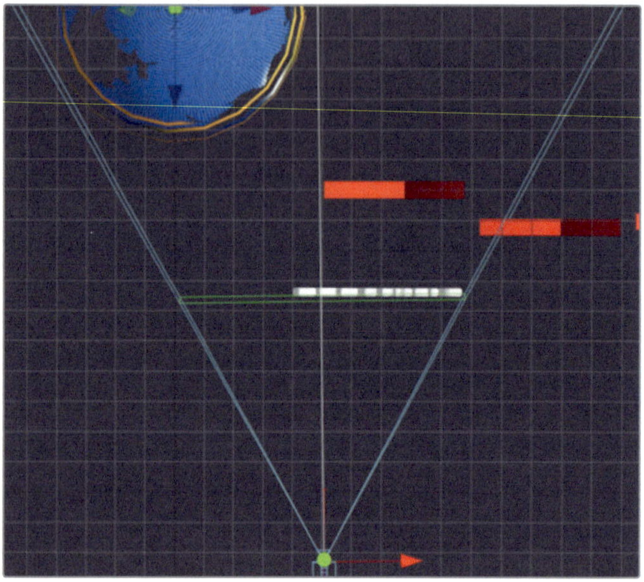

The focal plane is now fixed at the location where your text lands, causing the text to be in-focus in viewer 1.

Since the camera pulls back during the course of the comp, the text will eventually be out of focus again, so you'll need to keyframe the plane of focus.

8 Click the keyframe button to the right of the focal plane slider.

9 Move to the end of the render range and adjust the focal plane slider again to be directly over the text.

Since there is no text at the start of the composition, you can focus on the sphere and have it rack focus as the text comes into view.

10 Move to the start of the render range, and adjust the focal plane slider to be directly in front of the earth sphere.

11 Press spacebar to preview the now-in-focus animation.

TIP if you regularly want a node to appear with certain settings already configured, you can right-click over the node in the node editor and choose settings > save default. Any time you add that node to a node editor, it will come preset with the current configuration.

Depth of field adds a significant enhancement to the look of graphics. When creating a 3D motion graphics design, you cannot just depend on 3D processing within the merge node. The render 3D options, as well as other 2D image processing, can also enhance the design.

Adding 2D Look Design

The output of the renderer 3D node is a 2D image that you can use like any 2D source image. Once you configure the renderer 3D node, its output can be used like any image in your node tree. We'll add a background node to create a solid color background, and then add a standard merge node with our renderer 3D as the foreground.

1 Click in an empty area below the renderer 3D node, and then press Shift-spacebar to add a background node.

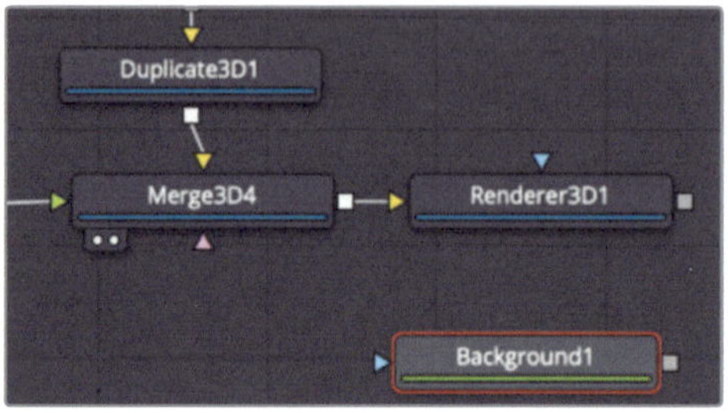

The background node will create a solid black background that we will use for our comp.

2 Drag the output of the renderer 3D node to the output of the background node.

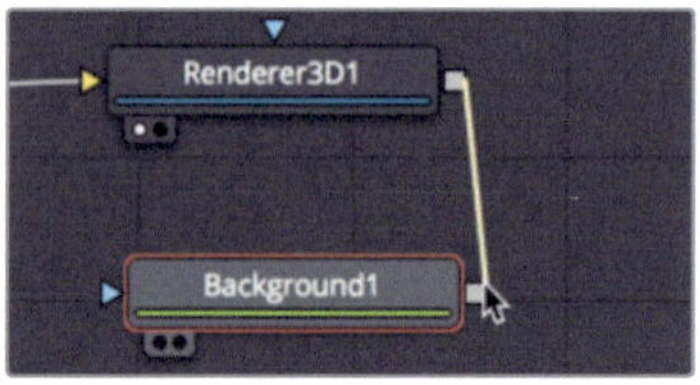

Dragging the two outputs together creates a merge node with the renderer 3D as the foreground.

3 Press 2 to see the merge node in the viewer.

With the composite now in 2D, you can add image-processing filters to enhance the look.

4 Click the effects library button to open the effects library.

The effects library includes all the Fusion page's compositing tools, as well as all the ResolveFX that you may have used previously in the edit and color pages.

5 In the effects library, click the OpenFX category, and then select the ResolveFX light category.

6 Drag the lens reflections from the effects library to the connection line between the renderer 3D and the merge nodes.

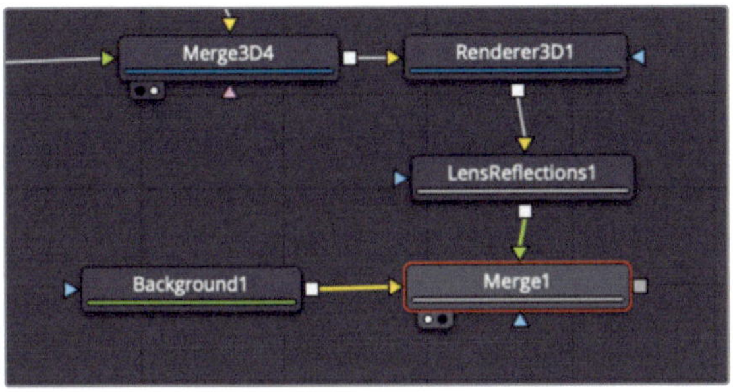

The lens reflections node is added to the comp, simulating flaring lens effects based on the shape and motion of highlights in the shot. The simulated reflections also move based on the animation of the elements, thereby creating an animated effect without adding keyframes to do so. However, the effect is a bit too strong, so let's lower its intensity.

7 In the Inspector, click the settings tab, and drag the blend slider to around 0.5 to lower the intensity of the lens reflections.

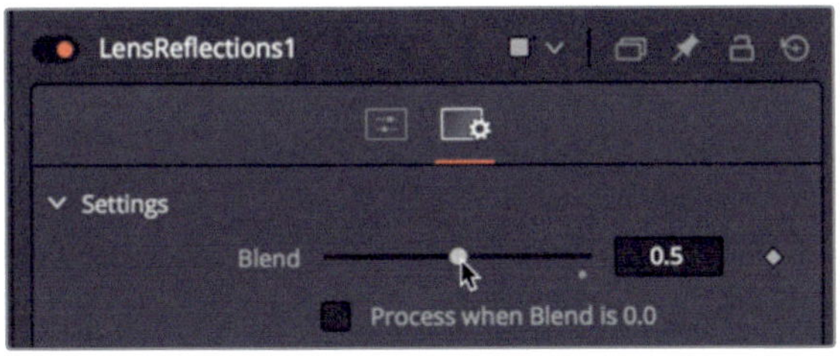

3D compositing is often too clean and sharp on its own, so adding a 2D image-processing pass can create a more organic looking composite.

8 Drag the output of the merge 1 node to the media out node.

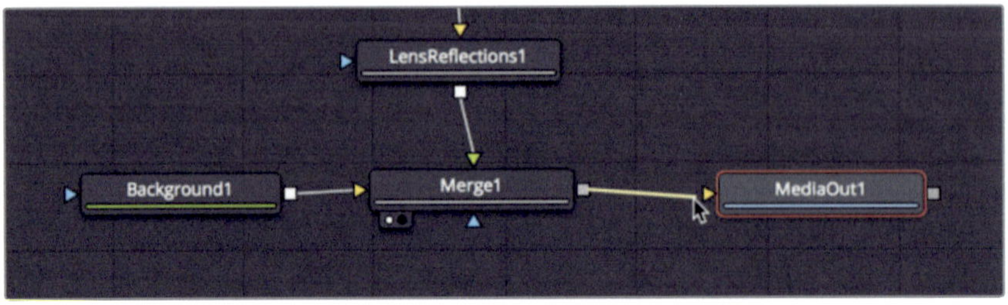

Using the combination of shape 3D, merge 3D, render 3D, and regular merge nodes, you can create multiple 3D and 2D sections within a single composite, which makes Fusion's 3D compositing extremely powerful.

That's it for this lesson, but you should go back and modify each parameter to your liking. Don't be afraid to mix and match new 2D filters to see how they affect each other, too.

Hopefully, you have a good understanding of building a 3D scene with text and objects and repurposing those as templates. Also, the idea of building animation first, developing the look second, and then finishing the final render with color correction, glows, and filter effects will help structure your projects. You can build upon a lot of the powerful techniques over these past two lessons and take your motion graphics to the next level using Davinci Resolve 16.

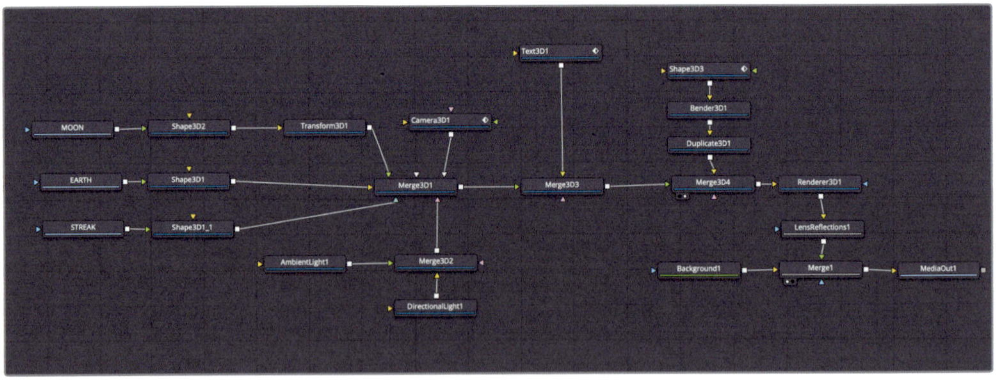

Completed 3D text node tree in Lesson 10.

Lesson Review

1 True or false: The 2D text+ node cannot be connected into a merge 3D node.

2 True or false: Depth of Field is enabled in the camera 3D node.

3 Which node must be added to change the material or shader on a text 3D node.

   a) a new shader node

   b) a 3D material node

   c) A replace material 3D node

   d) None of the above.

4 True or false: The last node after any 3D compositing must be a renderer 3D node.

5 True or false: Only a 3D foreground, background, a light, and a camera node can be connected to a merge 3D node.

Answers

1 True. The 2D text+ node cannot be connected into a merge 3D node. Only text 3d nodes can be used for text in a 3D composite.

2 False. Depth of Field is located in the renderer 3D node.

3 The answer is C. A replace material 3D node must be added to change the material or shader on a text 3D node.

4 True. A renderer 3D node is required at the end of any 3D compositing to convert the 3D scene to a 2D image.

5 False. Unlike the standard 2D merge node, there is no foreground or background limitation on a merge 3D node. An unlimited number of 3D elements can be connected.

Lesson 11

Exploring 3D Particle Systems

Fusion's very powerful 3D particle system generates large numbers of automatically animated objects. The objects can be any image, movie file, polygon shape, text, or Fusion generator.

Particles have infinite uses in just about every type of project. You can create practical effects such as fire and rain, or create more abstract motion graphics designs. Fusion also includes particle-specific tools to simulate physical phenomena—wind, gravity, and bounce—as well as forces that allow particles to be attracted to or repelled by other objects.

With so many ways to build and manipulate particles, it's easy to become overwhelmed with the possibilities. So, in this lesson, you'll focus on building some simple blowing smoke for the music video shot from Lesson 8. You'll learn about Fusion's essential particle tools and their many artistic options.

Time

This lesson takes approximately 40 minutes to complete.

Goals

Completed particles for Lesson 11.

Setting Up a Particle System

To start, you'll use a 3D composition that is already set up for you. Unlike the 2D green screen shot you worked on in Lesson 8, this shot from the same music video is set up as a full 3D scene. Since we will be working in 3D, the particles will be able to surround every element in the scene. You'll begin by adding the essential nodes to create particles and get them moving.

1. In the Fusion 16 Lessons part 3 project, open the **Lesson 11 START** timeline. Position the playhead over the Fusion clip and click the Fusion page button.
2. In the node editor, select the media out node, and press 2 to see it in the viewer, if necessary.

3 Select the merge 3D node, press 1 to see it in the viewer, and then set viewer 1 to display the perspective view.

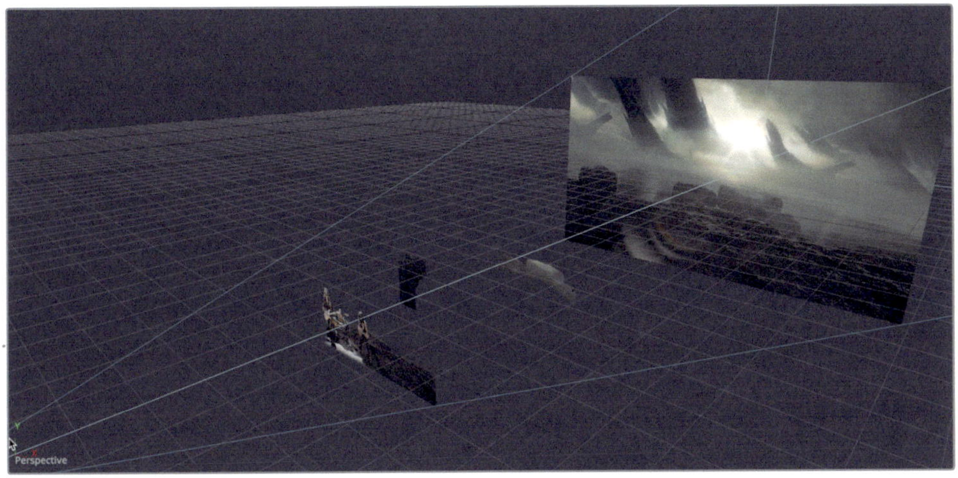

This node tree includes a 3D layer for the band, a purple underlay containing the background elements broken into different 3D layers, and a camera. All you need to do is add some apocalyptic blowing smoke to this scene.

Any time you create a particle system in Fusion, you must start with two nodes: a particle emitter node that generates the particles, and a particle render node that renders the results.

4 From the toolbar, drag the pEmitter tool into an empty area of the node editor, above the merge 3D node.

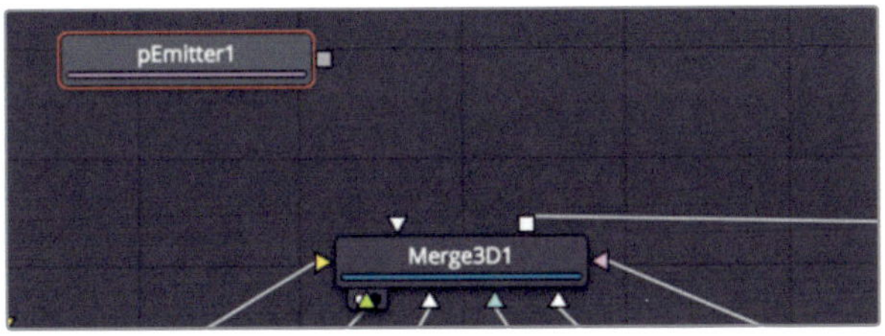

5 With the pEmitter selected, go back to the toolbar and click the pRender to add it to the node editor.

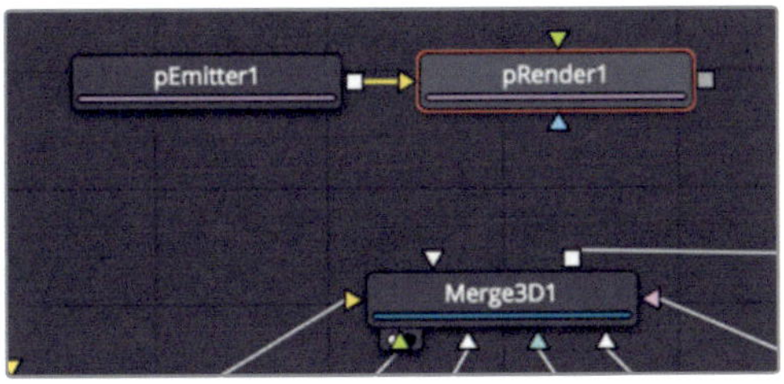

TIP You could use a pImage emitter node instead of a pEmitter node. The pImage emitter node is best used in special cases when you want to take an image and treat each pixel of the image as if it were a particle. Otherwise, in all other cases, you would begin with a pEmitter node connected to the pRender node.

When starting with particles, your first choice is whether to generate particles for a 2D or a 3D composition. For this exercise, you will create a 3D particle system because doing so will allow us to create a sense of depth for the particle smoke.

6 In the node editor, select the pRender node, and in the Inspector, ensure that the output mode is set to 3D.

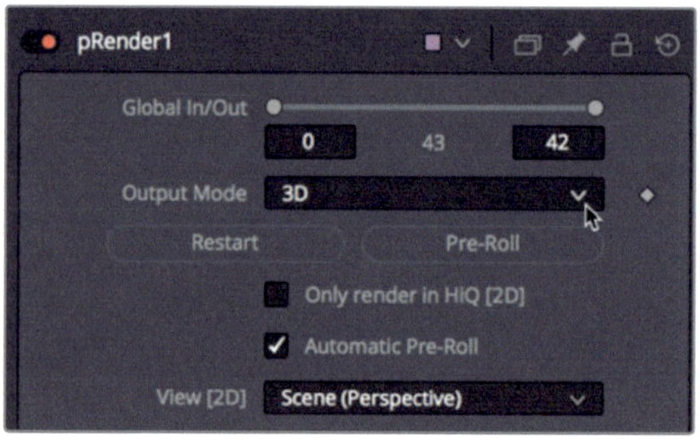

To view the particles, you need to view only the pRender node. All other particle nodes are only visible through the pRender node.

TIP The names of all particle-specific nodes begin with a p: pSpawn, pBounce, pTurbulance, and so on. This naming convention makes it easy to locate and identify particle-specific nodes.

7 Press 1 to view the pRender in viewer 1.

Because the pRender node is set to 3D, the viewer shows a 3D scene. The pRender is all you will view for a while, so you can open up more space in the interface by using just a single viewer.

8 In the upper-right corner of viewer 1, click the single viewer button.

You now have one viewer to work in and more room for the rest of the interface. You can improve performance by not viewing any nodes that are not particle nodes. Even though viewer 2 is hidden, viewing the media out in viewer 2 will still impact performance.

9 Select the media out node and press 2 to stop it from viewing in viewer 2.

Most of your actual particle setup begins in the pEmitter node.

10 In the node editor, select the pEmitter.

A particle system is divided into two primary parts: the particle emitter and the particle cells.

The particle emitter is the source of the particle cells. By default, the viewer displays this emitter as a green wireframe sphere, but you can easily change it into almost any shape, including a rectangle, a line, text, or a polygon. It defines the area that produces the particles.

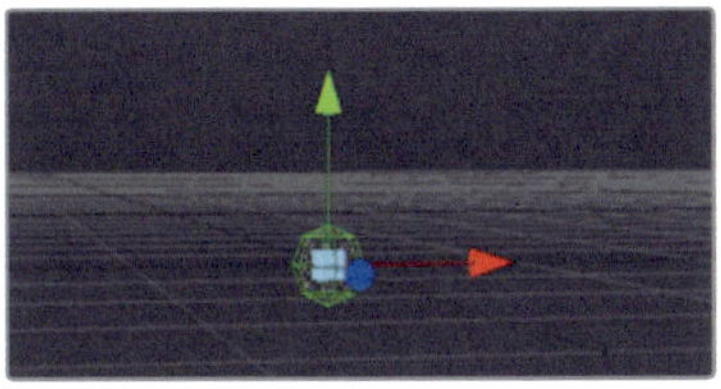

The particle cell is the object that is multiplied and animated by the particle emitter. By default, it is represented by small white points within the sphere, but as with the emitter, the particle cell can be any image or built-in particle cell object. The default small white points tend to be difficult to see clearly, so temporarily we will set the style of the particles to be a larger object.

11 In the inspector, click the style tab.

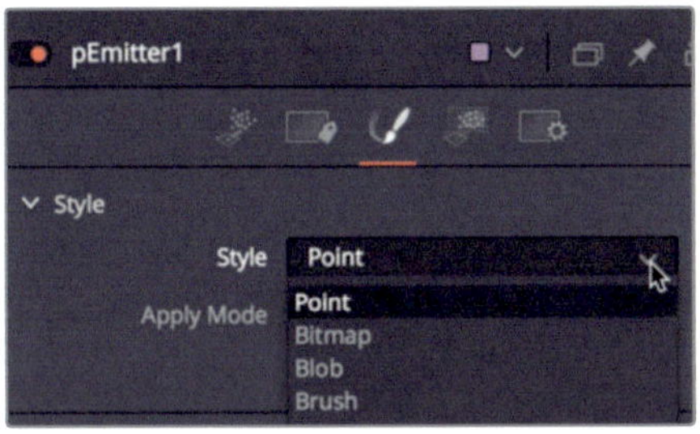

The style tab includes a menu in which you can choose the object to be used for the particle cells.

12 In the style menu, change point to blob.

Blobs are white soft circles that can be useful when creating glowing bokeh effects. We'll use them here temporarily, just to get our particles started. Next, let's work on getting these particle cells moving.

Adding Motion to Particles

The default tab in the inspector for the pEmitter is the controls tab. There, you can start to define how many cells are generated and in which direction they move. To give your particle cells some movement and a trajectory, you use the velocity and angle controls in the inspector. It is easier to perform motion-based adjustments on particles while the comp plays.

1 Click the play button to begin playing the comp.

2 Click back on the controls tab in the inspector.

Particle cells start with no motion. They require some force that makes them move. So, when you first click the play button, the particle cells appear to fill up the sphere emitter shape, but they do not go anywhere.

3 In the velocity section of the inspector, set the velocity value to about 0.2.

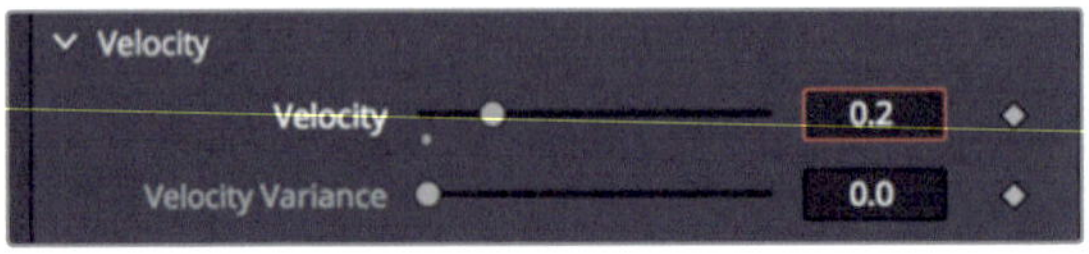

All the particle cells shoot to the right at a steady, constant rate. To change the angle of direction, you can use the angle control.

4 Below the velocity setting, adjust the angle to 180, causing the particles to emit to the left.

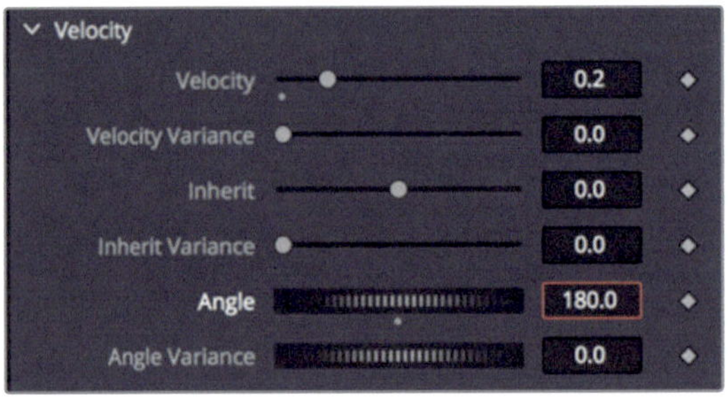

The angle setting determines a clockwise or counter-clockwise angle, so a setting of 180 causes all the particle cells to move in the opposite direction.

If you want to produce more natural-looking particles, you need to add variation to a setting. For instance, to impose a more natural look on your particle cells, they should not all move at the same constant speed. Adding variance to the velocity will cause some individual particles to move slower and some to move faster.

5 Set velocity variance to 0.1. Doing so causes some particles to move 5% faster and others 5% slower than the current 0.2 velocity setting.

The amount of particles generated on each frame is determined by the number setting. The default number setting of 10 generates 10 particle cells on every frame, which may be too many for what we are creating, so let's lower that number.

6 Set the number to 0.3 to create 1 particle roughly every 3 frames. That doesn't sound like a lot, but smoke doesn't require many, and remember that you can always change the number later.

The last step in your initial setup of a particle system is to determine how long the particles last. You control the duration of the particles through the lifespan value. Your composition is 42 frames long, as you can see by the render range end value to the left of the time ruler. You will want all your particles to end after frame 42 to ensure that they last for the entire composition and do not just disappear mid-way.

7 Set the lifespan to 50 to ensure that all your particle cells will last the entire duration of the composition.

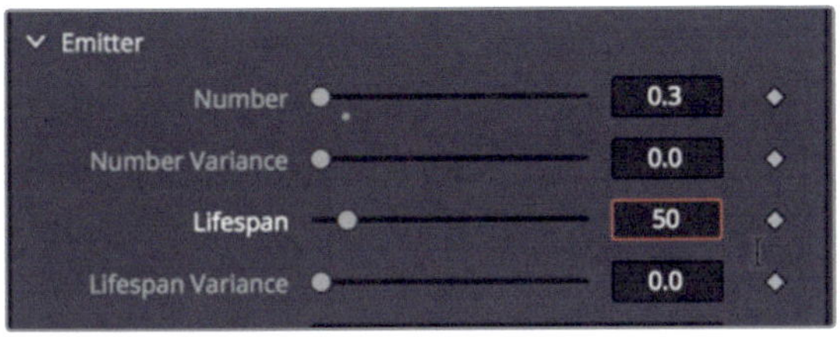

These settings are the essential parameters for the initial setup of any particle system. You set the number of particles, the general speed, and the direction in which they should travel.

The other essential set of controls determines size, position, and shape of the emitter. While you've been using the emitter's default sphere shape, it is now time to explore its other options.

Defining the Emitter Shape

The emitter is the source from which particles are generated. If you were creating fireworks, the emitter would be a small point on the ground. If you were making rain or snow, the emitter would be a rectangular shape that hovers over the scene. For the blowing smoke, you want the particles coming from the area off-screen to the right of the band and deep enough to cover the entire 3D scene. To redefine the emitter shape, you use the region tab in the pEmitter inspector.

1 In the inspector, click the region tab.

The region menu in the region tab is the primary control that defines the object used as the emitter. Once you decide on the basic shape, you can then use the transform controls to position and scale the emitter object how you need it.

2 In the region menu, change sphere to cube.

The viewer now shows the particle cells emitting from a cube shape. To make a larger emitter area for the particles, you can make the cube taller and deeper. You'll size it roughly for now, and you can return to refine it after you connect the particles into the larger 3D scene.

3 At the top of the inspector, set the height to 0.5 and the depth to 1.0.

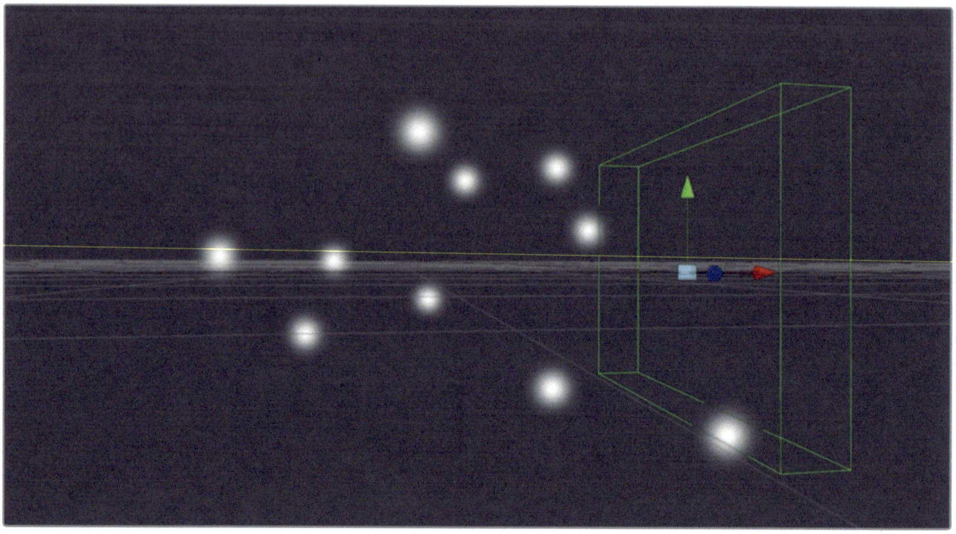

When you connect the particle system into the 3D scene, you'll need to position it based on the objects in the scene. For now, you'll leave it here.

Using Images for Particle Cells

It is unusual to stay with the small white blobs for your particle cells. In fact, the cells can be almost any object you choose. Often, you'll use one of the built-in shapes to get started and later switch to an image or small movie file when you have completely configured the cells' motion. You set the cells' appearance in the style tab.

1 From the particle bin in the media pool, drag the noise clip to an empty place in the node editor, near the pEmitter. Then press 1 to see it in the viewer.

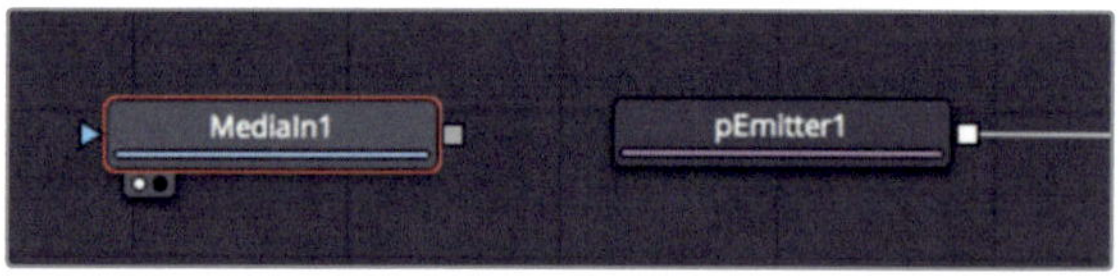

The file is a small 200 x 200 pixel movie that somewhat resembles some type of vapor like clouds, smoke, or fog. It was created using Fusion's fast noise generator.

NOTE You can learn to create the noise movie in Appendix C.

Although this movie is just a single puff of smoke, you can make it into larger plumes by connecting it into the pEmitter. The style tab where you previously selected the blob type also allows you to choose to use an image from the node tree for your particles.

2 Select the pEmtter node, and in style tab's style menu, choose bitmap.

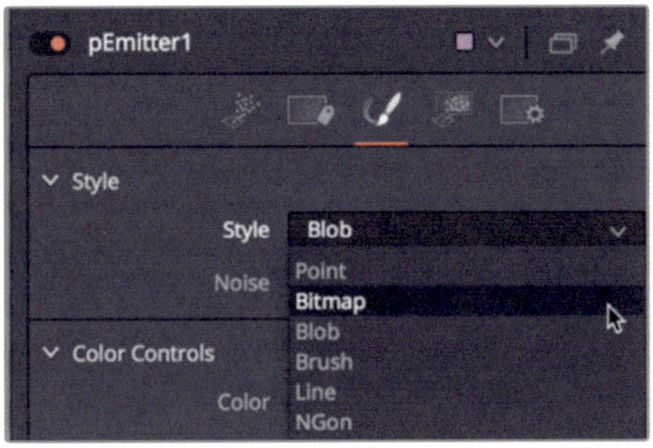

Bitmaps allow you to connect and use images as the particle cells.

When you set the style to bitmap, a yellow input is displayed on the pEmitter node that allows you to connect any image in the node editor.

3 Drag the output of the noise movie to the yellow input on the pEmitter.

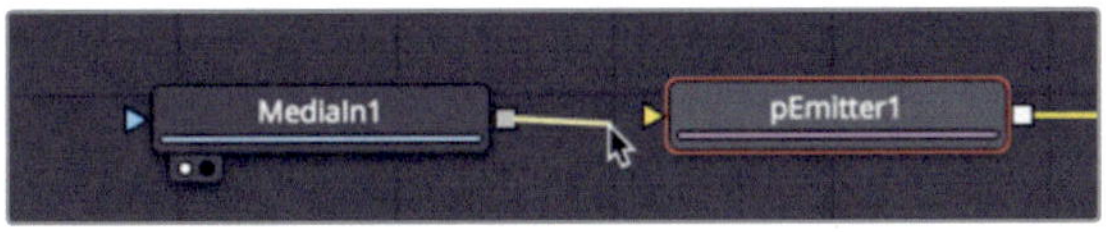

For the remaining exercises in this lesson, you'll need two viewers to see the final output and the 3D perspective view.

4 In the upper-right corner of the viewer, click the dual viewer button.

5 Select the merge 3D node and press 1, and then select the render3D node and press 2.

To make sure your smoke fits well into the 3D scene, you'll connect the particles into the merge 3D node.

6 Drag the output of the pRender node to the open input on the merge3D.

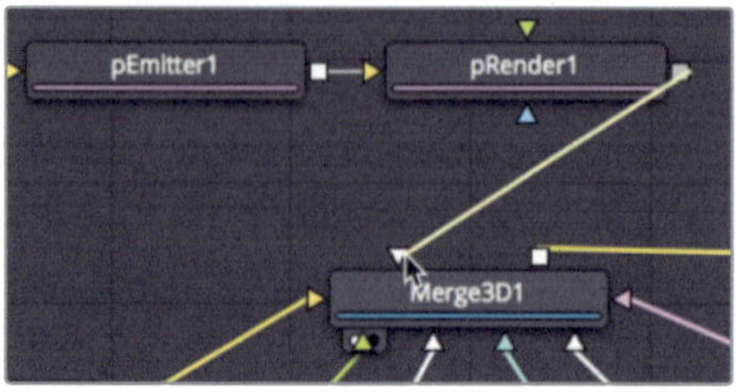

For the first time, you are able to see the particle smoke integrated into our scene.

7 Click the play button to view the smoke in the scene.

Yeah, not very impressive...yet. A few issues prevent the particles from looking like smoke. The particles are very small, sparse, and too light to match the light in the comp. All these are easy to fix, but before we dive in too deep, let's optimize the playback performance of our composition.

Optimizing Performance

The larger your composition grows, the longer it takes to load into RAM for playback. You can improve the performance for playback in several ways. One way is to limit the area of the viewer that gets updated when you play the composition.

8 In the upper right of viewer 2, click the drop-down menu next to the region of interest button.

9 From the region of interest menu, choose set.

10 Drag a wide rectangle around the three band members in viewer 2.

11 Click the play button to watch the composition.

TIP During playback, clearing one of the viewers of any image will also increase performance.

Only the portion of the frame outlined by the region of interest rectangle is updated as the composition plays. The smaller the rectangle, the better performance you will see playing back a composition.

Controlling the Timing and Look of Particles

OK, now that we have better playback performance, we can get back to creating our particle smoke. The most obvious problem is the small size, so we'll adjust the size first, and then match the color.

1 In the pEmitter Style tab, set the size to 0.5 and the size variance to .05.

Some of the smoke particle cells are now larger, and some are smaller. As with most other particle settings, adding variance to the size brings a more natural appearance to the particle system.

The emitter also must be corrected. First, the emitter is centered in the frame, so the particles are generated in the frame instead of off-screen to the right. Second, the particles should already be covering the frame at the start of the shot, but currently they are only starting to generate on frame 1. Let's deal with the positioning problem first. All we need to do is position the emitter off to the right side of the frame.

2 At the top of the inspector, select the region tab.

3 While looking at the perspective view, drag the X offset slider to the right until the green emitter outline in the viewer is positioned off to the right of the band.

The second issue requires you to give the particles a running head start before the composition starts. You can pre-generate particles to ensure that the smoke is already covering the frame on frame 1.

4 Select the pRender node, and then toward the bottom of the inspector, set the pre-generate frames slider to 40.

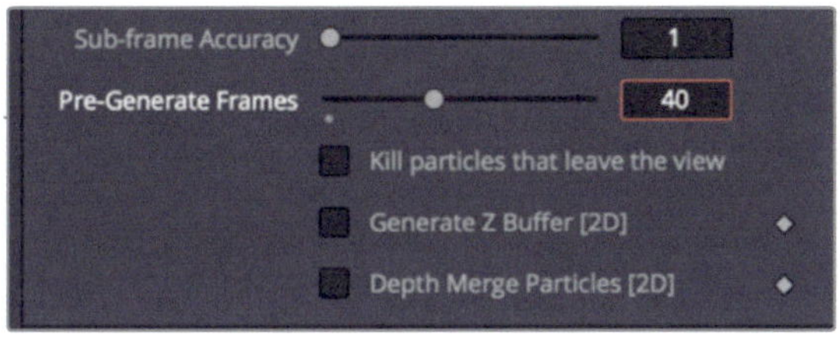

Setting the slider to 40 causes the particles to render on frame 1 in the state they would appear on frame 40 of the composition.

5 Click the play button to view the smoke in the scene.

Although better, the transparency for the particles isn't perfect and the particles still end too early. The problem with the timing is that earlier you set the lifespan of the particles to 50 frames. At the time that seemed long enough, but now the particles are pre-generating 40 frames, so their life must be extended.

6 Select the pEmitter node, and then select the controls tab and increase the lifespan to 100. This should be long enough to cover the pre-generated frames for the duration of our composition.

The particles now last long enough for the composition, but when they overlap, they appear to have very hard edges. This is a result of the fast Z-depth sorting Fusion performs by default. The behavior is corrected when you render the 3D particles using the render 3D node.

7 Select the render 3D node in the node editor.

8 At the bottom of the inspector, set the transparency menu to sorted (accurate).

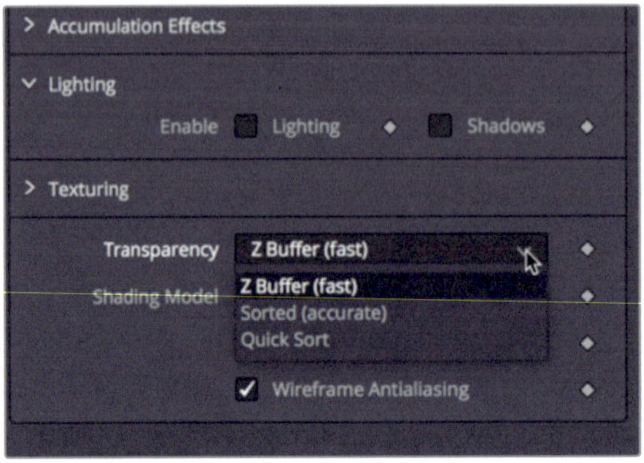

All these adjustments have created a much more realistic effect. You can further improve the integration of the smoke into the scene with a small amount of color correction.

9 Select the pEmitter node, and in the style tab of the inspector, open the color controls section.

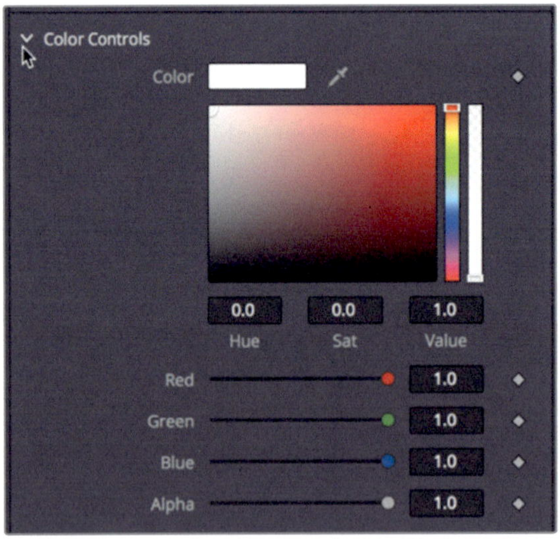

The color swatch in the color controls is for setting an overall color to of the particles.

10 Set the color swatch to a light yellow/green tint to match the background light.

The lack of transparency in the color makes it appear more like thick fog than blowing smoke. You can introduce more transparency in two ways. First, you can use the alpha slider under the color controls swatch to add transparency to the color.

11 Under the color swatch, drag the alpha slider down to around 0.5.

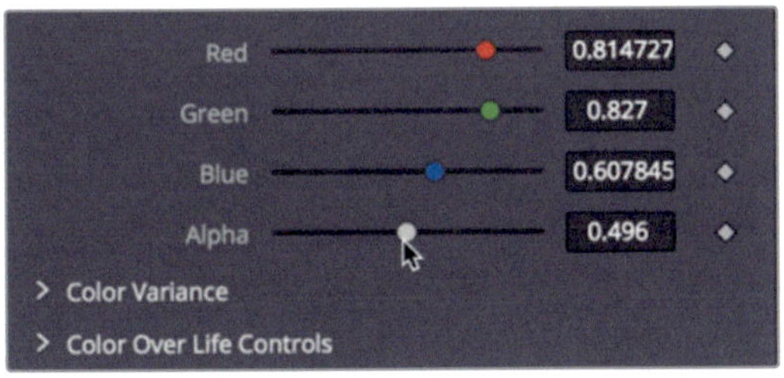

12 You can control the overall blend of the particles by going to the top of the inspector and adjusting the gain control to between 0.2 and 0.3

Lastly, to add more randomness to the motion, you'll add a spin animation to the particles.

13 Select the controls tab in the inspector, and set the Z spin to 1.

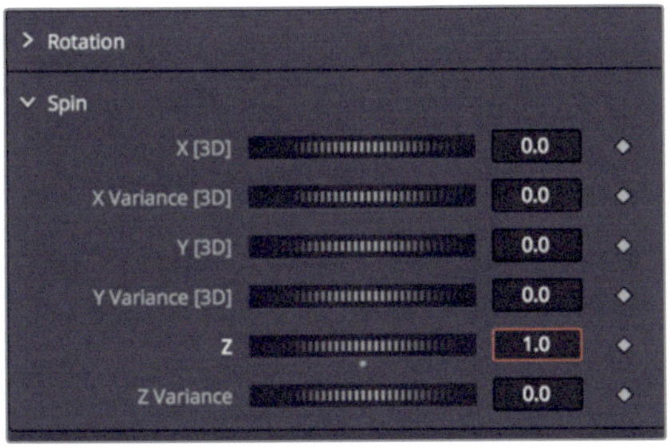

A 10-degree spin is added to each particle cell on every frame. This will add a tumbling animation to the particle cells. As with every aspect of particles, increasing the variance for a given parameter will make them appear much more realistic.

14 Increase the Z spin variance to 0.5.

So far, you have used just a few fundamental controls, but even these controls can realize a multitude of particle variations. As you have experienced, a small change to one of these parameters can yield big results. Beyond the basic controls are several additional nodes specifically designed for particles, that can add character and increase the realism in particles.

Introducing Particle Forces

The pEmitter and pRender nodes are the essential requirements for your particle system, and they offer a wealth of control, but you can apply a variety of particle forces or behaviors to shape the flow and speed of your particles. From re-spawning new particles, to bouncing off objects in the scene, customizing particles is just a matter of stringing together other particle nodes between the pEmitter and pRender nodes.

TIP The only nodes that you can add between the pEmitter and pRender nodes are particle-specific nodes with names that start with a lowercase p, such as pTurbulence or pSpawn. You cannot add standard image-processing nodes such as blur or color corrector between the pEmitter and pRender nodes.

1 Above viewer 2, click the region of interest button to disable it.

It will be easier to see the affect turbulence has on the particles if you view the entire frame, even at a diminished performance.

2 With the pEmitter node selected, open the effects library and select the tools > particles category, and then click the pTurbulence tool.

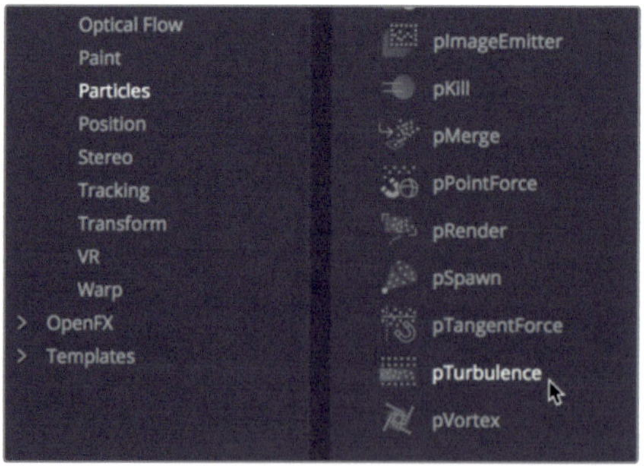

The pTurbulence node adds random chaos to the particle cells' motion, making it appear more unpredictable and more natural. Let's adjust the turbulence to increase the variety of movement among the particle cells.

3 Set the density to 75.0. This will cause steadier, less chaotic movement to our particles while still adding some turbulence.

4 Play the particle composition to view the results.

Congratulations! You just created your first particle animation!

You now have a clearer understanding of how to use Fusion to create realistic particles. It is an extremely powerful and fun aspect of DaVinci Resolve 16.

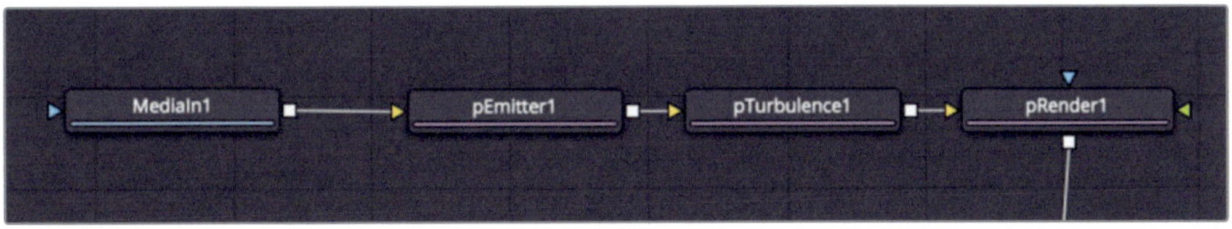

Completed particle node tree for Lesson 11.

Lesson Review

1 True or false: The end of every particle system requires a pRender node.

2 True or false: To make particles glow, you can add a soft glow between a pEmitter and a pRender node.

3 True or false: The pRender node controls how many particles are generated, the size, and the color of the particles.

4 True or false: Setting the velocity to 1.0 will cause the particles to move down in the frame.

5 True or false: Configuring particles to work in 2D or 3D occurs in the pRender node.

Answers

1 True. Every particle system must end with a pRender node.

2 False. Only particle nodes can be inserted between a pEmitter and pRender node.

3 False. The pEmitter node controls how many particles are generated, the size, and the color of the particles.

4 False. Velocity has nothing to do with direction.

5 True. The pRender node includes an output mode menu where you can set the particles to work in 2D or 3D.

Lesson 12

3D Camera Tracking

One of the most difficult tasks in budget filmmaking is set design. Without a lot of money—and a small army of carpenters—it's extremely difficult to make your set look like ancient Rome, or the command bridge of a high-tech starship, or an alien planet with three moons. Typically, efforts to turn your cousin's basement into a secret government research lab will end up looking like your cousin's basement.

Fusion changes all that. Using its powerful 3D camera tracker node, you can enhance set designs with surprisingly little effort. During this lesson, you'll learn how to set up, perform, and refine a 3D track in order to realistically add a pirate ship off the coastline of a simple beach shot.

NOTE This lesson uses the camera tracker node, which is available only in DaVinci Resolve 16 Studio.

Time

This lesson takes approximately 70 minutes to complete.

Goals

Completed composite for Lesson 12.

Rotoscoping for 3D Tracking

Camera tracking uses sophisticated photogrammetry to recreate a virtual 3D scene that corresponds to the physical set in your live-action scene. The position, movement, and focal length of your camera are recreated in Fusion's 3D space, along with the spatial locations of landmarks within your set.

While the intricacies of photogrammetry are *far* beyond the scope of this book, here's a simplified explanation of the process: The camera tracker uses the relative speeds and movement directions of items in your scene to determine where they are in space. When you ride in a car or train, you observe that objects closer to the car move more quickly than objects in the far distance. The camera tracker can use this motion *parallax* to calculate the position of each element in the physical scene and calculate where a virtual camera should be to replicate the same parallax within the computer.

Now this calculated parallax works convincingly as long as everything in your shot is "nailed down." Objects within the shot that exhibit independent motion—such as those pesky actors who always seem to find their way into visual effect shots—can confuse the calculations and produce poor results because their speed is not dependent solely on their distance from the camera. Therefore, before you can perform a 3D camera track, you need to use garbage mattes to identify which objects you want to track and which you want ignore.

1 In DaVinci Resolve 16, open the Fusion 16 Lessons part 3 project.

2 From the master bin, load the Lesson 12-START timeline and play over the first clip in this timeline.

As it plays, try to identify objects in the clip that will require garbage matting; that is, locate the objects in the shot that do not exhibit independent motion.

In this shot, the actors and ocean obviously have independent motion. The rock outcropping and the beach move only because the camera moves. You'll garbage matte these two elements separately.

3 Switch to the Fusion page, and move the playhead to the start of the clip.

4 Click in the gray space of the node editor to deselect any selected nodes.

5 Press Shift-spacebar to open the select tool dialog, and type **polygon** to add a polygon tool.

6 Create a very rough outline around the rock outcropping at the start of the render range. You may want to pan the viewer to allow you to draw the shape beyond the left edge of frame.

TIP Emphasize speed over accuracy. As long as you mostly keep the rocks within your garbage matte, the camera tracker algorithm can ignore stray pixels that fall on either side of the matte edge. Your garbage matte shape should require no more than a dozen control points.

7 Advance the playhead to last the last frame in the render range, frame 126.

8 In the viewer toolbar, click the select all points button.

9 Move and reshape the polygon to fit the rocks.

TIP Unless you need to move individual points, it is easier to keyframe the polygon using a shape box.

10 Press Shift-B to enable the shape box around the entire polygon, and then drag the control handles on the shape box to stretch and resize the polygon shape to encompass the rock.

11 Move to roughly the middle of the clip, around frame 60, and reshape the polygon to fit the rocks.

12 Scrub through the clip to make sure the shape consistently fits the rocks. Reshape the polygon using the shape box or individual control points in areas where it needs refinement.

TIP Pressing the Option-left or right arrow keys (macOS) or Alt-left or right arrow keys (Windows) will move the playhead to the previous and following keyframes, respectively.

13 When you are done keyframing the shape, select the polygon node, press F2, and rename the node **rocks_gshape**.

14 Press cmd-left arrow (macOS) or Ctrl-left arrow (Windows) to move to the start of the render range at frame 11.

15 Click again in the gray space of the node editor to deselect all nodes. Press Shift-spacebar, and then press enter or return immediately to repeat the previous tool selection, the polygon tool.

16 Repeat the previous matte process to draw and animate a shape around the foreground beach.

17 Select the new polygon node, press F2, and rename it **beach_gshape**.

18 Click in an empty space in the node editor and add two more unconnected polygon nodes.

19 Repeat this rotoscoping process to add and animate two smaller shapes for the pirates running up the beach. Starting at the end of the render range on frame 128 might be easier because the pirates are larger and clearer in this part of the shot.

20 Rename these new polygons **SR_pirate_gshape** for the pirate on the right and **SL_priate_gshape** for the pirate on the left (SR and SL for screen right and screen left).

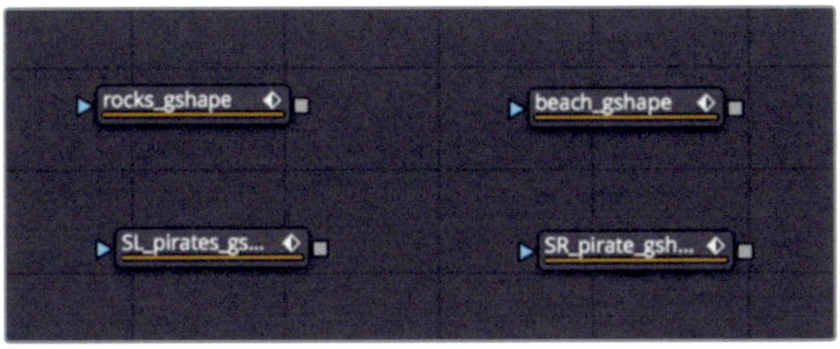

At the moment, these are just unconnected polygons. To make them into a mask that you will use with the camera tracker, you need to combine them.

Combining Garbage Mattes

To create a single mask for your camera track, you must connect all the masks and combine them so that the actors are subtracted from the beach and rocks mask. You'll start by combining the two main mattes, the rocks matte and the beach matte.

1 Connect the output of rocks gshape into the effect mask input of beach gshape.

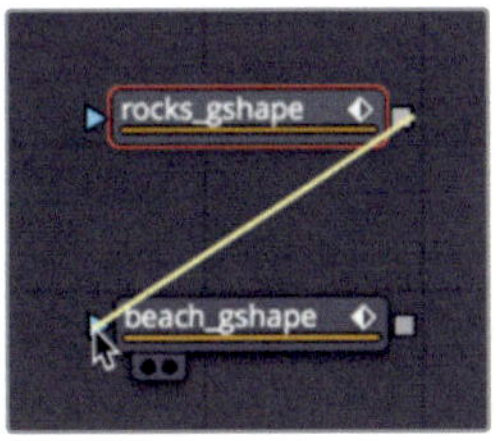

2 Select beach gshape, and press 1 to display its output in viewer 1.

In viewer 1, you can confirm that the two images are combined. The camera tracker tracks the white areas and ignores the black portions of the matte. So, it will track the rocks and the beach, but the pirates are not "nailed" to the set, so they must be excluded from the tracking because they overlap with the rocks.

3 Connect the output of beach gshape to the effect mask input of SL pirates gshape.

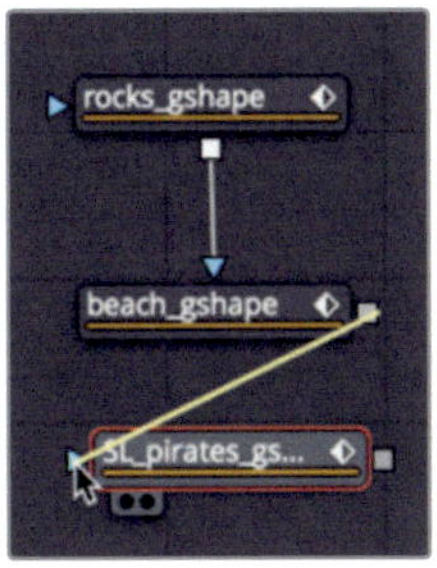

4 Select SL pirates gshape, and press 1 to display its output in viewer 1.

5 In the inspector, set the paint mode to subtract.

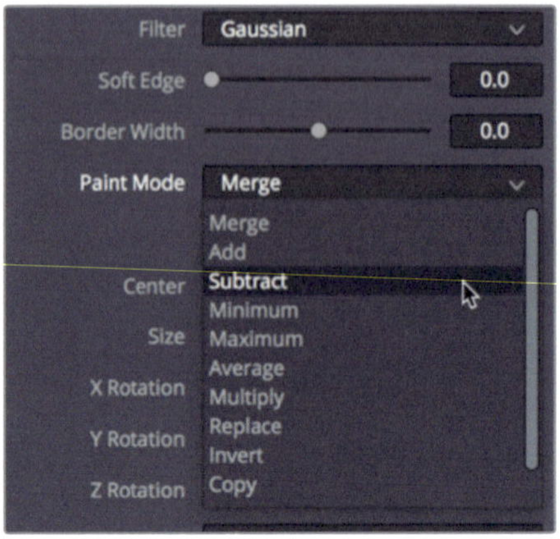

You can see that the SL pirates gshape is now cut away from the main garbage matte. You'll repeat this procedure for the SR pirates gshape and reposition nodes in the nodes editor to clean things up a bit.

6 Connect the output of SL pirates gshape to the effect mask input of SR pirates gshape, and display the SR pirates gshape in viewer 1.

7 In the inspector, set the paint mode to subtract.

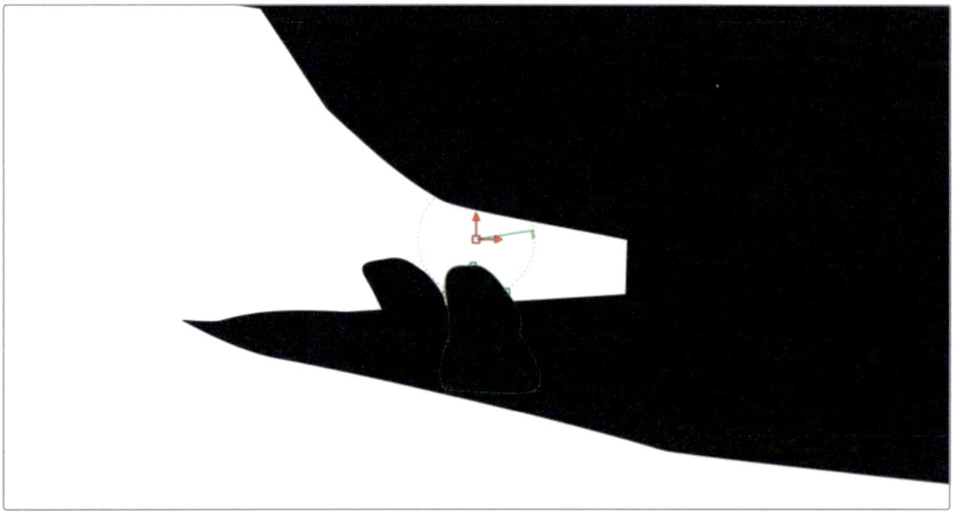

8 In the node editor, drag all the connected polygon nodes to rearrange them into a neat column.

9 Select the lined-up polygon nodes, and press Shift-spacebar to open the select tool window.

10 Type **underlay**, and press enter to add an underlay that surrounds the selected nodes.

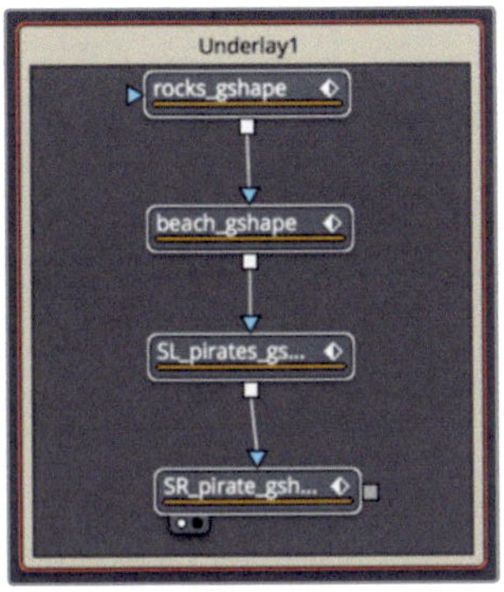

We'll use the underlay to group all the matte nodes visually.

11 Cmd-click (macOS) or Ctrl-click (Windows) each of the polygon nodes to deselect them. Press F2, and rename the underlay **garbage_matte**.

The underlay surrounds the selected nodes, making it easier to locate and move these nodes as your node tree grows.

Preparing the Camera Tracker

With all the manual labor out of the way, it's time to set up the tracker and let it do all the hard number crunching.

In case you were unable to complete the mattes for all the frames, we'll jump ahead and open a completed version of the previous exercise. You can access previously saved versions of a composite directly from the clip in the Fusion page.

1 In the upper-left corner of the DaVinci Resolve window, click the clips button to display a thumbnail timeline.

Along the bottom of the window, a thumbnail is shown for each edit in your timeline. The currently selected shot is highlighted with a red outline. Because each Resolve user tends to create and keyframe garbage mattes differently, you'll use a saved version of the garbage mattes.

2 Right-click the first thumbnail, and choose rotoscope done > load.

This new composite will appear similar to the one you just created, but it includes all the mattes created and animated.

3 Click the clips button to hide the thumbnail timeline.

4 Select the media in node, and press Shift-spacebar. Type **cam**, and from the list of tools, select the camera tracker. Click OK to add the tool to the node editor.

> **NOTE** The camera tracker is available only in DaVinci Resolve 16 Studio.

The camera tracker is added to the node editor with the media in 1 node connected to its input. To use all the garbage mattes that you created, you must connect them into the track mask input of the camera tracker.

5 Connect the output of SR pirates gshape to the track mask input of the camera tracker node.

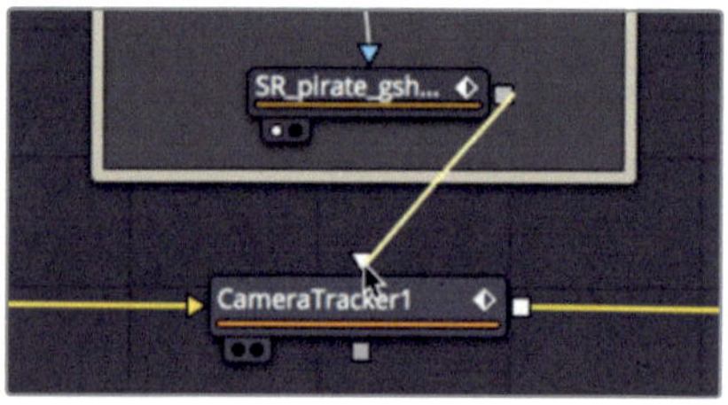

6 Press cmd-left arrow (macOS) or Ctrl-left arrow (Windows) to move to the start of the render range at frame 11.

The camera tracker starts by generating a cloud of trackers informed by areas of contrast in the image. The first job is to tune the settings for this particular shot to get a good set of tracks.

7 Select the camera tracker node, and press 2 to display it in viewer 2.

8 In the camera tracker's inspector, enable preview autotrack locations.

The small green dots indicate features that would be tracked if you were to begin tracking now, but you want to track a lot more features than these to improve the tracking data. You can always delete features later if they are inaccurately tracked.

9 Reduce the detection threshold to around 1.8 and minimum feature separation to around 0.01.

The detection threshold determines how much contrast an image feature must display for it to be considered trackable. Feature separation determines how close tracking features can be to be considered unique. By lowering these two sliders, you should have a significant number of trackable features displayed in the viewer.

The camera tracker uses an optical flow-based algorithm to follow pixels from frame to frame. You can further refine the tracking using a pattern recognition-based method similar to the normal tracker in Fusion: a planar tracker algorithm that works well when you have large areas of planar transformations in a shot. Or, you can continue using optical flow, which works well for a shot like this with very few crisscrossing objects or motion blur.

10 Click the new track defaults disclosure arrow, and in the tracker menu, choose optical flow.

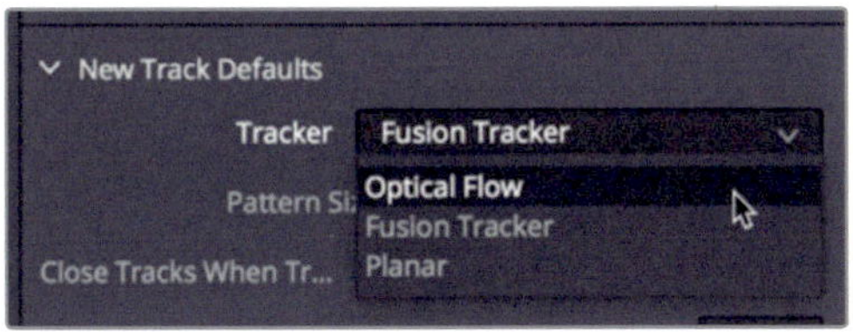

11 Click the auto track button to begin tracking.

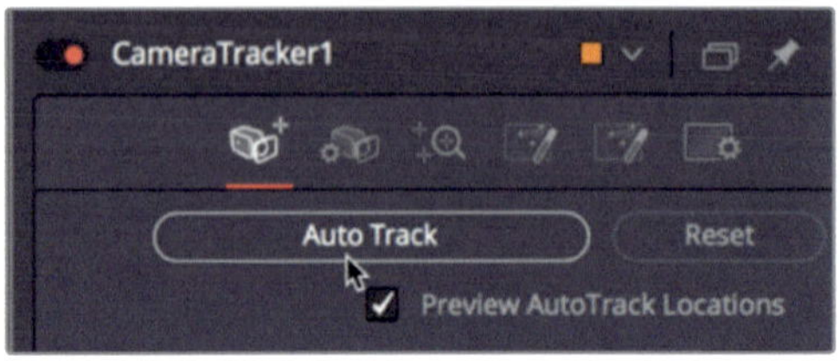

The camera tracker steps frame-by-frame, calculating the positions for all the tracking points. Obviously, the more tracking points you have created, the longer the process takes. When it gets to the end of the render range, the tracker moves backward frame-by-frame to refine the existing points.

Solving for the Camera

Once auto tracking is completed, you can begin to enter known camera parameters, and finally calculate the 3D representation of your live-action set. While Fusion can estimate much of the camera information, the more information you provide about the physical camera that captured the scene, the better the results will be. Often, the information is logged on set, but you can also examine a clip's metadata in Resolve's metadata inspector to see if useful information is listed there.

This footage was shot with a Blackmagic Ursa 4K using a focal length of 12.65 mm. At a minimum, you need to enter the focal length to solve the camera track.

1 Click the camera button to switch to the camera tab.

2 Enter a focal length of **12.65**, and set the film gate to Blackmagic URSA/Production 4K 16:9. Resolve automatically enters the correct aperture settings for that camera type.

> **TIP** You can often ignore mild lens distortion (as in this shot), but when working with lenses that have more distortion, you can select the refine lens parameters checkbox to automatically correct the distortion.

Time to solve the camera track!

3 Click the solve button to switch to the solve tab, and click solve. Depending on your computer, the solve may take between four to eight minutes to calculate.

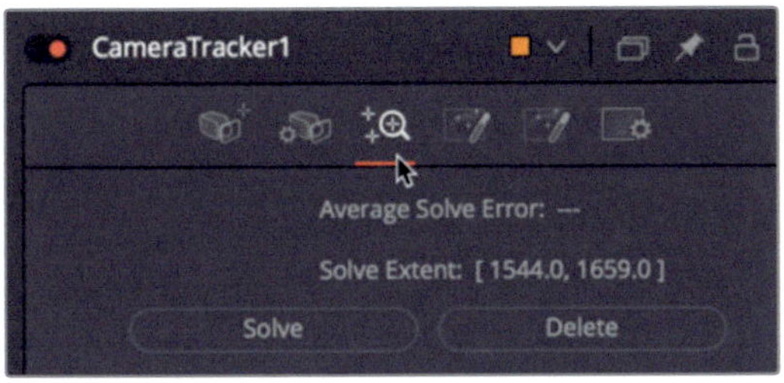

When the solve is completed, the reward for your efforts will be just a single line of information: the average solve error.

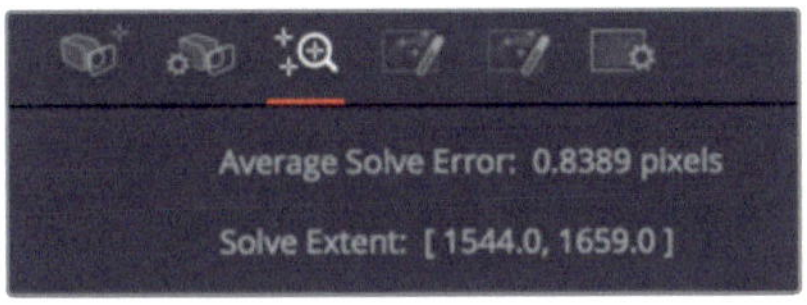

NOTE Depending on whether you are using the saved Rotoscope Done composition or the composition with your animated garbage mattes, your average solve error will be different from the value in the preceding figure.

An average solve error of around 0.5 means that, on average, any digital environment work should be just over a half a pixel off, at most. For high-resolution 4K footage, this is borderline workable, but for this lower-resolution 1080p footage, it's more than acceptable. Typically, you want to get this error value below 1 at a minimum, and ideally below 0.5. No matter where your average solve error ended up, let's refine the solve a bit more to see whether you can get the error below 0.5.

Refining the Solve

The average solve error is also called the *reprojection* error, because it measures how closely the computer model of your live-action set can predict and recreate 3D locations in the physical set.

Imagine replacing your live-action camera with a digital projector placed at exactly the same location as the camera, pointing in the same direction, and using the same lens. If your virtual set is perfectly *reprojected* back onto the scene using the virtual camera image, every projected pixel should line up perfectly with the object in the physical scene. If the reprojected pixels miss their marks, that's a *reprojection error.*

The solve error is measured in pixels and refers to, on average, how far pixels are misaligned from the original scene.

Looking at the viewer, most of the tracked features appear green to indicate a reprojection fit. You'll also see a few of the tracked features colored red to indicate that they have an unacceptable reprojection error. To improve the overall solve quality, you can delete the features with high errors.

TIP Solving is a computationally demanding and RAM-intensive process, but it is also an iterative process as you refine the calculation. Deleting too many tracking markers and resolving can cause problems when your computing power is not up to it. So if you plan on doing multiple iterations, it is advisable to make a copy of the original camera tracker node after each solve.

1 Select the camera tracker node.
2 In the viewer, drag a selection over a group of red trackers.

The trackers will highlight yellow to indicate that they are selected.

3 In the inspector, click delete to remove the high-error tracked features.

This shot doesn't have many red trackers clumped together, which makes it very difficult to select more than a handful of them at once. The Inspector provides for an easier way to select trackers when manual selection is difficult.

The maximum track error determines how poorly a feature was tracked during the solving phase. The maximum solve error determines how poorly a feature reprojects based on the final scene.

4 In the solve tab, set the maximum track error to 0.2 and the maximum solve error to 3.0.

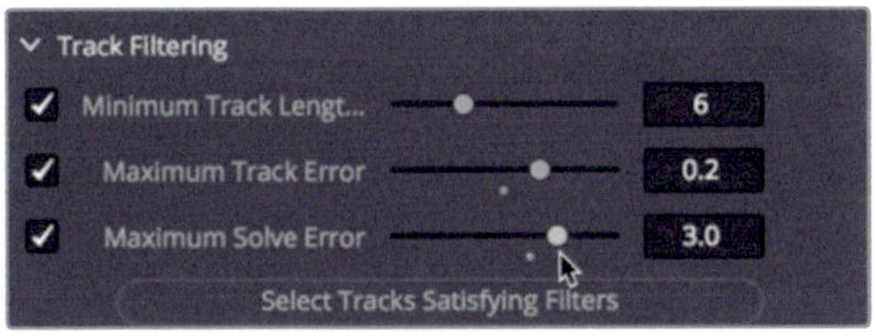

If you push these values too far, you may actually find the reprojection error worsening. By starting with a value of 3.0, you avoid deleting too many points at once.

5 Click select tracks satisfying filters to select for deletion all the tracks with errors worse than the ones you just set.

6 Click delete to remove the high-error tracked features.

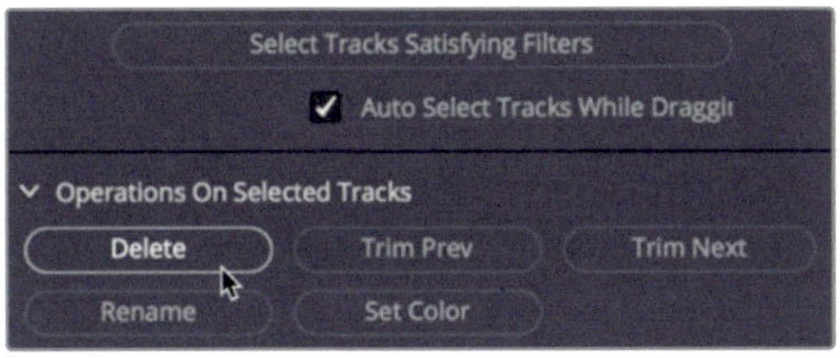

7 Click solve again to re-solve the scene with a leaner, more accurate sample of features.

At this point, you've achieved the goal of reducing the solve error a little lower than your initial results. It is probably not worth the effort to further reduce the error.

NOTE Your solve errors will almost certainly be different from the ones shown here. For example, slight differences in the positions of your garbage matte shapes will change the accuracy of your solve. These differences are negligible, though, and you'll almost certainly achieve similar or better results by following the steps described here.

On tougher shots, you'll often repeat the solving and error reduction process several times, lowering the error rate a little more each time until your solve fails (and you must return to the solve values that you previously saved as a backup), or you achieve your desired average solve error goal of less than 1.0.

Setting the Ground Plane and Exporting the Scene

At this point, the camera tracker node has computed a virtual 3D scene that matches the original live-action scene to within slightly more than half a pixel. But before you can play with this new scene, you need to establish some ground rules. In fact, you need to establish a ground.

The camera tracker has no access to camera accelerometer data, so it doesn't know if the camera was level, tilted, upside down, or on its side. So before you begin working with the 3D scene, you need to tell the camera tracker where the ground is located.

NOTE If you were unable to get a satisfactory track, there is a saved version of the composition with the camera track completed. Open the clips timeline, right-click the selected thumbnail, and choose solve done > load. You can now follow the lesson using the composition 3D node tree.

1 Click the export tab button at the top of the inspector.

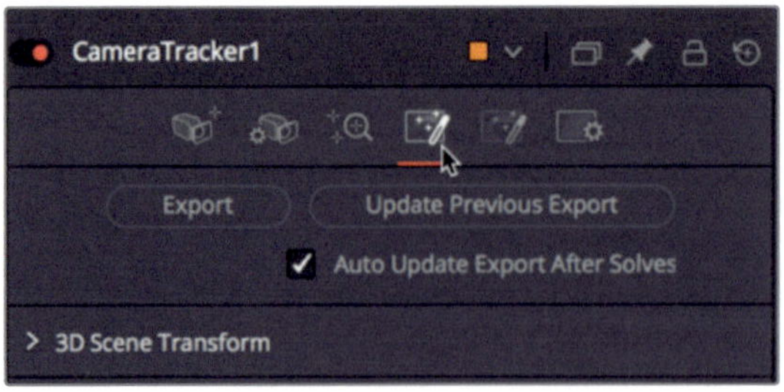

2 Click the disclosure arrow next to 3D scene transform.

The options for setting the ground plane aren't made visible until you switch to unaligned.

3 In the aligned menu, choose unaligned. The option to set the ground plane becomes visible in the orientation section.

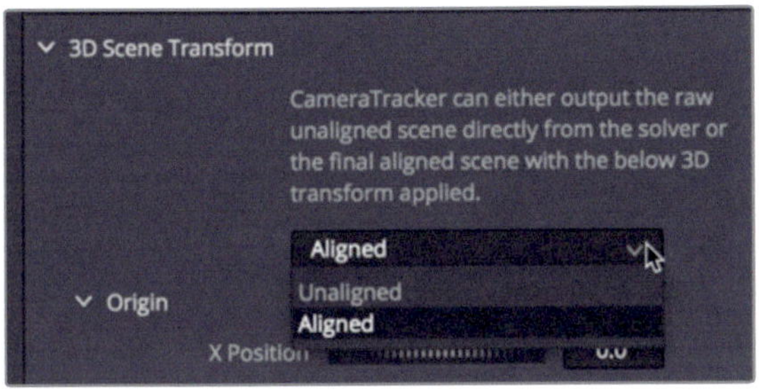

In this (and almost every) 3D coordinate system, X is the horizontal axis, Y is the vertical axis, and Z is the depth axis. So, the default XZ plane is the typical ground plane as defined by the horizontal (X) and depth (Z) directions.

> **TIP** In some shots, the ground may never be visible. In such a case, it might make more sense to use another plane to identify the ground. For example, if a green screen wall has clear tracking features, selecting the XY plane would allow you to lock camera direction to the green screen wall, even when the floor wasn't visible in the shot or didn't track well.

4 Move to the start of the render range.

Here you have a clear view of the "ground" of the beach. To set the ground plane, you first select all the tracking points that tracked features located on the beach.

5 In the viewer, drag through the trackers on the beach to select them.

6 Shift-click any beach track points that you may have missed with your selection.

7 In the orientation section, click set from selection. The camera tracker adjusts the scene rotation to align with the selected feature tracks.

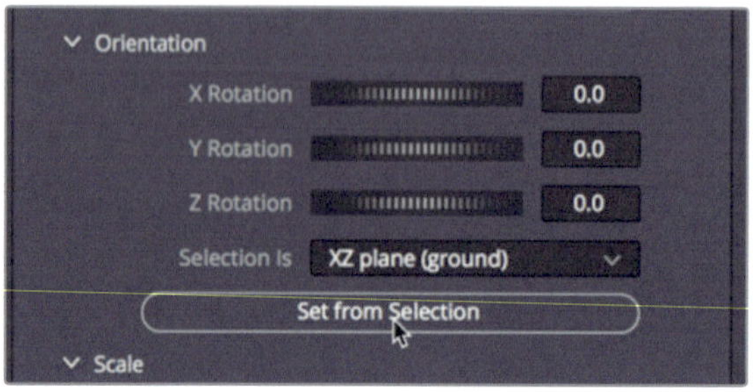

Finally, you need to tell the camera tracker where the origin point—the center of our 3D universe—should be. It can be anywhere that's convenient. We'll select a feature track point from the center of the beach and set it as the origin point.

8 Move to the last frame of the render range. Hover your mouse pointer over the viewer, and notice how Fusion offers a readout of the solve error for the specific feature beneath the pointer.

9 Select a feature with a reasonably low solve error located somewhere on the beach.

10 In the origin section, click set from selection.

The camera tracker sets the center of the internal 3D scene to the selected feature.

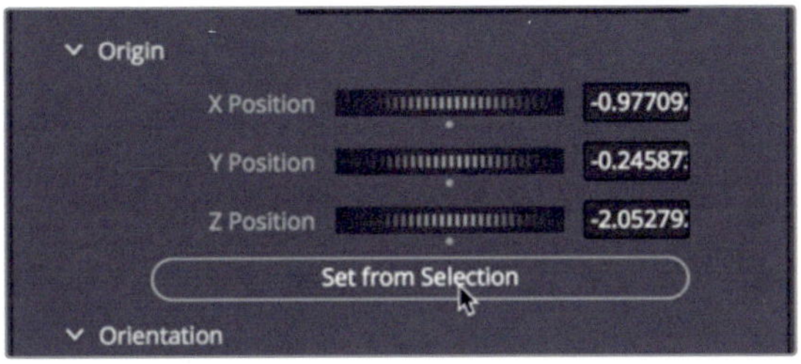

Now that the ground plane is set, it's finally time to export the solved 3D scene.

11 Reset the unaligned menu option to aligned to lock in the ground plane adjustments.

12 At the top of the inspector, click the export button.

A group of five nodes representing the created 3D scene is automatically created in the node editor.

13 Drag to reposition the new nodes, as desired.

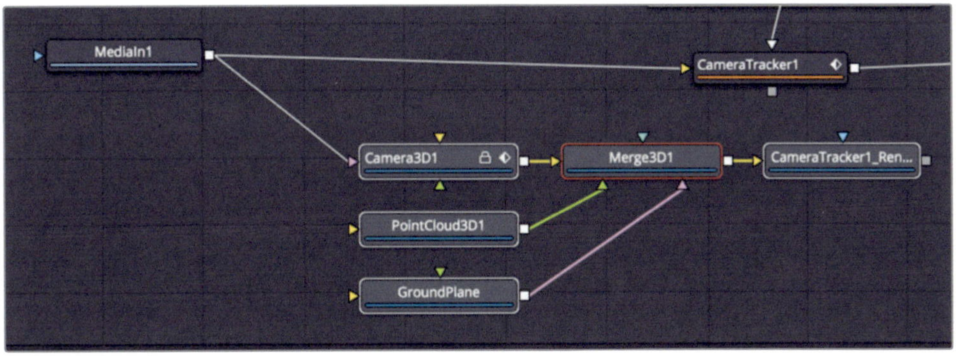

Congratulations! You've performed your first 3D track.

The new nodes create a 3D scene that includes a merge3D, camera, ground plane, point cloud3D node, and a camera tracker renderer.

With the 3D camera track performed, it's now time to take all that quality data and turn it into movie magic. This will be a fairly straightforward composite of adding the 2D pirate ship out on the horizon.

Positioning Objects in a 3D Set

To add a ship to the horizon, most of the heavy lifting is already done for us, and we can achieve the result with just six more nodes. Let's dive in.

1 From the 3D tracking bin in the media pool, drag **PirateShip.png** into the node editor.

2 Press 1 to see the PirateShip.png in viewer 1.

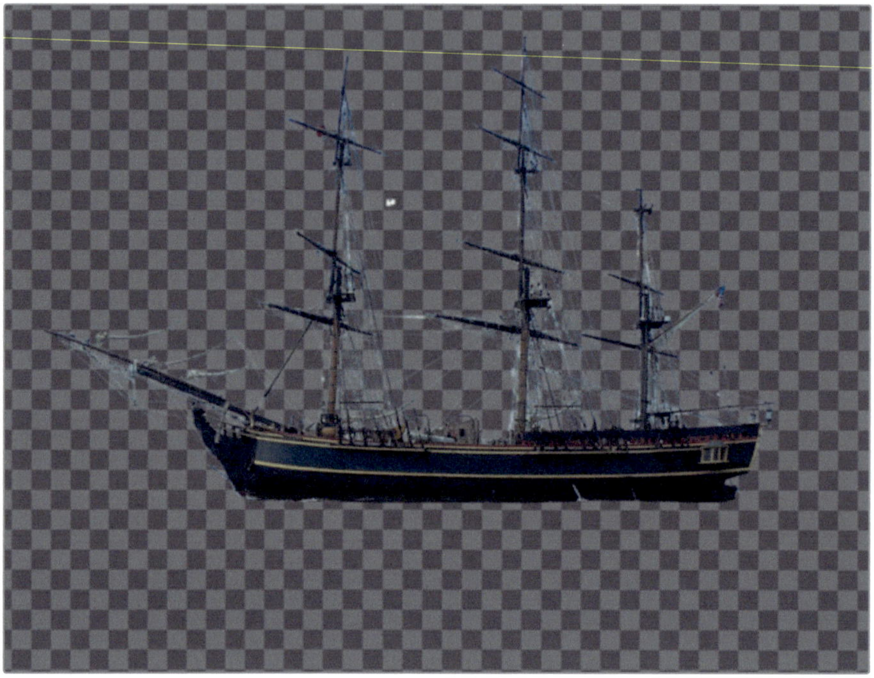

To composite this ship into our 3D scene, just like the earth and moon we used back in Lesson 9, the media in node must be placed on a 3D shape. We don't need a sphere this time, so we can just use an image plane 3D node.

3 With the media in selected, use the select tool dialog to add an image plane 3D node.

We can then connect the image plane into the merge 3D node.

4 Drag the output of the image plane 3D and connect it into the merge 3D node. Then press 1 to see the merge 3D in viewer 1.

> **TIP** If you do not see the ship in the viewer, click in viewer 1 and press F to fit the pirate ship in the viewer.

5 Press cmd-left arrow (macOS) or Ctrl-left arrow (Windows) to position the playhead at the start of the render range.

The ship is composited into the 3D scene but is clearly too large and not in the right location.

We want the ship to appear far off on the horizon, so we need to push it far back in the scene, and then scale it up. It is easiest to do this while viewing the output of the camera tracker 1 renderer node

6 In the node editor, select the camera tracker 1 renderer node, and press 2 to view the output in viewer 2.

7 Select the image plane 3D node, and then, in the inspector's transform tab, use the Z translation slider to push the ship out in Z until it's sufficiently distant but doesn't disappear behind the image of the beach. The Z translate value will be around -40.

8 At the bottom of the inspector, use the scale slider to increase the size of the ship to around 20.0.

9 Adjust its X and Y translation controls to position the pirate ship just to the right of the rock outcropping and just below the horizon line. The pictures in this book used the following values: X translate: 18, Y translate: 9.

10 Adjust the Z rotation to level the ship to the horizon (around value 6.0).

11 Play the composition to see how the ship matches the camera move.

NOTE In the clip timeline, the composition titled Ship placed in 3D has the ship positioned correctly if you are having difficulty finding the right settings.

That's all it really takes to add the ship and have it match the camera move from the 3D track. Now comes the compositing part, where you must make it appear like the ship fits naturally in the beach setting. That will require some lighting and color correction.

Matching Color and Light

Compositing isn't just about placing one object on top of another. At this point, you probably know that. Although the ship follows the camera motion perfectly, it still needs work look realistic. Because the ship looks a little dark, we'll add some ambient light to the scene. Ambient light would come from the sky, so it will need a light blue color to it.

1 Select the merge 3D node in the node editor.

2 Using the select tool dialog, add an ambient light node.

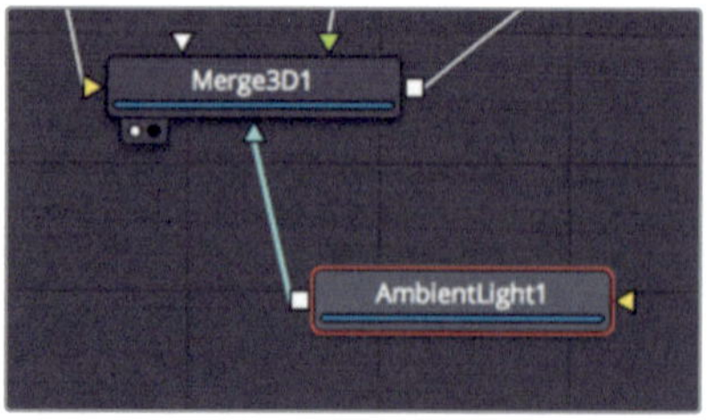

Just as you did in the render 3D node, you must enable lighting in the camera tracker renderer to view the results in viewer 2.

3 Select the camera tracker renderer node, and click the enable checkbox in the lighting section of the inspector.

4 Select the ambient light node, and adjust the intensity to 0.8.

Sky light is blue, so we'll also tint the ambient light a little to better simulate blue sky light.

5 Drag the eye dropper from ambient light 1's color swatch over the lightest part of the sky in viewer 2 to sample its color.

We've now completed the ambient lighting of the ship object. Don't panic; we still need to color correct it into the scene to make it blend well. Like the render 3D node, the camera tracker renderer is used to render from 3D to 2D, so we add a color corrector node after it.

6 Select the camera tracker renderer 3D node, and add a color corrector node.

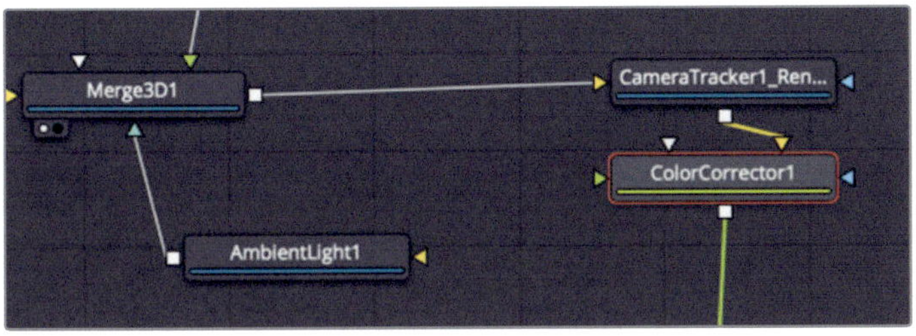

The camera tracker renderer renders this to a 2D image, and now the color corrector will allow us to correct the ship to match the beach shot.

Before we start making adjustments, we no longer need the media in 1 connected to the camera 3D 1, which was only required for the tracking. We can disconnect them and composite the media in 1 to a merge node.

7 Click the connection line near the camera 3D input to disconnect the media in node.

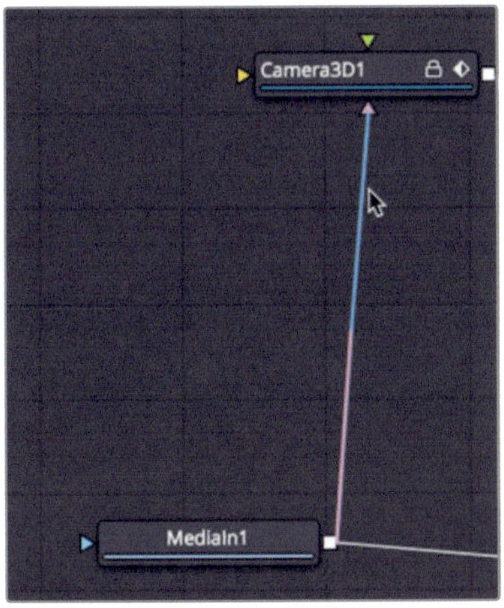

Now let's add a merge to the media in 1 and composite the color corrector node as the foreground.

8 Select the media in node, and add a new merge directly following it.

9 Connect the color corrector node into the foreground input of the new merge.

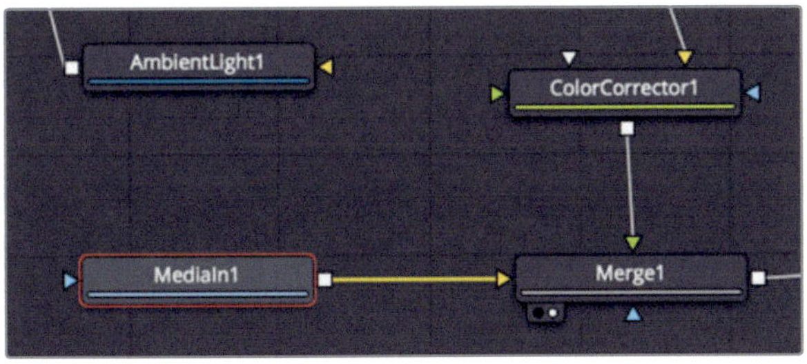

10 Select the new merge node, and press 2 to load the viewer.

11 Select the color corrector node for the ship and adjust the lift and gamma controls up to create an atmospheric wash (lift: 0.05, gamma, 1.3).

The obvious problem that we have come across before is that the background changes when you color correct the ship. If you recall, we had this same issue back in Lessons 4 and 8. The pirate ship includes a premultiplied alpha channel. So, before we can color correct, it we must first divide the alpha. Then, once we color correct it, we must multiply the alpha before it is connected into the merge node. All of that can be done from inside the color correct node.

12 In the inspector, click the options tab and enable the pre-divide/post-multiply checkbox.

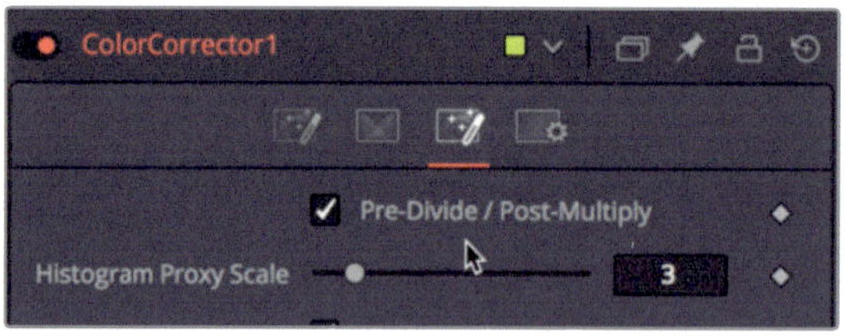

Now, the background is not affected, and we can continue color correcting the ship.

13 From the color corrector inspector, drag the color indicator in the color wheel slightly toward blue to account for the distant color shift.

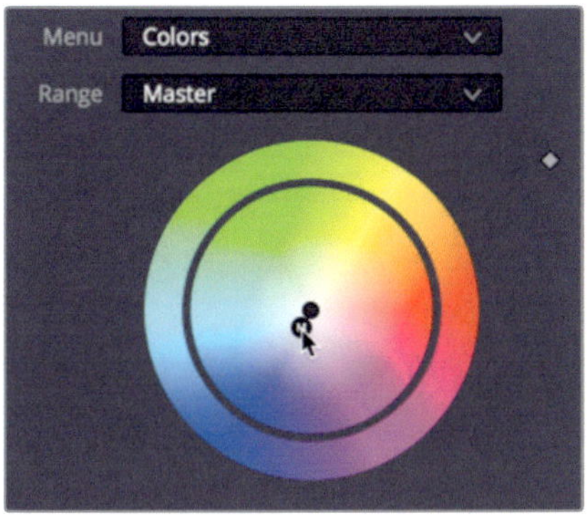

14 To compare the ship with and without the color correction, press cmd-P (macOS) or Ctrl-P (Windows) to disable the color correction node, and then then press it again to enable it.

You should have a fairly convincing match. The final touch for this ship will be to add some grain/noise so our CG ship better matches the live action background.

15 With the color corrector selected in the node editor, add a grain node.

16 To see the grain clearer, hover your mouse pointer over viewer 2 and scroll the middle mouse wheel to zoom in to the viewer.

The initial settings are fairly strong, so we will back them off to get a better match.

17 In the inspector, set the grain size to 0.5.

18 Click the play button to cache out the sequence and admire the menacing pirate ship in the distance.

You can see how easy it is to add elements into a 3D-tracked shot. Having 3D tracking built in to DaVinci Resolve 16 Studio means that with a bit more effort you can transform simple sets into grand period pieces or galactic starships. And best of all, there is no waiting for a separate studio to deliver the visual effects to you. As the composite progresses, you can instantly see how it works in the scene and make changes without causing massive delays.

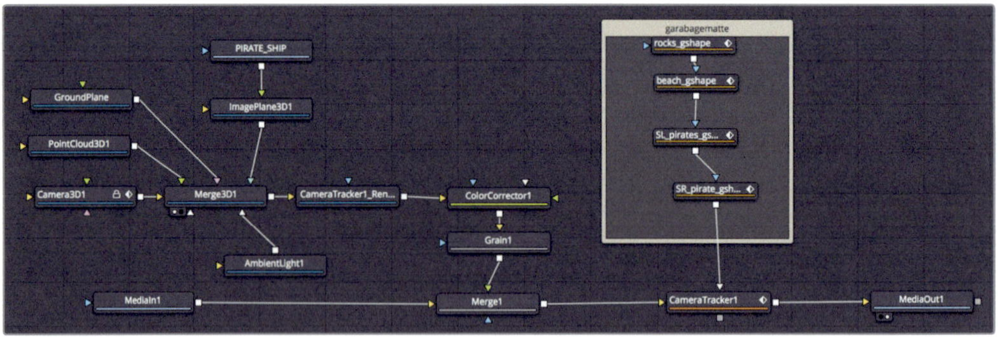

Completed node tree for Lesson 12.

Practice Exercise

For this exercise, you'll camera track a new beach shot. This one also requires some rotoscoping of the actors and ocean. Then, you'll take an image of a pirate's flag and place it convincingly (more or less) on the beach:

1 Select the second edit in the timeline.
2 Create mattes for the areas you want to avoid (pirates, swashbucklers, and ocean).
3 3D track and then solve the track.
4 Set the ground plane and export the scene.
5 From the media pool, add the flag.psd and composite it anywhere along the beach.

Lesson Review

1 True or false: 3D camera tracking is available only in DaVinci Resolve 16 Studio.
2 True or false: It is OK to have the camera tracker follow people and moving cars since they will help with calculating the parallax.
3 What piece of camera information must be entered into the camera tracker for solving to work?
4 When setting the ground plane for a floor in the frame, should the coordinates be set to XY, XZ or YZ?
5 True or false: Adjusting the maximum solve error and the minimum track length can potentially improve a high solve error.

Answers

1 True. 3D camera tracking is not available in the free version of DaVinci Resolve 16.

2 False. You only track objects that are "nailed to the set." All other moving objects should be eliminated using a garbage matte.

3 The lens focal length must be entered to solve the 3D track.

4 XZ are the appropriate coordinates for a ground plane because X represents the horizontal axis, and Z represents depth. The other two coordinates are more appropriate if the ground is not in the frame, such as for a wall.

5 True: Adjusting the maximum solve error and the minimum track length can potentially improve a high solve error.

Congratulations!

Congratulations! You just completed some high-end professional visual effects and motion graphics from the comfort of your own computer workstation. Hopefully, this training guide has opened your eyes to the power and flexibility of node-based compositing, as well as the awesome visual effects tools available as standard in DaVinci Resolve.

Test your skills by taking the online assessment: **www.blackmagicdesign.com/products/davinciresolve/training**

Appendix A

Creating a Light Wrap

This appendix is an optional exercise for the green screen you created in Lesson 8. Once you have completed the green screen lesson, the last color correction you should consider making on a green screen composite is the useful yet horribly overdone *light wrap* effect. The idea is to add a simulated spill light from the composited background. Had this guitarist really been playing on an apocalyptic rocky ledge, the eerie green/yellow sunlight would have spilled around him, tinting the edges just as the green screen spill did. Building a light wrap in a node tree is a simple process now that you have a foreground with a nice alpha channel.

Time

This lesson takes approximately 10 minutes to complete.

Goals

Overview of a Light Wrap

As an overview, a light wrap is essentially the creation of a new foreground image. It is made up of an edge matte of the foreground that has an inner blur.

Then, a blurred background image will act as the new, simulated spill.

All of that will be merged over the top of the current foreground image to add a subtle spill color around the edges that matches the background light directly behind your foreground subject.

Staying Organized with Color

Before we expand the green screen node tree you created in Lesson 8, you should consider organizing your tree before it becomes a tangled web of nodes and connection lines. The Fusion page includes a number of ways to help you organize your composites beyond simply naming the nodes. Before our node tree expands, let's start getting organized using color to outline the different parts of our tree.

1 Return to the Fusion 16 lessons part 2 project.

> **NOTE** You will need to enable the DaVinci YRGB Color Managed setting in the settings window, if you have disabled it since completing this lesson.

2 From the timelines bin, double-click the Part 2_START timeline, and then go to the fourth and final red marker, which is the green screen you created.

> **NOTE** If you did not finish Lesson 8, you can use the Lesson 8 completed timeline located in the timeline > completed bin.

3 Click in an empty area of the node editor directly above the Delta keyer and two mattes you created.

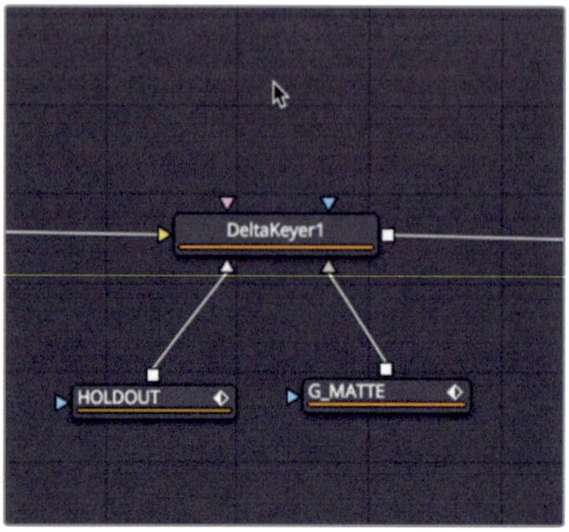

4 Open the effects library and choose tools > flow.

5 From the flow category, click the underlay tool.

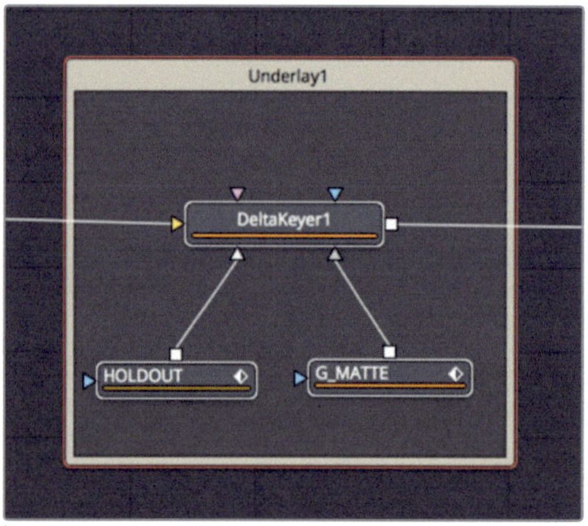

The underlay tool is added to the node editor and encompasses the Delta keyer, garbage matte, and hold out matte in a rectangular outline. The underlay tool is not an effect but an organizational tool that helps you group parts of your node tree using a background color. Let's rename the underlay and assign a color to make it stand out in the node editor.

6 Hold the cmd key (macOS) or Ctrl key (Windows) and click on the Delta keyer and two matte nodes to deselect them.

7 Right-click over the underlay title bar and choose rename from the menu. Then rename the underlay to **MATTES**.

8 Once you save the name, right-click over the title bar again and choose set color > orange.

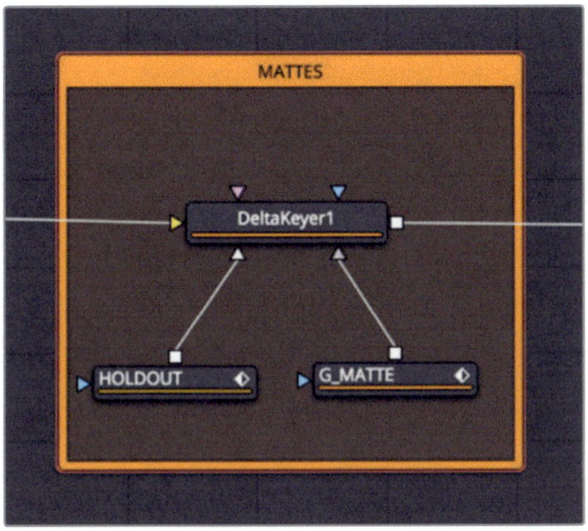

TIP Moving the underlay by dragging the title bar, moves all the nodes encompassed in the underlay rectangle.

This one underlay organizes our key and its auxiliary mattes into a group. As you expand the composite with the light wrap, you can add another underlay, using a different color to distinguish the different areas from each other.

9 Click in an empty area of the node tree between the color corrector 1 and the merge 1 node.

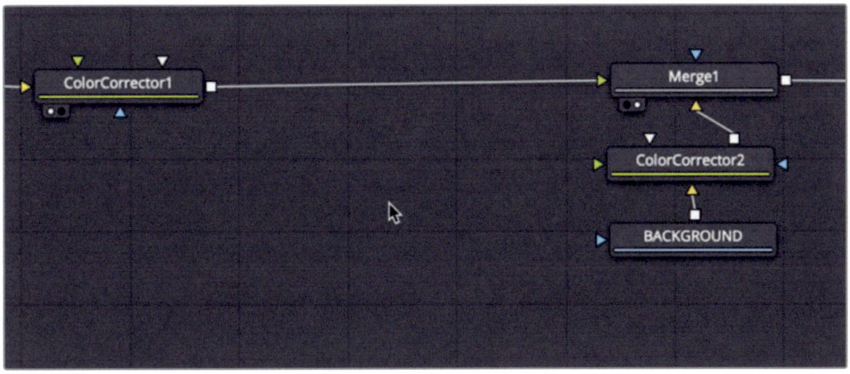

10 From the effects library, click the underlay tool to add it to the node editor.

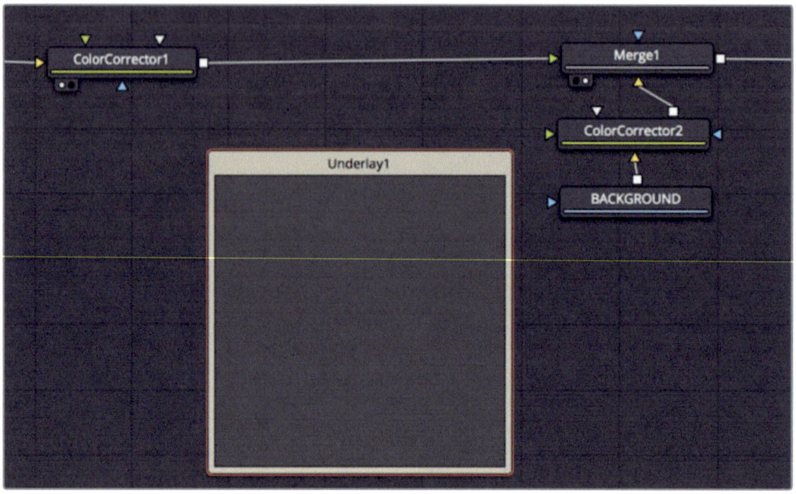

NOTE You may want to move nodes around to create more room between the underlay and the merge 1 node.

11 Rename the underlay to **LIGHT_WRAP** and set its color to blue.

Combining Mattes

Now you will make two copies of the matte from the Delta keyer to create an edge matte that will represent the matte for the light wrap. An edge matte is a matte that just masks around the edges of the foreground subject.

1 From the toolbar, drag two matte control tools inside the underlay.

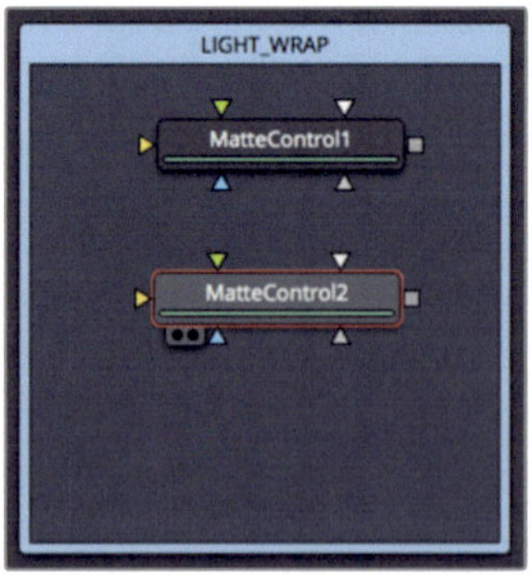

2 Connect two additional outputs from the Delta keyer, one to each of the matte control nodes.

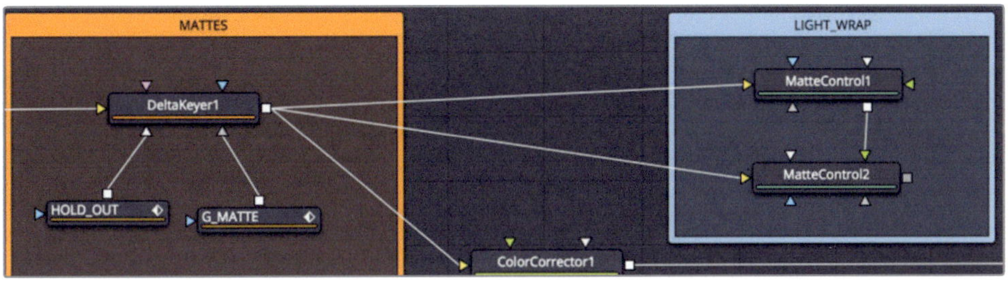

3 Select matte control 1 and press 1 to view the alpha channel. In the inspector, increase the blur slider to around 2 or 3.

4 Select matte control 2 and press 1 to view its alpha channel.

Matte controls are used for a variety of operations on mattes. You can use the matte from the foreground input to combine in any number of ways with the matte from the background input. For our purpose, we will connect the two matte controls and subtract one from the other.

5 Connect an output of matte control 1 to the green foreground input of matte control 2.

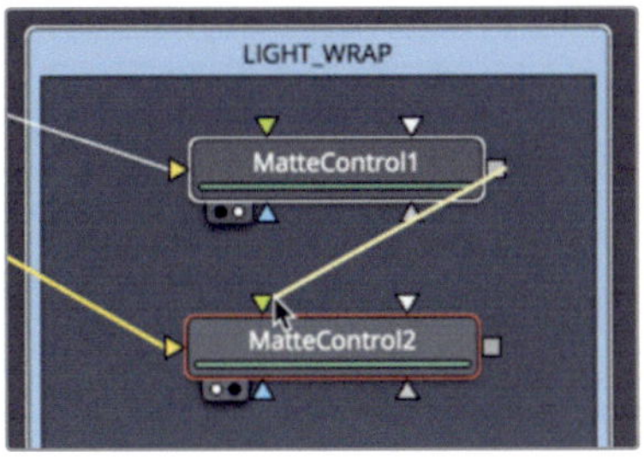

6 In the inspector of matte control 2, set the combine menu to combine alpha, and set the combine op menu to subtract.

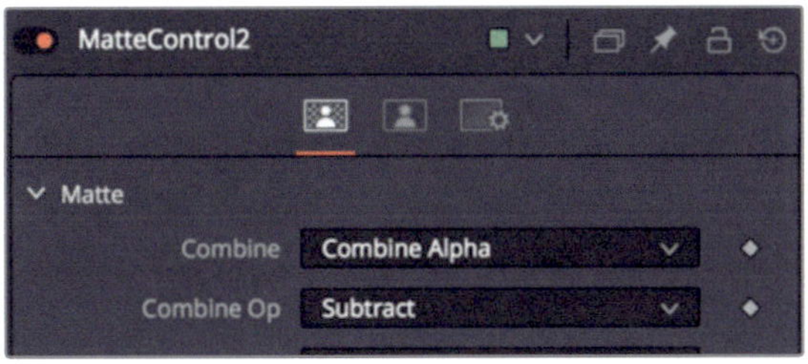

You have just created the edge matte you need for the light wrap. The blurrier you make the matte, the more the color from the background image will spread around the edges of the foreground subject.

Merging the Light Wrap and Foreground

The remaining task for the light wrap is a simple merge composite. You'll use the edge matte and fill it with the background, and then merge that over the top of our musician.

1 Drag a merge node from the toolbar, and place it under the matte control 2 node.

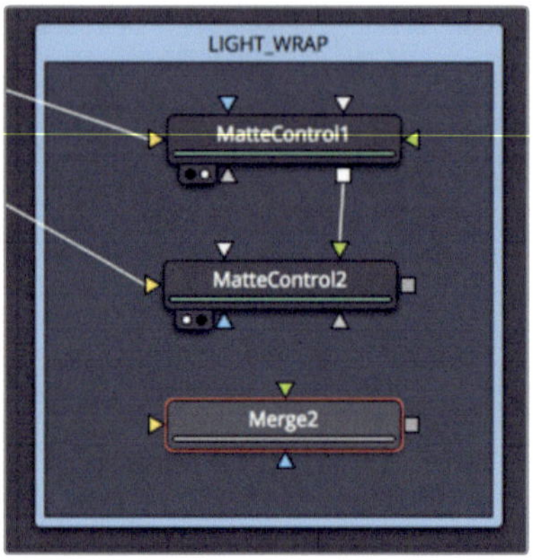

2 Drag the output of the matte control 2 node to the blue mask input of the merge 2 node to create the effects mask for your composite.

3 Disconnect color corrector 1 from the merge 1 node.

The final result of the light wrap will become the new foreground for this merge 1, so this connection is no longer needed.

4 Connect the output of the color corrector 1 node to the yellow background input on the merge 2 node.

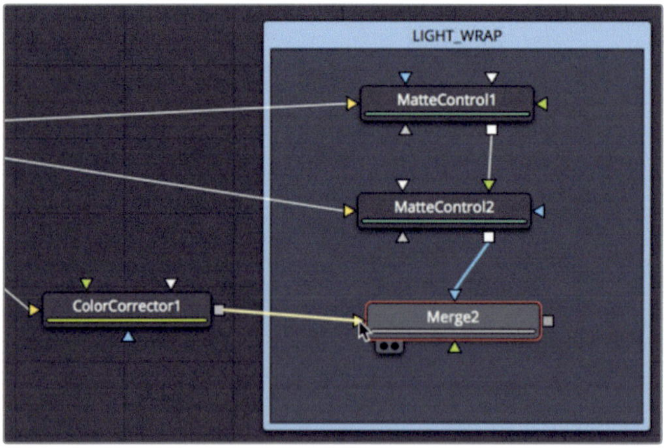

It may seem counterintuitive to connect the color-corrected guitarist to the background input, but remember that you are going to merge a blurred background over the top of him using the edge matte you created.

5 Drag a second output from the color corrector 2 node into the green foreground of the merge 2 node.

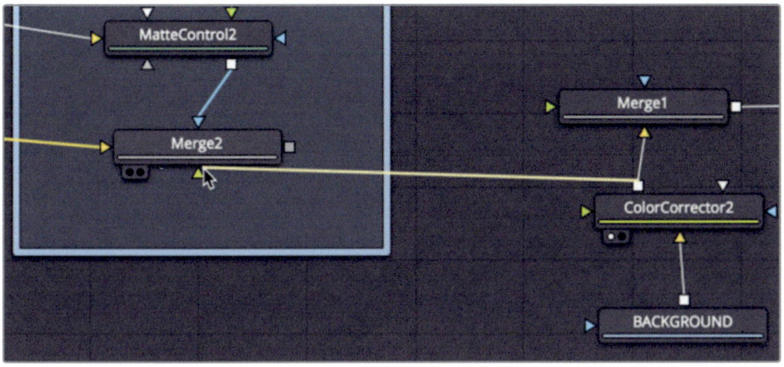

The light wrap is essentially complete. However, spill light from any light source is a diffused light. It should not have detailed, sharp edges. That being the case, you must soften the spill colors coming from the color corrector 2 node using a blur node.

6 To soften the spill from the background clip, insert a blur node between the color corrector 2 and the merge 2, and then, in the inspector, increase the blur slider to around 10.

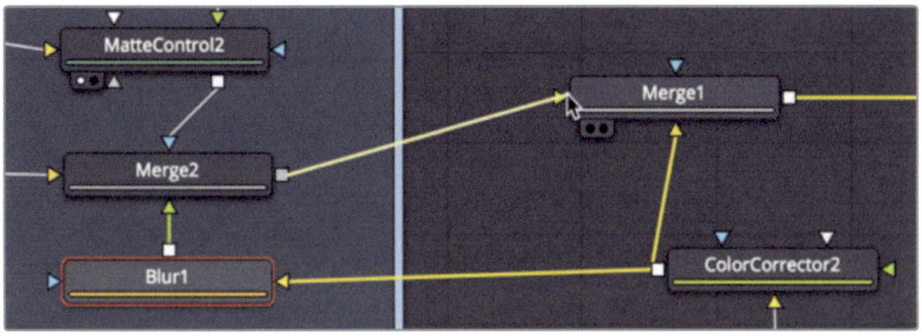

The output of merge 2 is now your new foreground image, and you can connect it to merge 1 as such.

7 Drag the output of merge 2 to the foreground input of merge 1, and then view merge 1 in either viewer to see the final result.

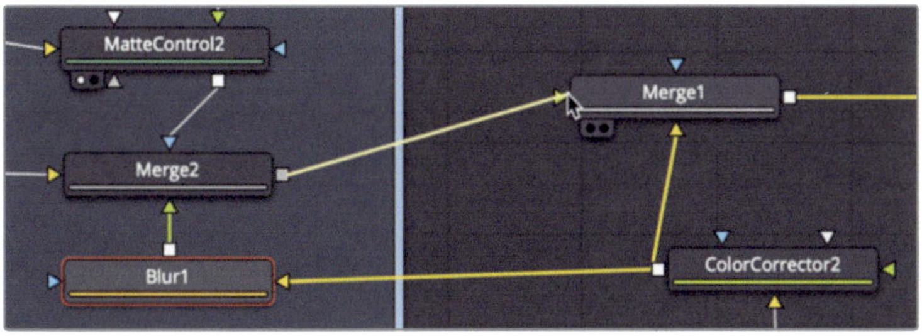

The light wrap is just one more piece you can add on to a green screen composite to make it appear that much more realistic. Since you have complete control over it, adjust the amount of blur and the gamma of the matte in the matte controls to see how it affects the strength and spread of the light wrap "spill."

Appendix B

Using Fusion Mattes in the Color Page

As you have experienced in Lesson 8, the Delta keyer is an amazing tool for green screen shots. And although the Fusion page also includes extremely adept color correction nodes, DaVinci Resolve offers world-renowned color grading tools on the color page. So, in some cases you may want to use the two pages together for compositing a shot. In this appendix, you'll use the green screen clip you started in Lesson 8. Using mattes created in the Fusion page, you'll tint the color of the guitarists' glasses in the color page. It's a simple process to show you how easy it is to bring mattes from Fusion into color.

Time

This lesson takes approximately 15 minutes to complete.

Goals

Adding a Second Media Out Node

The main requirement in sending the matte from the Fusion page to the color page is adding a media out node. You'll open the Fusion 16 lessons part 2 project and load the Part 2 Start timeline for this exercise.

1 Return to the Fusion lessons part 2 project.

> **NOTE** You will need to enable the DaVinci YRGB Color Managed setting in the project settings window if you have disabled it since completing this lesson.

2 From the timelines bin, double-click the Part 2 -START timeline.

3 Move the playhead over the fourth and final red marker, which is the green screen you created in Lesson 8.

4 Click the Fusion page button.

Even if you did not complete Lesson 8, we have a finished version saved for you in the thumbnail timeline.

5 At the top of the Fusion page, click the clip button, and then right-click over the last thumbnail and choose the extra matte version.

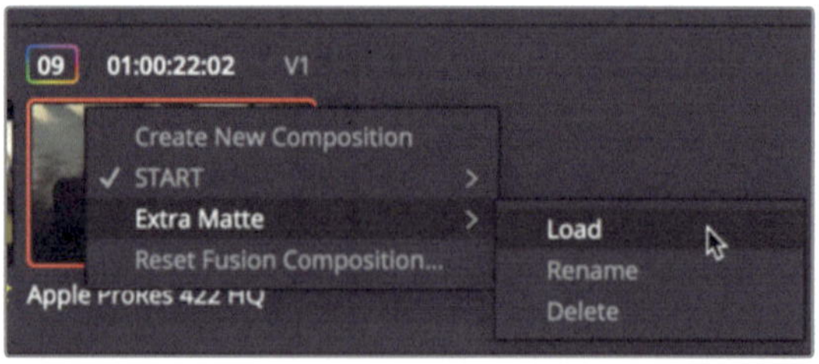

The orange underlay in the node editor displays the mattes you will use for the color page. These mattes isolate the sunglasses the guitarist is wearing.

6 Select the HAT_BRIM node, and press 1 to see the combination of the three mattes in the viewer.

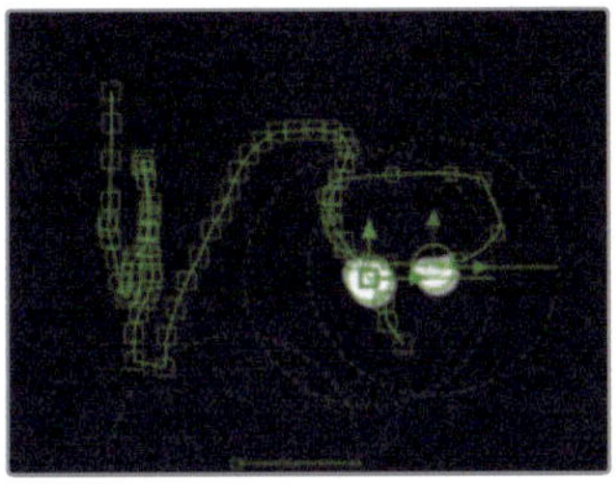

To bring these mattes into the color page, you must add another media out node at the end of the string of matte nodes

7 Press Shift-spacebar, type **MediaOut**, and then press enter to add the node to the node editor.

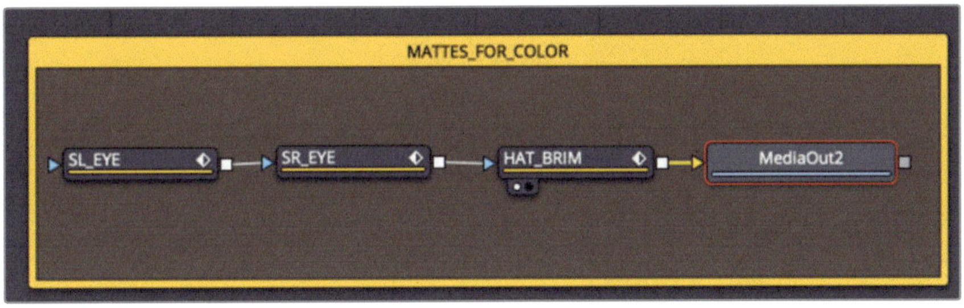

The first media out node in the node editor is used to go to the edit page. A second media out goes to the color page. You can add as many media outs as you have mattes. We are bringing the combination of these mattes into the color page, but we could just as easily bring all three mattes individually.

Setting Up the Color Page

Most of the time you use tools in the color page, you are using just the color data. Occasionally, you will bring an external matte into the color page node editor. That external matte source is a clip located in the media pool. In our case, we need a source from the Fusion page, similar to the source that comes from the edit page.

1 Click the color page.

The green screen clip you were working on in Fusion is now the currently selected clip in the color page. You now need to add a source that represents the media out 2 node (the mattes) from the Fusion page.

2 In the node editor, right-click in an empty gray area and choose add source.

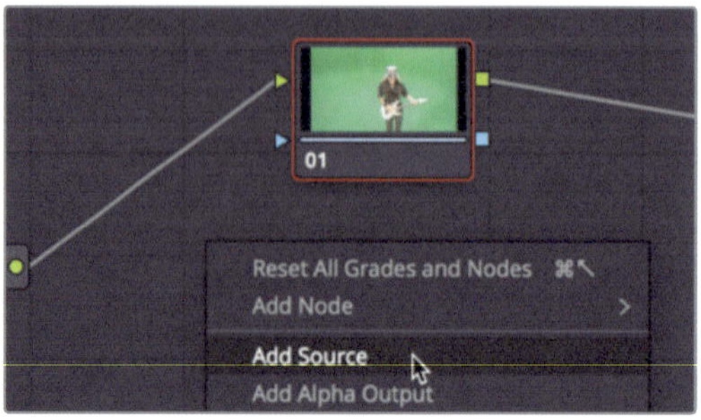

The source is added as a green icon on the left side of the node editor, directly under the RGB source from the edit page. This second source is from the media out 2 from the Fusion page. You can use it as either RGB data or an alpha channel since it includes both. For this exercise, we'll use the second source as an alpha matte source.

3 Drag the output of the second source to the key input of node 1.

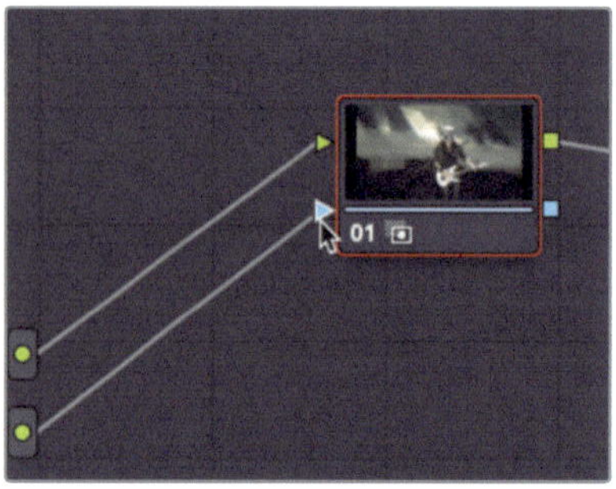

You now have a mask connected to node 1, so that any color correction you do will be limited by the mask from the Fusion page.

4 Using the color page color wheels, drag the lift and gain color wheels toward a yellow/green color to tint the glasses.

Using Fusion's drawing tools, you can build more detailed mattes for secondary color corrections.

Appendix C

Creating Smoke with Fast Noise

The fast noise tool is a very flexible, organic-looking noise generator. It is a common technique when generating particles like clouds and water caustics. The following steps will take you through the process of creating the single smoke movie you used to create the particles in Lesson 11.

Time

This lesson takes approximately 10 minutes to complete.

Goals

Using Fast Noise for Particle Cells

Fast noise is particularly good at creating vapors like smoke or clouds. However, since fast noise can be very demanding on your computer because it must generate the image in real time as you create the particles, it is often easier just to render out the fast noise as a short, small movie. You can then use that small, less taxing movie as the image for your particles.

1 Open a new project in DaVinci Resolve.

2 Add a Fusion composition effect to the timeline and leave it at the 5-second duration.

3 Click the Fusion page button.

4 From the toolbar, drag the fast noise tool to an empty place in the node editor.

The fast noise tool is a very flexible, organic-looking noise generator. It can be useful for a wide range of effects from clouds to water caustics, and in our case, smoke. The first thing to do when making an image to be used as a particle cell is to set a resolution as small as possible. Most particle cell images should be under 200 x 200 pixels so they use very little graphics card memory.

5 In the inspector, click the image tab and uncheck the auto resolution checkbox.

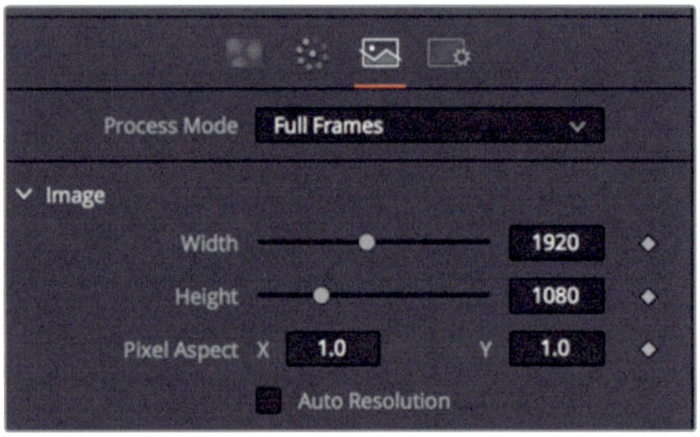

6 In the width and height fields, enter 200 x 200.

The image tab is primarily used for setting the resolution. The majority of controls for the noise are located in the noise tab. Here, we will control the tonal quality of the noise as well as the auto-animation feature.

7 Click back on the noise tab, and set the detail to 5 to give more texture to our smoke.

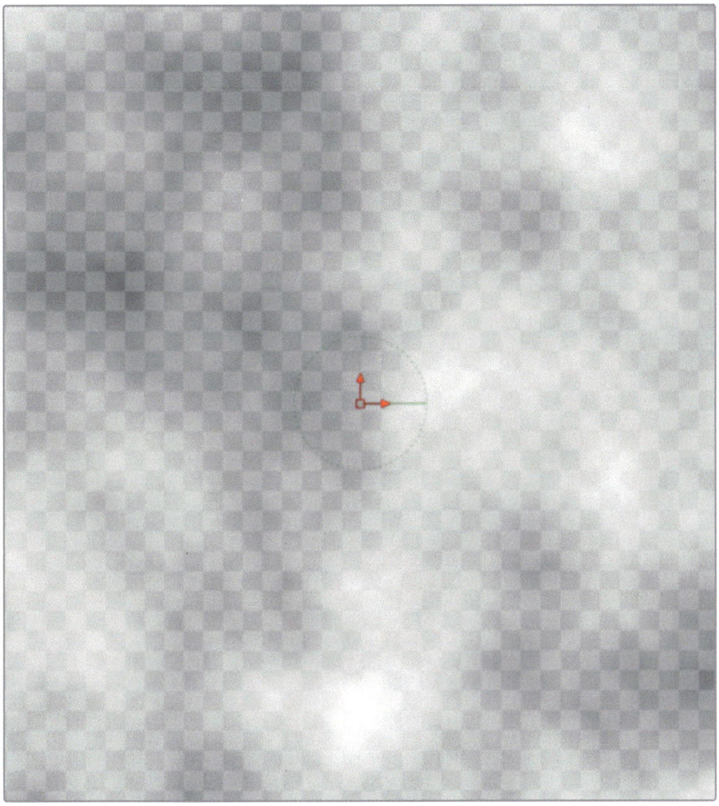

8 Increase the seethe rate to 0.1 to create a slow smoldering animation for the noise.

Now that the noise texture looks right, we need to shape it. We cannot have the noise go to the edges of the frame boundary or those sharp edges will appear in each particle cell. To fix this, you will make the noise use a radial gradient and set the outer edges of the gradient to be transparent.

9 Click the color tab, and set the type to gradient.

10 From the gradient type menu, choose radial.

Switching to a radial gradient displays a gradient line in the viewer. Using the green gradient line, you will set the center point of the gradient and the outer edge.

11 Using the handle on the left end of the green gradient line, drag the point to the center of the viewer.

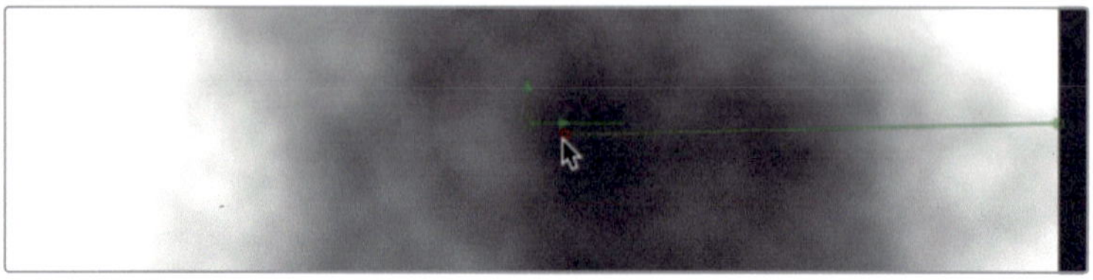

12 Using the handle on the right end of the green gradient line, drag the point in toward the center of the viewer, about one third of the way from the edge.

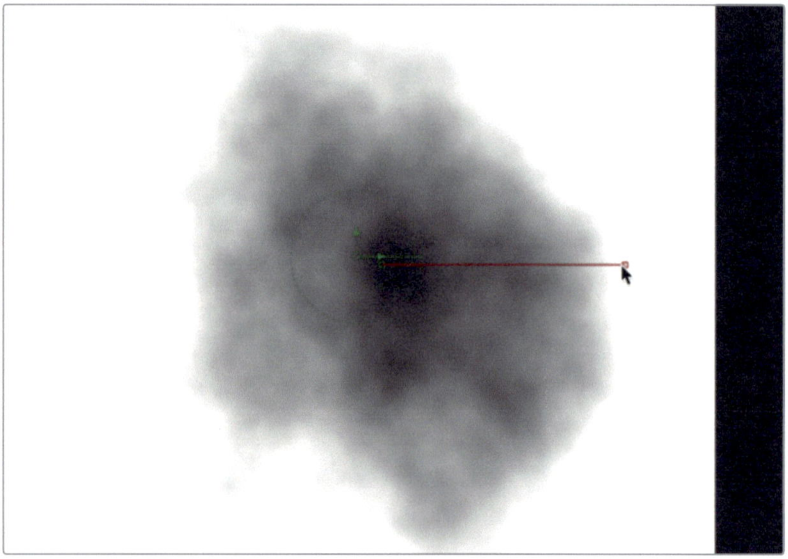

Now you can add some transparency using the alpha control in the inspector.

13 In the inspector, click the first color stop on the left end of the gradient bar and set the color to a middle gray.

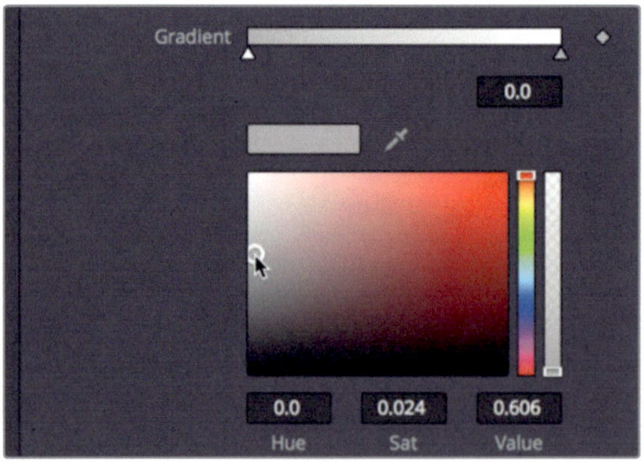

14 Click the last color stop on the right end of the gradient bar, and set the color to black and the alpha slider to 0.

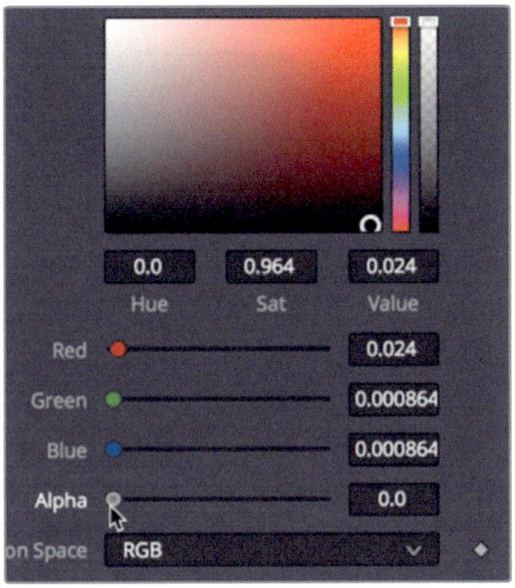

15 Play the fast noise effect in the viewer to see your puff of smoldering smoke.

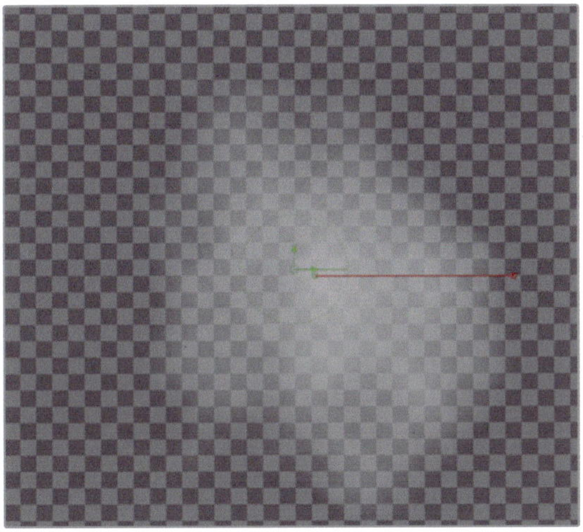

Now you can render this movie out and use it in your particle system.

Rendering from the Fusion Page

If you render this fast noise generator using the media out node, it will render using the timeline resolution. You need to render it using the 200 x 200 resolution you set in the image tab of the fast noise. You can do this by adding a save node.

1 Open the effects library and go to the Tools > I/O category.

2 Drag the Saver node to an empty location in the node editor.

3 Drag an output from the fast noise into the yellow input on the Saver node.

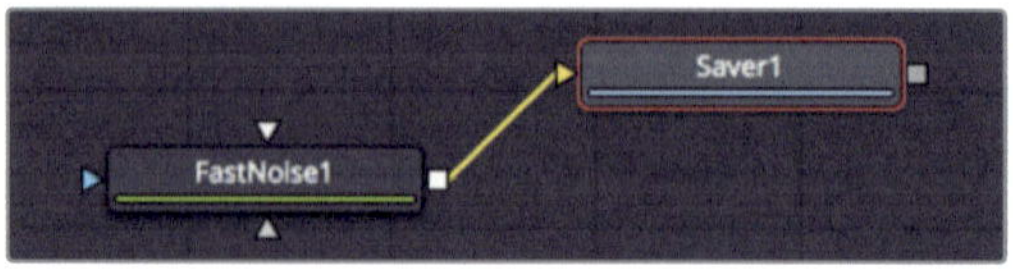

The saver node is unique because it allows you to render out any part of your node tree as EXR image files. It will use the resolution based on the image that is connected to its input. In your case, it will render out at 200 x 200 pixels, since that is the resolution of the fast noise generator.

4 Select the saver node, and in the inspector, click the browse button.

5 Navigate to a location where you want the image files to be rendered and create a folder for them to reside within.

6 At the top of the window, name the image files noise, and then click save to close the window.

The file field at the top of the inspector is now populated with the location of the file and the file name with an .exr extension.

7 In the menu bar, choose Fusion > render all savers to begin the rendering process.

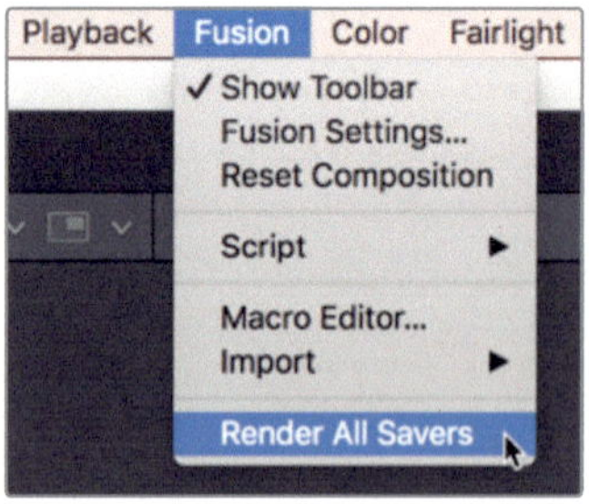

8 Once the rendering is complete, click OK in the dialog.

You can now test your render by importing the EXR files into the media pool. The fast noise generator is a fantastic tool to experiment with when creating particle effects. It is the perfect tool to turn to when trying to create organic vapor effects.

Appendix D

Fusion Keyboard Shortcuts

Many Fusion-specific shortcuts can help speed up working in the Fusion page. However, since the Fusion keyboard shortcuts are not available in the keyboard customization setting, as they are for other DaVinci Resolve pages, Fusion shortcuts are listed here for convenience.

General

Function	macOS	Windows/Linux
Go to Fusion page	Shift+5	Shift+5
Undo/Redo	Cmd+Z/Cmd+Shift+Z	Ctrl+Z/Ctrl+Shift+Z

Node Editor

Function	macOS	Windows/Linux
Open Select tool window	Shift+Spacebar	Shift+Spacebar
View/clear selected tool on the left viewer	1	1
View/clear selected tool on the right viewer	2	2
View/clear selected tool on the SDI/HDMI monitor	3	3
Clear all viewers	~ (tilde)	~ (tilde)
Rename a node	F2	F2
Create a group	Cmd+G	Ctrl+G
Swap background/foreground connections	Cmd+T	Ctrl+T
Lock/Unlock selected node	Cmd+L	Ctrl+L
Pass through (disable) selected node	Cmd+P	Ctrl+P
Update mode (Suspends viewer updates until you toggle back. Can be helpful when tweaking nodes takes too long to update in viewer.)	Cmd+U	Ctrl+U
Open Find dialog window	Cmd+F	Ctrl+F
Display navigator window	V	V
Scale node tree to default size	Cmd+1	Ctrl+1
Zoom in/out of node tree	+ or – key	+ or – key
Cut, copy, paste nodes	Cmd+X/C/V	Ctrl+X/C/V
Paste an instance of the copied tool	Cmd+Shift+V	Ctrl+Shift+V
Select all nodes	Cmd+A	Ctrl+A

Navigation

Function	macOS	Windows/Linux
Play	Spacebar	Spacebar
Go to next frame	]	]
Go to previous frame	[	[
Go to start of render range	Cmd+[	Ctrl+[
Go to end of render range	Cmd+]	Ctrl+]
Go to next keyframe	Option+]	Alt+]
Go to previous keyframe	Option+[	Alt+[
Set render range to frames enclosed by mouse drag	Cmd+drag in time ruler	Ctrl+drag in time ruler
Set render range to viewable frames in time ruler	Cmd+double-click in time ruler	Ctrl+double-click in time ruler

NOTE To reset render range to the comp length, right-click in render range > Auto render range. Yes, I know it's not a keyboard shortcut, but you wanted to know right?

Inspector

Function	macOS	Windows/Linux
Display Modifier tab	F11	F11
Display Tools tab	F9	F9
Move to next/previous number field	Tab/Shift+Tab	Tab/Shift+Tab
Open expression/pick whip field for selected number field	=, then Return	=, then Enter
Cut, copy, paste numbers in fields	Cmd+X/C/V	Ctrl+X/C/V
Lock/Unlock active Inspector/node	Cmd+L	Ctrl+L
Pass through (disable) active Inspector/node	Cmd+P	Ctrl+P
Pass through (disable) active modifiers (with modifier tab active)	Cmd+P	Ctrl+P
Rename modifier	Click modifier name, then F2	Click modifier name, then F2
Delete all modifiers (with modifier tab active)	delete	Backspace

Viewers

Function	macOS	Windows/Linux
Display red, green, blue, or alpha channel	R,G,B, or A	R,G,B, or A
Display Z depth channel	Z	Z
Display composite RGBA	C	C
Zoom in/out of viewer	+ or -	+ or -
Scale image to fit viewer	Cmd+F	Ctrl+F
Show image at actual size (100%)	Cmd+1	Ctrl+1
Display subview	V	V
Swap subview and main view	Shift+V	Shift+V
Show/hide on-screen controls	Cmd+K	Ctrl+K
Show/hide aspect ratio guides	Cmd+G	Ctrl+G
Display A viewer buffer	, (comma)	, (comma)
Display B viewer buffer	. (period)	. (period)
Display A/B buffer split screen wipe	/	/
Place selected viewer into single viewer mode (toggle, pressing again switches back)	F4	F4
Display Region of Interest (ROI)	Cmd+I	Shift+I

3D Viewers

Function	macOS	Windows/Linux
Same keyboard shortcuts as 2D viewer		
Center selected element in viewer	Cmd+F	Ctrl+F
Quad view toggle	Shift+Q	Shift+Q
Switch to position, rotate, scale controls for selected object.	Q,W,E	Q,W,E

Splines

Function	macOS	Windows/Linux
Set the active spline to click-append mode	Shift+C	Shift+C
Set the active spline to draw-append mode	Shift+D	Shift+D
Set the active spline to insert & modify mode	Shift+I	Shift+I
Set the active spline to modify only mode	Shift+M	Shift+M
Set the active spline to Done	Shift+N	Shift+N
Close the spline	Shift+O	Shift+O
Select all points on a selected spline	Shift+A	Shift+A
Delete selected point	Shift-delete	del
Smooth selected control points	Shift+S	Shift+S
Make control points linear	Shift+L	Shift+L
Hide Bezier handles on splines	Shift+H	Shift+H
Hide control points on splines	Shift+K	Shift+K
Display shaping box around selected points	Shift+B	Shift+B
Publish selected control point on spline	Shift+P	Shift+P
Set the selected points to follow existing published points.	Shift+F	Shift+F
Disable rendering until mouse up	Shift+R	Shift+R
Display point editor dialog	E	E

Advanced Spline Transforms

Function	macOS	Windows/Linux
Scale selected points around mouse pointer	Hold S+drag	Hold S+drag
Twist selected points around mouse pointer	Hold T+drag	Hold T+drag
Scale horizontally only	Hold X+drag	Hold X+drag
Shrink or expand shape based on selected points	Hold O+drag	Hold O+drag

Spline Editor

Function	macOS	Windows/Linux
Same keyboard shortcuts as Splines (unless noted below)		
Open the spline editor (not a toggle)	F8	F8
Zoom in on rectangle (region zoom)	Cmd+R	Ctrl+R
Zoom in/out of spline editor	+ or – keys	+ or – keys
Fit all keyframes in window	Cmd+F	Ctrl+F
Select all keyframes	Cmd+A	Ctrl+A
Cut, copy, paste control points	Cmd+X/C/V	Ctrl+X/C/V
Delete control point	delete	Backspace
Display ease in/ease out control in toolbar	T	T
Create a keyframe on active splines at playhead position	Cmd+K	Ctrl+K
Set selected points to Step In/Step Out mode	I/O	I/O
Reverse selected points	V	V
Limit dragging points horizontally or vertically	Option+drag	Alt+drag
Copy select points when dragging	Cmd+drag	Ctrl+drag
Flatten spline handles horizontally	F	F
Show Guide List	Shift+G	Shift+G

Index

SYMBOL

NUMBERS

A

B

F

G

H

R

S

T

U

V

X

Y

Z

About the Authors

Damian Allen is a visual effects and animation consultant, developer, and supervisor in Hollywood. He is the owner of VFX company Pixerati LLC, with a focus on picture-lock visual effects emergencies and VR and animation tool development. Damian is also a core contributor to the moviola.com training site for filmmakers.

Tony Gallardo, ever since picking up his first VHS camera, has been hooked and cut his teeth at a very early age making short films and promo videos for his school and church. A story editor from the start, Tony quickly expanded to all aspects of post production and production. From designing award-winning motion graphics to directing tear jerking real-life stories, his passion for the craft and tools is endless. After co-running an award-winning production facility in San Antonio, Texas for a little over 14 years, he branched out and now runs his own post boutique, Tomiga. Tomiga is a hybrid creative boutique focusing on short form content from brand commercials to informative PSAs to promotional media. When he's not creating brand commercial campaigns or social media ads, Tony is learning and educating about all his favorite creative tools—Davinci Resolve and Fusion being at the top of his list.

Dion Scoppettuolo is a Certified Blackmagic Design Master Trainer and co-author of The Beginner's Guide to DaVinci Resolve 16. He has taught classes on DaVinci Resolve in Hollywood and New York City, as well as across Europe and Asia.

Mr Scoppettuolo has extensive industry experience in editing and visual effects, having held the position of Senior Product Manager for Shake and Motion at Apple Inc.

Made in the USA
Las Vegas, NV
10 November 2026

57851953R00226